D1528865

A General Interpreted Modal Calculus

A GENERAL INTERPRETED MODAL CALCULUS

by Aldo Bressan

Foreword by Nuel D. Belnap, Jr.

New Haven and London, Yale University Press, 1972

Library of Congress catalog card number: 77-151568

International standard book number: 0-300-01429-5

Designed by Sally Sullivan

and set in IBM Bold Face One type.

Printed in the United States of America by

The Murray Printing Company

Forge Village, Massachusetts.

Published in Great Britain, Europe, and Africa by Yale University Press, Ltd., London.
Distributed in Canada by McGill-Queen's University Press, Montreal; in Latin America by Kaiman & Polon, Inc., New York City; in Australasia and Southeast Asia by John Wiley & Sons Australasia Pty. Ltd., Sydney; in India by UBS Publishers' Distributors Pvt., Ltd., Delhi; in Japan by John Weatherhill, Inc., Tokyo.

to Anna

Contents

SYNTACTICAL SECTION

Foreword

The book in your hands, written by a professor of physics working in Padua in the very shadow of the chair from which Galileo first preached the marriage of mathematics and nature, is the most important contribution to date concerning the introduction of quantifiers into modal logic. It surpasses any article or book in the generality of its conceptions, the degree of their development, and the profundity of the attendant analysis. Perhaps one should credit the author's near total isolation from the logical community for allowing him to proceed with the elaboration of his fresh ideas unobstructed by premature criticism, and doubtless one must credit his uncompromising insistence on "usability" to the fact that his enterprise arose from and has been continually nourished by the felt needs of a practicing physicist.

The prefatory notes to follow are intended for two groups, to each of which I should like to address a preliminary remark. In the first place, some professional logicians may be put off by the fact that Bressan does not—except in the last part—advert to the contemporary "possible world" treatments of quantified modal logic by Kripke, Hintikka, Montague, etc. (see the descriptive accounts and bibliography of [18]). The historical reason is of course that he worked in complete independence of "the tradition" (which dates only to 1959!), knowing nothing of it until the present work was substantially complete. But even more important, the concepts involved in Bressan's work are sufficiently different from those of the tradition for him to have been unable in any event to base his work on it with profit.

In the second place, the less than professional logician will find the work hard going, with much notation. I am convinced, however, that this is due to the nature of the topic. Handling the mixture of quantification and modality—"modal involvement of the third degree," as Quine says—just _is_ extraordinarily difficult, especially when the whole is type-theoretical. But there is in Bressan's work an underlying clarity of conception which perseverance makes manifest: it's worth it. I have had the privilege of using the typescript as text in a course for philosophy graduate students, some with but a term's previous work in logic, and can testify that it is eminently usable in such a capacity. My own introductory device was to have the students begin by reading a few chapters in Hughes and Cresswell's An Introduction to Modal Logic ([18]), after which background they were altogether capable of profiting from Bressan.

The remainder of this preface is divided into three sets of remarks: on the General Considerations underlying Bressan's logic, on its Distinctive Semantic Features, and on the New Directions in which he points us. But lest you read this preface no further, I should like to point out here the feature in Bressan's ML^{ν} distinguishing it from all its predecessors: a new analysis of predication which does not compel all predication to be extensional (see paragraph 11, below).

I. General Considerations

Among the considerations guiding Bressan's work I wish to mention eight: Types, Cases, Extension and intension, Semantic composition, "Case-intensional semantics," Uniformity, Simplicity, and Generality.

1. Types. Bressan's ML^{ν} is a many-sorted logic (where many = ν), which is a useful but not so important feature; the fact, however, that ML^{ν} is—uniquely among modal logics—a complete type theory, with no upper limit on its types, is extremely important. For in the first place, the availability of higher types allows natural constructions which would be ad hoc on the first level (e.g. arithmetic as in 14, below, and quantification over possible cases as in 18; and in attending to the requirements of higher types one finds one is forced to deepen significantly the semantic analysis (e.g. of predication as in 11, below) in ways which then turn out to reflect back on the lowest level. (Montague, in "Universal Grammar" forthcoming in Theoria, has independently moved to higher types, but without the new analysis of paragraph 11. Gallin reports various completeness and independence results concerning Montague's system in an abstract forthcoming in the Journal of Symbolic Logic.)

2. Possible cases. Bressan bases his analysis of modality on the concept of a set Γ of "possible cases" γ—what the tradition calls "possible worlds." (The "case" terminology is better since it is less overpowering.) This generalization over state descriptions (Carnap [12]) or models is of course one of the key ideas of the tradition; it was conceived independently by Bressan in his work on the foundations of physics which was presented at the

1960 International Congress of Logic, Mathematics, and Philosophy of Science at Stanford, and in his 1962 monograph [4]. Bressan does not employ, however, the "relative possibility" relation of Kripke, etc.; necessity just means "true in every possible case" simpliciter, and possibility "true in some possible case." So the underlying propositional logic is S5 (see [18]), which is all anyone needs for the kinds of applications to physics which guide Bressan's work.

3. Extension and intension. Bressan's semantics, like those of the tradition, is based on Carnap's Method of Extension and Intension [12], so that every meaningful expression is conceived as having both an extension and an intension. Paradigmatically, the extension of a sentence is a truth-value, while its intension might be called a "proposition," and the extension of an individual constant is an individual in some preselected domain, while its intension might be called an "individual concept."

4. Semantic composition. Bressan's semantics, like all modern formal semantics, is based on a version of Frege's analytic principle that "the meaning" of compound expressions shall be a function of "the meaning" of the parts. (For an account and references, see A. Church's Introduction to Mathematical Logic, Princeton University Press, 1956.) Bressan chooses also to satisfy the additional principle that no grammatical expression shall be without meaning, an article not of faith but of convenience.

5. "Case-intensional semantics." I made up this term for the joining of ideas 2, 3, and 4 according to the following recipe: in the first place, extension is always relativized to cases, so that expressions never have extensions bare, but only relative to a case. This means that the fundamental semantic locution in the vicinity is the extension of Δ in γ, for Δ a piece of notation and γ a case. (The extension of Δ can be identified with the extension of Δ in the real case.) In the second place, the intension of an expression Δ is identified with, or is made isomorphic with the pattern of its extensions as the cases vary through the set Γ of all cases. More mathematically: the intension of an expression can for purposes of semantic analysis be identified with a function from the set Γ of possible cases into the collection of objects appropriate to serve as its extensions. (Bressan calls these "quasi-intensions," as one might say that the von Neumann numbers are "quasi-numbers.") Paradigmatically, then, the intension of a sentence—a proposition—can be identified with a mapping from possible cases into truth values, while the intension of an individual

constant—an individual concept—can be identified with a mapping from possible cases into individuals in the pre-given domain. Third and lastly, case-intensional semantics is based on the following specification of the Fregean idea: the intension of a composite expression should be a function of the intensions of its parts. That semantics should be case-intensional in the above sense is a regulative principle guiding alike the work of Bressan and the tradition, though there are wide variations in how it is applied.

6. Uniformity. Another principle guiding Bressan's work, but not the tradition, is that all expressions of a given logical type should be treated semantically in exactly the same way, satisfying a principle of uniformity. Thus, consider the three chief categories of expressions of the type of individuals: individual constants, individual variables, and definite descriptions; according to the principle of uniformity, since individual constants are assigned an intension (an individual concept), exactly the same should hold true of variables and of definite descriptions. The tradition, however, does not treat variables and definite descriptions uniformly with constants. Variables are often assigned only a special kind of individual concept (perhaps a "substance" as in [34, 35]) or else assigned an individual in the domain instead of an individual concept (Kripke; see [18]); and definite descriptions, when treated at all (which is not very often or very well), are sometimes given no intension at all, à la Russell (again see [18]). Now it may be that treating expressions of these three categories each in a different way is in some respects closer to English or to extensional logic; but there is no doubt that it greatly complicates both the semantics and the proof theory. So one who is interested in really applying the logic, and not just constructing a model of something, will do well to be guided by the principle of uniformity: treat expressions of the three categories alike, letting each have an individual concept as its intension. Another application of this principle emerges at the higher types: Bressan demonstrates, by his usage, that there is no need for two styles of variables, one for (say) classes and one for (say) attributes (Lemmon in Acta Philosophica Fennica, fasc. 16, 1963). In fact I now believe that the whole class-attribute intuition, while prima facie appealing, was seriously misguided. (In terms of ideas defined later, we might wish to identify classes with those modally constant attributes which are extensional in the real world. Then—and this is the point—a class is a particular kind of attribute.)

7. Simplicity. ML^{ν} was not designed to "model" this or that feature of discourse, but as a calculus to be used in the axiomatic

development of physics. Hence, that its proof theory is simple counts heavily in its favor. In fact its propositional modal part is the simplest of all, i.e. S5, while unlike the logics of the tradition, its quantificational part is just exactly ordinary quantificational logic (as in say [32]). In particular, ML^{ν} has the principles of universal specification and existential generalization in full generality: with respect to variables, individual constants, and definite descriptions. Further, ML^{ν} contains an unfettered principle of lambda abstraction, and the full thrust of the simplifying Barcan and converse Barcan formulas (from [2]; see [18]).

There is, however, one complication in Bressan: the logic of identity is *not* classical, and in particular ML^{ν} cannot tolerate the indiscernibility of identicals (Leibniz's Lie): necessarily (in every case), if $x = y$ then for every attribute F, x has F just in case y has F. But two close cousins are available: the thing is true *if* (*mere*) identity is replaced by necessary identity, and also if identity is kept contingent but the range of F is restricted to extensional attributes (see below, end of 11). Because prior to Bressan's introduction of absolute concepts (see 15) it was easy to suppose that *all* "genuine" attributes were extensional, it is little wonder that the Lie got turned into the Law.

Upshot: one has to unlearn hardly anything.

8. *Generality*. Of the construction of a logic as simple as ML^{ν}, "they said it couldn't be done," with arguments; but they were wrong.

Of course the arguments were not silly; indeed it is possible to cater to the intuitions underlying them *in* ML^{ν} whenever it is appropriate to do so, thus keeping the global structure of the language as simple as possible while leaving room for introducing local complexities when they are wanted. Examples: (*a*) Bressan shows how to handle in ML^{ν} those cases which have led the tradition to doubt the truth of the Barcan formula (see [18]). (*b*) He shows how to get the effect of quantification over "substances" (Thomason and Stalnaker [35]) via his deployment of absolute concepts. (*c*) It is easy to see that one could get the effect of a "free logic" (e.g. Lambert, "Notes on E!: I–IV," *Philosophical Studies* 10 [1959]: 1–5; 12 [1961]: 1–5; 13 [1962]: 51–59; and 15 [1964]: 85–88), admitting nondesignating singular terms and empty domains in ML^{ν}. (*d*) If one is wedded to Russell descriptions, one could of course add them to ML^{ν} by way of Russell's own definition as in the *Principia Mathematica*—though I cannot see why anyone, having seen the beauty and ease of Bressan's theory of descriptions, should want to. And (*e*) Bressan shows how Kripkean "relative possibility" relations (see [18]) can be built into ML^{ν}, to be called on when needed. And all of this without corrupting the elegant simplicity of the logic itself.

II. Distinctive Semantic Features

Of these features I shall remark on the following: Range of variables, Contingent identity, Nonextensional predication, Lambda abstraction, and Definite descriptions.

9. Range of variables. In ML^{ν}, for a fixed domain D and set of cases Γ, the range of variables is the set of all intensions of the appropriate type; for the lowest type the range would consist of all individual concepts. (Principal alternatives: range directly over individuals in domain, or range over only a limited subset of all the intensions; see [18].) Bressan's alternative has been rejected by much of the tradition for at least two reasons. First, it appears to lead to the logical truth of "if necessarily (in every case) something has F, then there is something which has F necessarily (in every case)" (see [18]). But this appearance dissolves on the new analysis of predication (see 11, below). Second, we have been taught that variables constitute the privileged mode of "reference," so that we shall wind up "referring" to (mere) individual concepts instead of (real) individuals (e.g. Quine in [12]). But this doctrine is twice faulty. In the first place, it treats variables as data on a par with the given features of natural languages; when in fact variables are constructed, not given, so that they mean what—and how—we want them to mean. Second, while the doctrine is plausible and helpful with respect to a wide range of extensional languages, its nature there is caught by describing it as a kind of "correspondence rule"; and the rule simply does not transfer to modal languages based on case-intensional semantics. Why should it, especially since the rule uses the one notion of "reference" where case-intensional semantics has both "intension" and "extension in γ"?

10. Contingent identity. Much of the tradition has allowed identity statements involving individual constants or definite descriptions to be contingent, while making noncontingent those identity statements involving variables—a clear breach of the principle of uniformity. Bressan's identity is in contrast everywhere contingent: regardless of whether Δ_1 and Δ_2 are variables, constants, or definite descriptions, $\Delta_1 = \Delta_2$ is true in γ just in case Δ_1 and Δ_2 have the same extension in γ. The intuitions underlying the mentioned portion of the tradition can be recaptured, however, by a judicious

use of absolute concepts in the sense of paragraph 15, below: if x and y both fall under the same absolute (substance) concept, then they are either necessarily identical or necessarily distinct.

11. Nonextensional predication. We come now to Bressan's new semantics for predication, which I take to be the cardinal innovation in ML^{ν}. (Bressan tells me that he himself takes the notion of absoluteness of paragraph 15 to be more important and that, historically, his analysis of predication was designed to accommodate this notion.)

First some background: Every functor in everyone's case-intensional semantics is intensional in the trivial sense that the intension of the compound is a function of the intensions of the parts; indeed that, according to 5, above, is one of the three leading principles of case-intensional semantics. But let us further define a functor as extensional if, in each case γ, the extension in γ of the compound depends only on the extensions in γ of the parts. Then one should clearly expect negation (say) to be extensional, since the truth value in γ of $\sim p$ (not p) depends only on the truth value in γ of p; and one should equally expect necessity to be nonextensional since the truth value in γ of Np (necessarily p) does not depend just on the truth value in γ of p; instead in computing the truth value in γ of Np one must know the whole intension of p, not just its extension in γ.

So much for the background. Now the question about predication can be put as follows: should it be extensional like negation or nonextensional like necessity? Oddly enough, almost the entire tradition (save Scott in his "Advice on Modal Logic," in Philosophical Problems in Logic, ed. K. Lambert [Humanities Press, 1969]; and even he weakened in his postscript) has conspired either not to ask the question or to plump for purely extensional predication (see survey in [18]). And just here, I conjecture, is where types come in, for whether an attribute F has in (say) the real case a higher level attribute G should clearly depend in general not just on the extension in the real case of F, but on its whole intension. Example: whether or not F has the attribute of being contingent depends on its whole intension, for contingency of F means that the extension of F has some variation from case to case; therefore contingency of F cannot be determined by inspecting only the extension of F in the real case.

Hence attention to higher types calls for a more general analysis of predication than any available prior to ML^{ν}. So by uniformity one applies the same analysis to the lower levels, and then one sees by hindsight that this more general analysis is also useful at the lower levels.

Let me come at predication in another way. In extensional logic the meaning of a predicate is given by a "class," while in modal logic its meaning, in the sense of its intension, is given by an "attribute." Now we know a class can be thought of as a mapping from individuals into truth values, so that at first glance it looks as if we become properly case-intensional if we think of an attribute as a mapping from individuals (not into truth values but instead) into propositions. Since propositions sound so nonextensional, it makes the whole thing sound nonextensional. But in fact what has just been described is equivalent to insisting that all predication be extensional! In order to be really and truly nonextensional we must think of an attribute not as a mapping from individuals—remember, individuals are extensions—into propositions, but from individual concepts into propositions. And that is the step Bressan has taken. His set-theoretical way of putting it is for good and sufficient reasons somewhat different, but isomorphic to the above. With each (one place) predicate he associates a set of ordered pairs, each pair containing an individual concept as left entry and a case as right entry. Then, to see whether Fx is true in γ, one first finds the intension i of x (an individual concept) and determines whether or not the pair $\langle i, \gamma \rangle$ is in the set of pairs associated with F. And this is seen to depend not merely on the extension in γ of x, but on the whole intension of x.

So an attribute can be identified with a mapping from individual concepts into propositions, or equivalently with a subset of the Cartesian product of the set of individual concepts with the set of cases. A third equivalent way of looking at an attribute (the intension of a predicate) is to think of it as a mapping from the set of cases into the set of subsets of the set of individual concepts. This third way is useful for visualizing (and drawing pictures of) the extension in γ of a predicate; its extension in γ is always a set of individual concepts (not individuals!).

Note incidentally that it is perfectly sensible to talk about the extensions in γ of not only pieces of notation but also of their intensions; e.g., "the extension in γ of an attribute" or "the extension in γ of an individual concept."

Some attributes F are of course extensional, which in accordance with the discussion above is just to say that the truth value in γ of the result of applying F to an individual concept i is a function of (just) the extension in γ of *i*. That *F* is extensional can be said *in* ML^{ν} by an appropriate use of identity: for each *x*, *F* holds of *x* just in case it holds of everything identical to *x*. Note that *F* might well be extensional in one case and not others; so to say without relativizing to a case simply that *F* is extensional should presumably mean either that *F* is everywhere extensional or that it is extensional in the real case; a distinction not possible prior to Bressan. What Bressan shows in his discussions of physics and other examples is that employment of *non*extensional attributes is altogether essential. For example, he shows that one cannot make sense of physics without the *non*-extensional concepts of "mass point" and "real number." (He points out that one also needs extensional versions of these concepts.) So much the worse for those who would have us believe that every "genuine" attribute must be extensional.

12. *Lambda abstraction*. Given a context (. . . *x* . . .) expressing an attribute, Church's lambda operator (see his *Introduction to Mathematical Logic*) allows one to make up a name (λ*x*) (. . . *x* . . .) of that attribute (and similarly for functions—we disregard these). Since the context *NFx* is obviously nonextensional, a logic limited to naming only extensional attributes cannot allow unfettered formation of lambda abstracts, for (λ*x*) (*NFx*) would have to name a nonextensional attribute. Furthermore, a logic which treats variables, constants, and definite descriptions differently with respect to the rule of universal instantiation will also treat them differently with respect to their interaction with lambda abstraction; such a logic will not have the full introduction-elimination equivalence according to which for arbitrary Δ (of the right type), Δ has the property (λ*x*) (. . . *x* . . .) just in case (. . . Δ . . .) (see [35]). But just because Bressan allows nonextensional attributes and treats all terms uniformly, he is able to introduce full-blooded (although of course type-theoretical) lambda abstraction without any annoying restrictions: he gives us both formation of lambda abstracts from arbitrary contexts, and the introduction-elimination equivalence for arbitrary terms. I should add that two more features of ML^{ν} conspire to make this possible: the unrestricted range of the variables, and the presence of higher types so that in the ontology there will be entities for the lambda abstracts to name.

Let me append that many, many other kinds of *useful* notations are introduced by way of definition, a sure sign of a fruitful formal systematization.

13. Definite descriptions. One of the finest features of ML^{ν} is its unparalleled treatment of definite descriptions; the theory is just right—if what one wants is a simple, transparent, and manageable theory. It is based on Frege's device of allowing definite descriptions like "the man at the corner of State and Madison" to denote some arbitrary entity—"the nonexisting object" if you like—whenever we might otherwise say that it does not denote at all because either there are not enough or too many entities (e.g. men) satisfying the descriptive condition (e.g. being at the corner of State and Madison). Of course this gimmick of Frege's is not close to English and should be eschewed by those who are trying to copy natural language, but its introduction is quite obviously simplifying if one wants to do serious work with the logic, and there is no real trouble connecting it with English via appropriate correspondence rules—to the extent that we are clear on English.

Given the foregoing, plus the desideratum that the construction of the definite description $(\imath \underline{x})(\ldots \underline{x} \ldots)$—the sole $\underline{x}$ such that $(\ldots \underline{x} \ldots)$—should be extensional, one can pretty well figure out what the semantics should be; just this naturalness is one of the blessings of Bressan's construction. Let us see how this works out in practice. We are given a condition $(\ldots \underline{x} \ldots)$ which we may think of as expressing a certain attribute as its intension, and we must construct the intension of the term $(\imath \underline{x})(\ldots \underline{x} \ldots)$. By case-intensional considerations, the latter must be a function of the former; and by the desideratum of extensionality, the extension of $(\imath \underline{x})(\ldots \underline{x} \ldots)$ in a case γ must be a function of the extension of $(\ldots \underline{x} \ldots)$ in γ. If we say what this function is, we shall be done, because by case-intensional semantics the intension of $(\imath \underline{x})(\ldots \underline{x} \ldots)$ is or can be thought to be just the pattern of its extensions as γ varies. Now the extension in γ of $(\ldots \underline{x} \ldots)$—call it $\underline{E}$—is, by the end of 11, a set of individual concepts, and we must come up with an individual (not individual concept) as the extension in γ of $(\imath \underline{x})(\ldots \underline{x} \ldots)$. Which individual? The only plausible suggestion is to throw away those portions of $\underline{E}$ which deal with cases other than γ. Let us therefore define $\underline{E}'$ as the set of those individuals which are the extensions in γ of some individual concept in $\underline{E}$. (Note that by concentrating on $\underline{E}'$ we will make the definite description in some sense doubly extensional, since its value in γ will depend on the extensions in γ of the members of the extension in γ of $(\ldots \underline{x} \ldots)$.) Now we need to define the extension in γ of $(\imath \underline{x})(\ldots \underline{x} \ldots)$ in terms of the set $\underline{E}'$ of individuals. How it is to be done is perfectly straightforward: if $\underline{E}'$ has exactly one individual as member, then we let that individual be the extension in γ of $(\imath \underline{x})(\ldots \underline{x} \ldots)$. In other words, if $\underline{a}$ is the sole individual in the domain $\underline{D}$ such that for at least one individual concept $\underline{i}$, $\underline{a}$ is the extension in γ of $\underline{i}$ and $(\ldots \underline{x} \ldots)$ comes out true in γ for the value $\underline{i}$ of $\underline{x}$, then the extension of $(\imath \underline{x})(\ldots \underline{x} \ldots)$ in γ is precisely $\underline{a}$. But if $\underline{E}'$ is either empty or contains more than one member, we use the Fregean device: the extension in γ of $(\imath \underline{x})(\ldots \underline{x} \ldots)$ is defined to be the "nonexisting object" or some arbitrarily (but pre-) selected individual in the domain. What else?

Of course the above sounds complicated, and in a way it is; but no more complicated than it has to be. It turns out it works properly (easily) only when the usual existence and uniqueness assumptions are available—a limitation shared with the accounts of definite descriptions in extensional logic—and also only when the context $(\ldots \underline{x} \ldots)$ is extensional (e.g., being at the corner of State and Madison). But in the third memoir Bressan shows how to introduce other description operators, of a nonextensional sort, which are useful also in connection with nonextensional contexts. These latter are exceedingly interesting, and I recommend their study.

III. New Directions

Of the new directions in which Bressan points, I should like to mention only five: Arithmetic, Absolute concepts, Physics, Homely puzzles, and Internal cases.

14. Arithmetic. It is the availability of higher types that permits Bressan to develop modal arithmetic in what is obviously the right way: one just takes over the Frege-Russell definitions of the natural numbers as they stand; e.g., the number 1 is the (higher type) attribute which an attribute has if it applies to exactly one thing. On this basis Bressan erects a full theory of modal arithmetic, including definitions not only of each natural number (they turn out to be extensional) but also of the attribute of being a natural number (it is quite importantly nonextensional, and indeed absolute in the sense of 15), below, and the usual arithmetic operations. With arithmetic built into the modal language rather than added ad hoc, one can now sensibly ask whether or not the number 9 is necessarily greater than 7, or whether, in a natural formalization of astronomy, it is necessary that the number of planets exceed 7. And it goes without saying that the type-theoretical machinery allows the construction of the rationals and the reals in the usual ways, after which one can investigate the subtle interplay between modal and nonmodal ideas in a natural setting.

15. *Absolute attributes.* An attribute *F* is said to be *absolute* if it is both modally constant (applies to same things in every case) and in every case modally separated (whenever *x* and *y* both fall under *F*, then *x* and *y* are either necessarily identical or necessarily distinct). The articulation and deployment of this notion is extremely important to Bressan's enterprise; he rightly claims it is a significant enrichment of the concept of "substance." Bressan both explains and illustrates absoluteness at length, and I shall not say much about it beyond mentioning that it does the work which others do by talk of "identity through possible worlds." But Bressan's conceptions are both more general and more flexible than those of others; and they go more deeply to the root of the matter. A single illustration: Bressan shows that it is crucial that the concept of real number be absolute, and by parity of reasoning also the concept of natural number. But it is not required and indeed it is not true that the number 1 be "identical through worlds" in the sense of having the same extension in each case. And in general what is required in actual reasoning is information about the substance *concept* ("secondary substance"—Aristotle) rather than information about the (primary) substances themselves. Let me add only my belief that Bressan's notion of absoluteness is the proper foundation for an adequate understanding of essentialism, essential predication, and the *de dicto*/*de re* distinction.

16. *Physics.* One of the most exciting aspects of Bressan's work is its potential in providing a rigorous foundation for certain parts of physics. One can tell from the book that the task of getting clear on the world guides Bressan's work. On more than one occasion he has explained to me that a certain suggestion would not do because of its lack of fit with the conceptualization of physics. This volume, of course, does not itself contain an axiomatization of physics (Bressan has written on this elsewhere in a somewhat different form; see [4]), but the outline is there, with a few details and a cornucopia of motivating considerations. In particular it becomes clear how the details of the language ML^{ν} are related to the problems one faces in formalizing physics.

17. *Homely puzzles.* Bressan sheds considerable illumination on such homely puzzles as how to make sense out of "If I had more money than I do I would buy a car." All of these turn out to involve absolute concepts in interesting ways, and many of them to require reference to the real world as in 18, below. Bressan's illustrations along these lines are invariably enlightening and almost by themselves worth the price of admission; but to my mind

it is the potentiality of the language for real and extended formal work as in 16, above, which is responsible for its true worth.

18. Internal cases. By an ingenious but by no means farfetched type-theoretical construction, Bressan brings the cases into the language ML^{ν} itself. Or better, Bressan shows that the cases can be already found in ML^{ν} without adding anything at all. (It goes without saying that the cases in ML^{ν} are type-theoretical constructs serving as surrogates for the cases themselves.) Because ML^{ν} can talk about its own cases, it can get the effect of saying that p is true in case γ by one of its own sentences, and by means of quantification can start counting the number of cases in which p holds (D. Kaplan, "S5 with multiple possibility," Journal of Symbolic Logic 35 [1970]: 355–56). One could also take subsets of internal cases as propositions, and quantify thereover (ibid.). Or, as Bressan shows, one can use the presence of the cases in ML^{ν} to impose relations on them à la Kripke, and hence define other sorts of modalities within ML^{ν} itself. Their presence in ML^{ν} is also an important ingredient of Bressan's relative completeness-and-consistency proof for ML^{ν}, and is in other ways of great mathematical interest. Of chief philosophical interest, however, are (a) Bressan's introduction, by means of the internal cases, of a case-dependent description operator, and (b) Bressan's extension of ML^{ν} by adding a constant denoting (the surrogate for) the real world. The developments based on these considerations make possible, I think for the first time, a formal evaluation of arguments like Plantinga's ("De Re et De Dicto," Noûs 3:235–58) which involve mention of the real world. For since the cases are already there in ML^{ν}, one knows that nothing ad hoc is going on in one's talk about them; they are the way they are. And exploitation of this apparatus should, I think, throw considerable light on whether there can be a truth theory for modal logic which satisfies Tarski's Convention T in the way in which Davidson and others would wish it to (see J. Wallace, "On the frame of reference," Synthese 22:117–50). Nor do I think that either these examples or those provided by Bressan constitute more than the tip of the iceberg of possible developments in philosophical logic utilizing internal cases.

Upshot: extraordinary riches.

Nuel D. Belnap, Jr.

Pittsburgh, Pennsylvania
May 1971

Preface

This monograph consists of three memoirs:

1. a semantical analysis of a general modal language ML^{ν} with an application to an axiomatization problem concerning classical mechanics;
2. a modal ν-sorted logical calculus MC^{ν} valid in the general modal language ML^{ν};
3. elementary possible cases, intensional descriptions, and completeness in the modal calculus MC^{ν}.

They were conceived and written as parts of the same work. The first belongs to semantics, and the other two belong to syntax.

The goal of the first two Memoirs is to construct an interpreted modal calculus, MC^{ν}, which on the one hand is general—in particular it has (1) constants and variables (of logical types) of infinitely many levels, (2) a contingent identity, and (3) a description operator, ι, such that $(\iota \underline{x})\underline{p}$ is meaningful for every matrix $\underline{p}$ and every variable $\underline{x}$—and on the other hand allows us to solve an axiomatization problem concerning the foundations of classical mechanics (according to E. Mach and P. Painlevé).

In Meaning and Necessity (1947)—see [12] in the Bibliography—Carnap says that "the problem of whether or not it is possible to combine modalities and variables in such a way that the customary inferences of the logic of quantification—in particular, specification and existential generalization—remain valid is, of course of greatest importance (pp. 195–96). The calculus MC^{ν} solves this problem, unlike most of the existing modal interpreted calculi, which often contain only variables of low levels—cf. e.g. Montague's "Pragmatics and Intensional Logic,"—see [28].[1] This makes MC^{ν}

* In item [18] in the bibliography, p. 210, G. E. Hughes and M. J. Cresswell say that "the topics of identity and description are among the most difficult in modal logic, and in the present state of the subject are still full of obscurities and unsolved problems," and that this situation affects the treatment of classes.

Incidentally in MC^{ν} all these problems have a solution, and to a large extent this is due to the fact that the semantical theory of attributes in MC^{ν} is developed from a more general point of view than those considered in [18].

efficient in connection with both physical axiomatization problems and other problems of traditional modal logic [Ns53, 55].

In the semantical system set up for the modal language ML^{ν} in the first memoir, a strong axiom [AS12.19] which has no extensional analogue holds. Incidentally, similar axioms hold in many other modal semantical systems; however as far as I know they were never explicitly stated.

The strong axiom mentioned above, which is hardly considered in the second memoir, has important consequences that are developed in the third memoir. Among other things, they allow us to prove a theorem of relative completeness for MC^{ν} [Ns63, 64], and to define an intensional description operator, ι_{u}, in contrast with the primitive one, ℩, which has a rather extensional character.

The introduction of a constant, ς, related to the real (possible) world and satisfying a simple axiom [N52], turns MC^{ν} into a calculus, MC^{ν}_{ς}, which proves useful in solving certain philosophical puzzles of traditional modal logic considered by Thomason and Stalnaker, as well as some generalizations of them [N55].

The author, an Italian mathematical physicist, is indebted to various professors of logic, in particular to R. Carnap, who strongly encouraged him to write this work and N. Belnap, Jr., who helped him to present this volume (and recommended a great deal of literature), as is said in more detail in the source footnote to the first memoir. In Mr. Belnap's fall 1970 seminar a prepublication version of the monograph was used as the text. His students, in particular I. Kvart, A. Gupta, J. Norman, J. Broido, and T. Siciliano, also deserve thanks for numerous improvements, most of which are linguistic, and because the first four of these students corrected a flaw in each of the following formulas, respectively: Def. 27.3, Theor. 51.1(IV), $(43)_2$ in Memoir 1, and Theor. 52.2(V).

The author wishes to thank Mrs. Jane Isay for her efforts in proposing the present monograph to the Yale University Press for publication, and the copyeditor, Mrs. Judy Maclay, whose work on the monograph was unusually difficult; in particular she offered many useful linguistic changes.

I am deeply grateful to my wife for her close participation in the difficult preparation of this book, particularly for typing, from 1964 on, several versions on the basis of my manuscripts, which are very difficult to decipher, and in addition to caring for our four children.

The research that led to this monograph was supported in part by the Air Force Office of Scientific Research under Grant AFOSR 728-66.

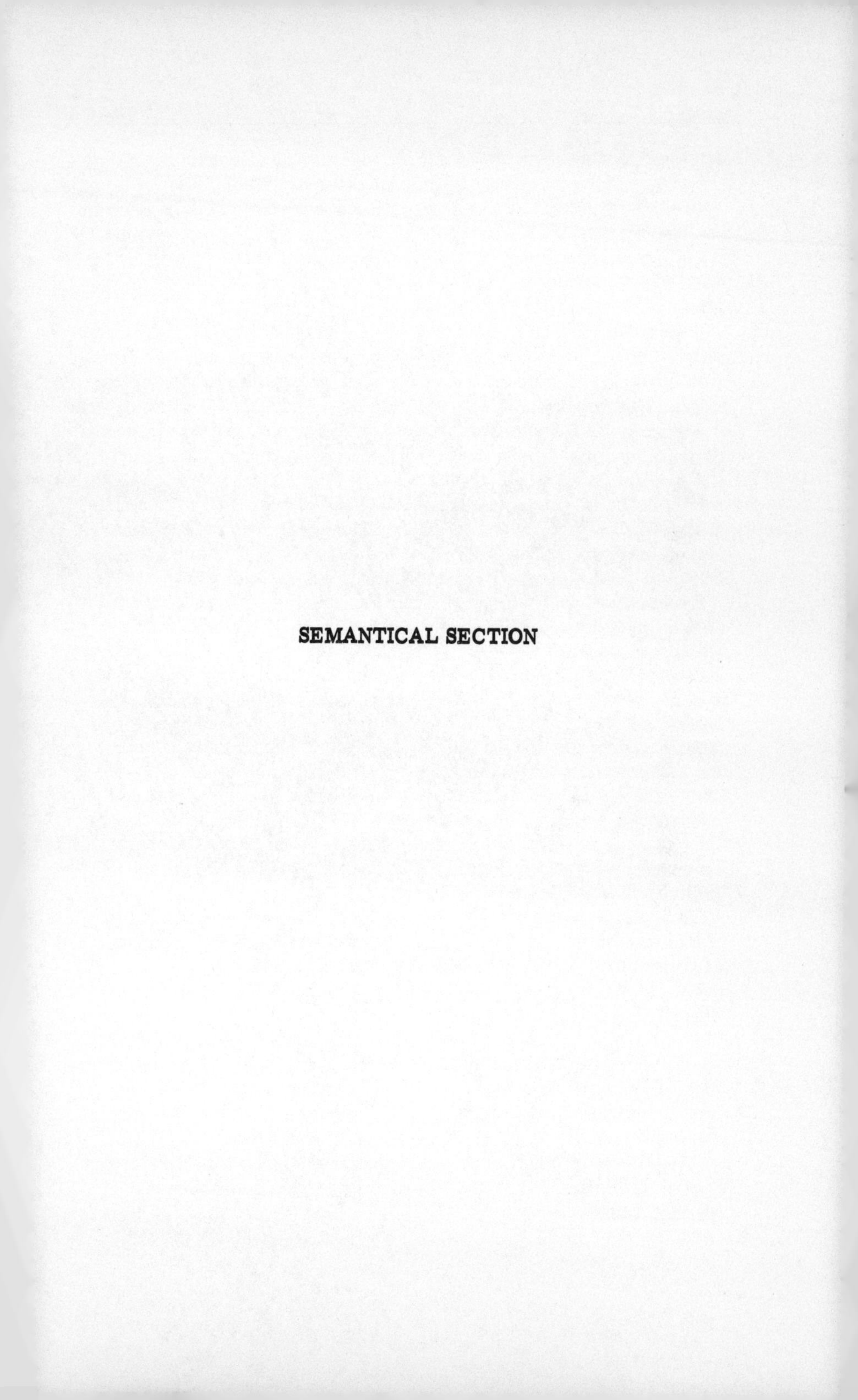

SEMANTICAL SECTION

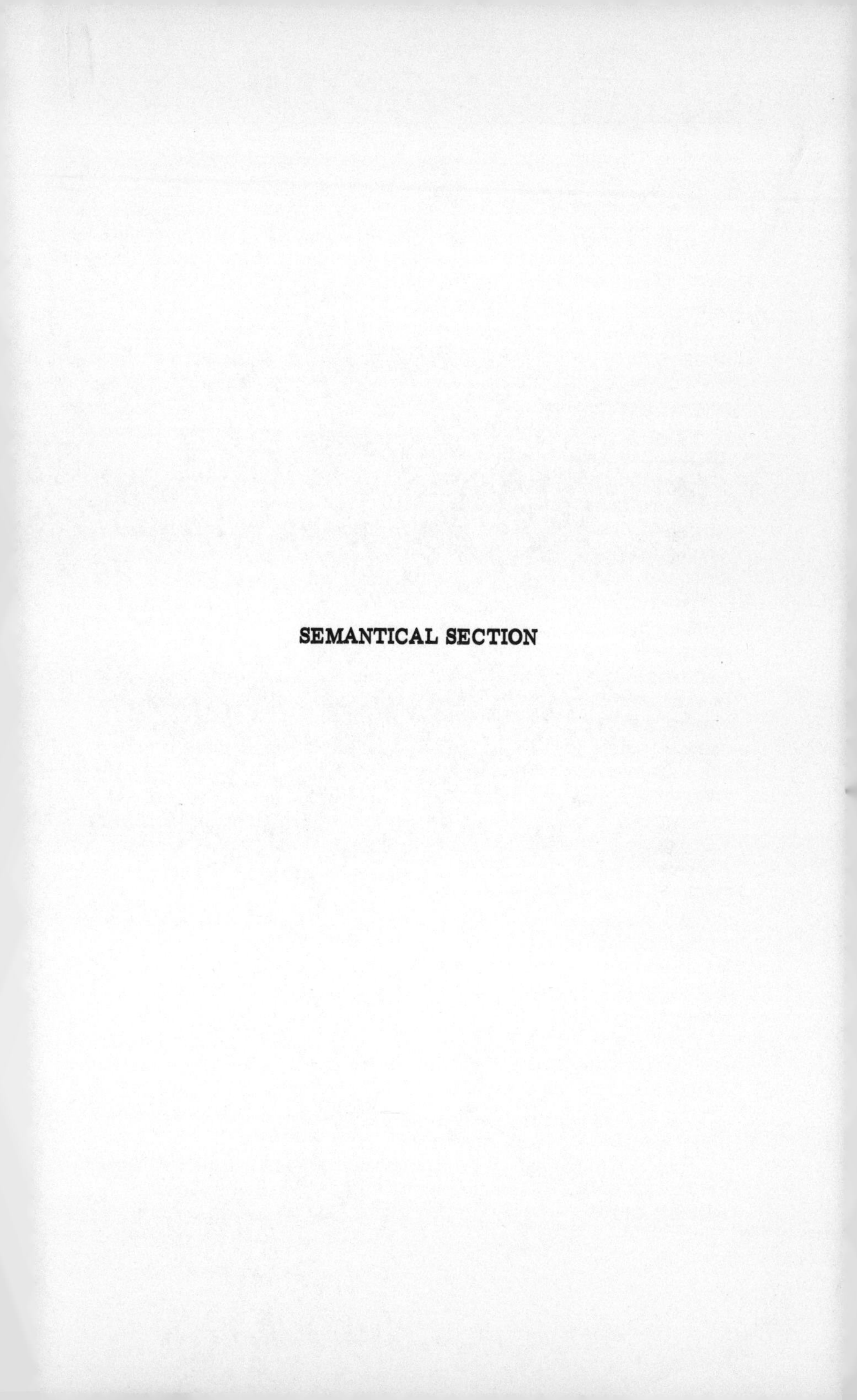

SEMANTICAL SECTION

Memoir 1: A Semantical Analysis of a General Modal Language ML^{ν} with an Application to an Axiomatization Problem concerning Classical Mechanics

N1. Introduction*

First we briefly describe the contents of this memoir and state why it was written. Second, the contents of part 1 are described in more detail. The contents of part 2 will be described in more detail in N17.

In part 1 a semantical analysis of a general ν-sorted modal language ML^{ν} is done using extensional means[1]—more precisely, suitable quasi intensions—in accordance with Carnap's method of extension and intension (see [12] and [13]). In addition, a set of axioms is proposed for a general modal calculus MC^{ν}[2] and its validity in ML^{ν} is shown.

In part 2 we introduce in ML^{ν} the concept of absolute attribute, and according to it and the results of part 1 we give a solution, based on ML^{ν}, to the problem of casting Mach's definition of mass into a rigorous form.[3] This problem includes the main difficulties arising from casting into a rigorous form the foundations of classical particle mechanics according to Painlevé—see [29]—who follows certain ideas of Mach and Kirchhoff and particularly accepts Mach's definition of mass.[4] We discuss our new solution to that problem and show some of its advantages with respect to previous solutions, based on extensional logic either directly—cf. [15] and [16]—or indirectly—cf. [4].[5]

In part 2 we also explain why, in accordance with a critical remark by Rosser—see N21 and fnn. 66, 67—for most systems of classical particle mechanics, the theories directly based on extensional logic are incompatible with nearly all of the initial conditions that can hold for them [N21]. We also show that to avoid this incompatibility it is natural to look for means enabling us to speak about possible facts explicitly. This aim is substantially reached in [4] on the basis of extensional logic and through a device suggested by certain extensional semantical rules used by Carnap for analyzing some modal languages. Let us remark that the language used in [4] is unusual,[6] and that in particular it differs from the ordinary modal language used by Painlevé in his well-known work [29] and followed by certain well-known textbooks, such as [33].[7]

The above considerations are sufficient, I think, to induce us to try and set up a general theory of modal logic such as the one for ML^{ν} in order to be able, among other things, to formalize in a direct way the intuitive modal language used by Painlevé, and to present in a rather conventional way the rigorous and general theory of foundations of classical particle mechanics according to Painlevé that was set up in [4]—cf. fnn. 7 and 53. Another motive for setting up such a theory is that it generalizes certain interesting theories of Carnap—see [12] and [13].

In [12] Carnap performs a semantical analysis of certain modal languages by means of his method of extension and intension—according to which every designator has both an <u>extension</u> and an <u>intension</u>—in contrast to the method of the name relation which had been used for instance by Frege, Quine, and Church. On pp. 181 and 182 Carnap represents the intensions by certain extensional entities which are now sometimes called <u>quasi intensions</u>

or QIs. By means of these QIs he also lays down extensional semantical rules for his modal languages—see §4.

In [13] Carnap briefly mentions two modal languages, $\underline{L}_2$ and $\underline{L}_3$. In $\underline{L}_2$ the modal sign $\underline{N}$ is not admitted within the scope of the iota or the lambda operators,[8] so that $\underline{L}_2$ admits only extensional properties [Def. 6.11]—cf. [13], III, 9.II, p. 892.[9]

In part 1, for $\nu = 1,2, \ldots$ we consider a ν-sorted modal language ML^ν of a general kind. One feature of ML^ν is that the necessity sign $\underline{N}$ is admitted within the scope of the lambda and iota operators;[10] another is that in ML^ν the levels of the logical types include all natural numbers.

In conformity with Carnap's work [13] we say that the extension of the designator Δ (in ML^ν) in the possible case (or state) γ is what Δ denotes if γ takes place; furthermore, we consider the intension of Δ to be known if and only if, for every possible case γ, the extension of Δ in γ is known.

Besides the formation rules [N3] determining the wffs, or typed expressions, we set up for ML^ν suitable QIs (discussed in N7) and corresponding semantical rules of an extensional kind—cf. Ns7, 11. So we construct a (particular) quasi-intensional semantical system for ML^ν, which characterizes a (modal) way of understanding ML^ν.

Since the language ML^ν conforms with Carnap's method of extension and intension, it fulfills the following requirements:[11]

(a) In connection with every logical type $\underline{t}$, only one kind of variable and one kind of constant is used; in particular, no distinction between class variables and property variables is made.

(b) In every possible case (or state) γ (and at every model and value assignment, i.e. whichever the QIs assigned to the variables and constants may be), every wff is meaningful, i.e. it is a designator.

(c) Modalities can be combined with quantification.[12]

(d) In ML^{ν} the usual inferences of the lower predicate calculus remain valid, particularly specification $[(\underline{x})\ \Phi(\underline{x}) \supset \Phi(\underline{a})]$ and existential generalization $[\Phi(\underline{a}) \supset (\exists x)\ \Phi(\underline{x})]$, cf. [12], §44, p. 196.

(e) The axioms for Lewis's propositional calculus S5 hold.

I believe that the above requirements fulfilled by ML^{ν} are advantageous; the more so as the language ML^{ν} has been constructed to be applied to mechanics and physics. More precisely, requirements (b) and (c) provide for, I think, good possibilities of application. Requirements (c), (d), and (e) serve for the construction of an efficient and essentially modal (logical) calculus MC^{ν} valid in ML^{ν}. Such a calculus is dealt with in Memoir 2 in this book. Requirement (a) conforms to the widespread tendency to reduce the primitive concepts and is, I believe, very useful, especially with regard to the relative simplicity of the calculus MC^{ν} mentioned above. As far as I know, ML^{ν} fulfills (b) in a stronger form than previous effectively interpreted modal languages do in that in ML^{ν} suitable analogues of the extensional axioms for quantification, identity, and the iota operator (and the lambda operator, cf. fn. 10) hold. These analogues, whose logical truth in ML^{ν} is shown in N12, are precisely related to Rosser's theory of extensional logic.[13]

As can be predicted, the axioms for identity and the iota operator proposed in MC^{ν} are different, in their ensemble, from their extensional analogues (as well as with regard to their number).[14]

It is interesting, I think, to show that in ML^{ν} a certain (essentially modal) axiom [AS12.19], which is new, as far as I know, holds.[15] To prove that the axioms proposed for MC^{ν} and mentioned above hold in ML^{ν} is useful for exhibiting essential features of ML^{ν}.

Again with the purpose of illustrating ML^{ν}, in N14 we present (without any explicit proof) some logically true matrices in ML^{ν} that have no extensional analogues and which concern the iota and lambda operators, existential quantification, and the uniqueness operator;[16] we also prove some nonevident fallacies by means of examples.

The QIs built for ML^{ν} [N6] are more complex than those indicated by Carnap in [12] and [13].[17] This is due to requirement (b) and the fact that in ML^{ν} the sign $\underline{N}$ is admitted within the scope of the lambda operator. I believe that an effective simplification of the above QIs is not compatible with the preceding requirements.

As to our semantical rules (which are extensional in that they assign a QI to every wff), let us remark that those for full sentences of predicates and functors [N8] are suggested by requirements (a) to (e)—especially (b) and (c)—and by the general way in which the sign $\underline{N}$ is used in the language ML^{ν}. Our semantical rules for descriptions—based on a suitable existence and uniqueness theorem [Theor. 11.1]—are suggested by requirement (b) and the condition that descriptions corresponding to extensional matrices fulfill the same requirements as do those theories of extensional logic in which Frege's method is accepted—e.g. Rosser, [32].[18]

Our semantical rules for molecular matrices deal with the necessity sign $\underline{N}$ as though it were a universal quantifier $(\forall \underline{x})$—cf.

[30]. This known analogy between N and $(\forall x)$ is emphasized in presenting the axioms proposed for the modal calculus MC^{ν}.[19]

In N14 we extend Carnap's concept of an L-determined intension—see [12], p. 88—to ML^{ν} and show its usefulness for representing extensions for ML^{ν}; cf. fn. 17.

Our semantical rules for ML^{ν} characterize a translation in a strong sense of the ν-sorted modal language ML^{ν} into a $(\nu+1)$-sorted extensional language $EL^{\nu+1}$.[20] This translation is interesting in the first place because it constitutes a proof of the so-called thesis of extensionality—see [12], p. 141—in connection with the general modal language ML^{ν}, particularly since it is a translation in a strong sense; the possibility of such a result is not, according to an assertion by Carnap, obvious.[21] In the second place, ML^{ν} will be applied to foundations of mechanics [part 2]. There is a rather widespread repulsion among physicists against the use of modal languages in axiomatic treatments, and the translation of ML^{ν} into extensional language $EL^{\nu+1}$ may reduce the antipathy of physicists to ML^{ν}.[22]

What can be said in ML^{ν} is also expressible in $EL^{\nu+1}$; but in conformity with a statement by Carnap—see [12], p. 142—the expressions in ML^{ν} are simpler than their translations into $EL^{\nu+1}$, as are the corresponding deductive manipulations. I think this will be apparent in Memoirs 2 and 3 in this book. Thus ML^{ν} is technically more efficient than $EL^{\nu+1}$.

Since we believe that the above translation is interesting and useful e.g. for proving a certain completeness theorem on MC^{ν}—cf. fn. 15—we have made it explicit [chap. 3]. Moreover a semantical theorem [Theor. 16.1] relates the meaning of every designator Δ in ML^{ν} to the one of its translation Δ. This theorem, together with the semantical rules for $EL^{\nu+1}$, provides in a new way the semantical rules for ML^{ν}.

Lastly let us note that the physical possibility of a proposition p is sometimes identified with the logical compatibility of p with the physical axioms,[23] but that such a procedure would be circular in the foundations of mechanics according to Painlevé, because in such foundations possibility appears explicitly in the physical axioms themselves. Therefore, in dealing with such a subject—see e.g. N19—by the physical possibility $\Diamond p$ of p we mean something to be tested by experience, for example something like an idealized technical possibility of realizing p up to a space-time translation [N5].[24] The semantical system for ML^{ν} fits both these concepts of possibility.[25]

PART I

AN EXTENSIONAL SEMANTICAL ANALYSIS OF A GENERAL ν-SORTED MODAL LANGUAGE ML^ν

1

Formation Rules for ML^ν and EL^ν

N2. The type system.

Unless otherwise noted, the letters $\underline{m}$, $\underline{n}$, $\underline{r}$, $\underline{s}$ will be used as metalinguistic variables ranging over positive integers. By $\langle \underline{a}_1, \ldots, \underline{a}_{\underline{n}} \rangle$ and $\{\underline{a}_1, \ldots, \underline{a}_{\underline{n}}\}$ we mean the ordered $\underline{n}$-tuple and the unordered $\underline{n}$-tuple respectively, formed from $\underline{a}_1, \ldots, \underline{a}_{\underline{n}}$.

For $\nu = 1, 2, \ldots$ we shall consider two ν-sorted languages, ML^ν and EL^ν, ML^ν being modal and EL^ν being extensional.

In ML^ν there are various well-formed formulas, or wffs. They are either matrices or terms. The terms are of various kinds. There are ν sorts of individual terms, which we shall refer to as wffs of the types $1, \ldots \nu$. There are in addition various kinds of $\underline{n}$-ary predicates and $\underline{n}$-ary functors (consisting of one or more signs).

In order to describe these expressions and, more generally, to define ML^ν, we now introduce some entities to be used as the types of the wffs of ML^ν:

DEF. 2.1. We define $\underline{t}$ as a term type for ML^ν, briefly $\underline{t} \epsilon \tau^\nu$, recursively by the following conditions:

(a) If $\underline{t}$ is any of the integers $1, 2, \ldots, \nu$, then $\underline{t} \epsilon \tau^\nu$.

(b) If $\underline{t}_1, \ldots, \underline{t}_{\underline{n}} \epsilon \tau^\nu$ and $\underline{t}$ is the ordered ($\underline{n}$+1)-tuple $\langle \underline{t}_1, \ldots, \underline{t}_{\underline{n}}, 0 \rangle$, then $\underline{t} \epsilon \tau^\nu$.

(c) If $\underline{t}_0, \underline{t}_1, \ldots, \underline{t}_{\underline{n}} \epsilon \tau^\nu$ and $\underline{t} = \langle \underline{t}_1, \ldots, \underline{t}_{\underline{n}}, \underline{t}_0 \rangle$, then $\underline{t} \epsilon \tau^\nu$.

We call 0 the sentence type and we call the term types of the forms $\langle \underline{t}_1, \ldots, \underline{t}_{\underline{n}}, 0 \rangle$ and $\langle \underline{t}_1, \ldots, \underline{t}_{\underline{n}}, \underline{t}_0 \rangle$ with $\underline{t}_0 \neq 0$ relator types and functor types respectively.

We call a ν-type or a type for ML^ν either 0 or any term type for ML^ν. We also set

(1) $$\bar{\tau}^\nu = \tau^\nu \cup \{0\}.$$

In order to comply with Carnap's notations for types,[26] which are shorter than those introduced above, we set

(2) $$(\underline{t}_1, \ldots, \underline{t}_n) = \langle \underline{t}_1, \ldots, \underline{t}_n, 0\rangle,$$
$$(\underline{t}_1, \ldots, \underline{t}_n : t_0) = \langle \underline{t}_1, \ldots, \underline{t}_n, t_0\rangle \quad (\underline{t}_0, \underline{t}_1, \ldots, \underline{t}_n \in \tau^\nu).$$

Naturally, for $\underline{t} \in \bar{\tau}^\nu$ we define the level of t, briefly, lev[t], recursively by the following four conditions:

(3) $$\text{lev}[0] = 0, \quad \text{lev}[\underline{r}] = 0, \quad (\underline{r} = 1, \ldots, \nu),$$
$$\text{lev}[(\underline{t}_1, \ldots, \underline{t}_n)] = 1 + \max(\text{lev}[\underline{t}_1], \ldots, \text{lev}[\underline{t}_n]),$$
$$\text{lev}[(\underline{t}_1, \ldots, \underline{t}_n : t_0)] = 1 + \max(\text{lev}[\underline{t}_0], \ldots, \text{lev}[\underline{t}_n]).$$

N3. The modal ν-sorted language ML^ν and the extensional language EL^ν.

Let the signs of our modal language ML^ν be the connectives '$\sim$' (negation) and '$\wedge$' (conjunction), the all-sign '$\forall$', the equality sign '=', the necessity sign 'N', the reversed iota '℩', the left parenthesis '(', the right one ')', the comma ',', the variables $\underline{v}_{tn}$, and the constants $\underline{c}_{tn}$ ($t \in \tau^\nu$; $\underline{n} = 1, 2, \ldots$).

Obviously $\underline{v}_{tn}$ and $\underline{c}_{tn}$, thought of as functions of t and n, are one-to-one functions, and the intersections of the counterdomains of these functions with one another and with the nine-member set {'$\sim$', '$\wedge$', '$\forall$', 'N', '=', '℩', '(', ')', ','} are empty. We say that $\underline{v}_{tn}$ and $\underline{c}_{tn}$ are a variable and a constant, respectively, of type t (and index n).

We call formulas or expressions the n-tuples of signs ($\underline{n} = 1, 2, \ldots$). If $\Delta_1 = \langle \underline{s}_1, \ldots, \underline{s}_n\rangle$ and $\Delta_2 = \langle \underline{s}_{n+1}, \ldots, \underline{s}_p\rangle$ are two expressions, then we simply denote by $\Delta_1\Delta_2$ the p-tuple

$\langle \underline{s}_1, \ldots, \underline{s}_p \rangle$. Here we define the wffs (or typed expressions) in ML^ν:

DEF. 3.1. For $\underline{t} \in \bar{\tau}^\nu$ we define the expression "Δ has the type $\underline{t}$"—briefly $\Delta \in \mathscr{E}^\nu_{\underline{t}}$—recursively by the following conditions, where $\underline{n}$ is understood to run over the positive integers:

(f_1) $\underline{v}_{\underline{tn}} \in \mathscr{E}^\nu_{\underline{t}}$ and $\underline{c}_{\underline{tn}} \in \mathscr{E}^\nu_{\underline{t}}$ for $\underline{t} \in \tau^\nu$.

(f_2) If $\underline{t} \in \tau^\nu$, $\Delta_1, \Delta_2 \in \mathscr{E}^\nu_{\underline{t}}$, and Δ is $\Delta_1 = \Delta_2$, then $\Delta \in \mathscr{E}^\nu_0$.

(f_3) If $\underline{t}_1, \ldots, \underline{t}_n \in \tau^\nu$, $\Delta_1 \in \mathscr{E}^\nu_{\underline{t}_1}, \ldots, \Delta_{\underline{n}} \in \mathscr{E}^\nu_{\underline{t}_n}$, $\underline{R} \in \mathscr{E}^\nu_{(\underline{t}_1, \ldots, \underline{t}_n)}$, and Δ is $\underline{R}(\Delta_1, \ldots, \Delta_{\underline{n}})$, then $\Delta \in \mathscr{E}^\nu_0$.

(f_4) If $\underline{t}, \underline{t}_1, \ldots, \underline{t}_n \in \tau^\nu$, $\Delta_1 \in \mathscr{E}^\nu_1, \ldots, \Delta_{\underline{n}} \in \mathscr{E}^\nu_{\underline{t}_n}$, $\Phi \in \mathscr{E}^\nu_{(\underline{t}_1, \ldots, \underline{t}_n : \underline{t})}$, and Δ is $\Phi(\Delta_1, \ldots, \Delta_{\underline{n}})$, then $\Delta \in \mathscr{E}^\nu_{\underline{t}}$.

(f_5) If $\Delta_1 \in \mathscr{E}^\nu_0$ and Δ is $\sim\Delta_1$, then $\Delta \in \mathscr{E}^\nu_0$.

(f_6) If $\Delta_1, \Delta_2 \in \mathscr{E}^\nu_0$ and Δ is $(\Delta_1 \wedge \Delta_2)$, then $\Delta \in \mathscr{E}^\nu_0$.

(f_7) If $\Delta_1 \in \mathscr{E}^\nu_0$ and Δ is $(\forall \underline{v}_{\underline{tn}})\, \Delta_1$, then $\Delta \in \mathscr{E}^\nu_0$.

(f_8) If $\Delta_1 \in \mathscr{E}^\nu_0$ and Δ is $\underline{N}\Delta_1$, then $\Delta \in \mathscr{E}^\nu_0$.

(f_9) If $\underline{t} \in \tau^\nu$, $\Delta_1 \in \mathscr{E}^\nu_0$, and Δ is $(\imath \underline{v}_{\underline{tn}})\Delta_1$, then $\Delta \in \mathscr{E}^\nu_{\underline{t}}$.

We call the elements of $\mathscr{E}^\nu_0$ the matrices in ML^ν, and for $\underline{t} \in \tau^\nu$ we call the elements of the class $\mathscr{E}^\nu_{\underline{t}}$ the terms in ML^ν of type t. In particular we say that these terms are individual terms, relators, or functors, according to whether $\underline{t}$ is a nonnegative integer or has the form $(\underline{t}_1, \ldots, \underline{t}_n)$ or else the form $(\underline{t}_1, \ldots, \underline{t}_n : \underline{t}_0)$ where $\underline{t}_0, t_1, \ldots, \underline{t}_n \in \tau^\nu$.

The signs of ML^ν will be freely (autonomously) used, and the metalinguistic symbols $\underline{x}$, $\underline{y}$, $\underline{z}$, $\underline{x}_1, \ldots$ will be used as (not necessarily individual) variables but only in argument positions; F, G, H, F_1, . . . will be used as relator variables; and f, g, h, f_1, . . . will be used as functor variables, having at least one nonargument position in the expressions being considered. In metalinguistic (or abbreviating) definitions, the metalinguistic symbols will be used in the same positions but for terms that are not necessarily variables.

The signs of the extensional ν-sorted language EL^ν are defined as those of ML^ν except for N.

For every $t \epsilon \bar{\tau}^\nu$—cf. (1)—the expressions of type t in EL^ν are those elements of $\mathscr{E}_t^\nu$ where N does not occur. Let E_t^ν be the class formed by them. Clearly the function E_t^ν of t ($t \epsilon \bar{\tau}^\nu$) can be characterized by the analogue of Def. 3.1 obtained by omitting condition (f_8) from it.

N4. Some conventions. Church's lambda operator in ML^ν and EL^ν.

We presuppose that the connectives $\vee$, $\supset$, $\equiv$ for disjunction, material implications, and material equivalence, and the existence sign $\exists$ and the possibility sign $\Diamond$, are introduced in the usual way. Now we lay down some conventions on the all-sign $\forall$ by the following metalinguistic definitions where, as in the remainder of the monograph, the condition p, $q \epsilon \mathscr{E}_0^\nu$ is presupposed.

DEF. 4.1. $(x)p \equiv_D (\forall x)p$.

DEF. 4.2. $(\forall x_1, \ldots, x_n)p \equiv_D (x_1) \ldots (x_n)p$.

For example, Def. 4.1 is understood to state that if p is a matrix of ML^ν, then $(x)p$ is the same expression as $(\forall x)p$.

Here and throughout we use the rules for dropping parentheses stated by Carnap in [11], pp. 9, 132, except that we take $\wedge$ to be more cohesive than $\vee$. We assume that the cohesive powers of $\sim$, $(\forall x)$, $(\imath x)$, $\wedge$, $\vee$, $\supset$, and $\equiv$ decrease in the order written, and if no confusion arises, the sign $\wedge$ will be dropped. Hence if p, q, $r \in \mathscr{E}_0^\nu$, then we may write for instance:

(4) $pq \equiv p \wedge q$, $\quad p \wedge q \vee r \equiv (p \wedge q) \vee r$.

Obviously an occurrence of a variable x in the typed expression Δ is said to be bound in Δ if it occurs within the scope of either of the operators $(\forall x)$ or $(\imath x)$; otherwise x is said to be free in Δ. If no occurrence of a variable is free in Δ, then Δ is said to be closed. Closed matrices are called sentences.

Let the substitution of the term Δ for the (free occurrences of the) variable x in the matrix p cause no confusion of bound variables—cf. [32], p. 209. Then we say that Δ is free for x in p. We often write e.g. $\Phi(x)$—or $\Phi(x, \Delta)$—for p, and $\Phi(\Delta)$—or $\Phi(\Delta, \Delta)$—for the result of the substitution of Δ for x in p, understanding that Δ is free for x in $\Phi(x, \Delta)$.

DEF. 4.3. A matrix p in ML^ν will be called modally closed if there are n matrices $p_1, \ldots, p_n$ such that p is constructed from $Np_1, \ldots, Np_n$ by means of $\sim$, $\wedge$, $(\forall x)$, and N. (N may be called the modal universal quantifier.)

Hence $(\exists x)Np_1 \vee Np_2$, for instance, is modally closed.

In accordance both with the way Rosser introduces the class notation $\hat{x}p$ in [32], p. 219—corresponding to Church's notation $(\lambda x)p$—and with the customary convention of considering n-ary functions as $(n+1)$-ary relations, let us introduce the lambda sign λ in connection with matrices and terms by the following metalinguistic definitions:

DEF. 4.4. $(\lambda \underline{x}_1, \ldots, \underline{x}_n)\underline{p} \equiv_D (\imath F)(\forall \underline{x}_1, \ldots, \underline{x}_n)[\underline{F}(\underline{x}_1, \ldots, \underline{x}_n) \equiv p]$,

DEF. 4.5. $(\lambda \underline{x}_1, \ldots, \underline{x}_n)\Delta \equiv_D (\imath f)(\forall \underline{x}_1, \ldots, \underline{x}_n)[f(x_1, \ldots, \underline{x}_n) = \Delta]$,

where (1) $\underline{p} \in \mathscr{E}_0^\nu$ and $\Delta \in \mathscr{E}_{\underline{t}}^\nu$ with $\underline{t} \in \tau^\nu$, (2) $\underline{x}_1, \ldots, \underline{x}_n$, $\underline{F}$, and $\underline{f}$ are distinct variables of the respective types $\underline{t}_1, \ldots, \underline{t}_n$, $(\underline{t}_1, \ldots, \underline{t}_n)$, and $(\underline{t}_1, \ldots, \underline{t}_n : \underline{t})$, and (3) on asserting Defs. 4.4 and 4.5 $\underline{F}$ [$\underline{f}$] is understood to be the variable of the type $(\underline{t}_1, \ldots, \underline{t}_n)$ [type $(\underline{t}_1, \ldots, \underline{t}_n : \underline{t}_0)$], which does not occur free in $\underline{p}$ [Δ] and has the smallest index (among such variables).

We accept the following common abbreviating definition:

DEF. 4.6. $\underline{x}_1, \ldots, \underline{x}_n \in \underline{F} \equiv_D \underline{F}(\underline{x}_1) \wedge \underline{F}(\underline{x}_2) \wedge \ldots \wedge \underline{F}(\underline{x}_n)$ where $\underline{x}_1, \ldots, \underline{x}_n$ are terms of type $\underline{t}$ and $\underline{F}$ is a term of type $(\underline{t})$.

2
Semantics for ML^ν

N5. On possibility and how we shall deal with it.

Modal languages enable us to speak about the possible truth of propositions. In order to study (more or less general) languages of this kind and to lay down semantical rules for them in an extensional metalanguage—which is practically equivalent to translating them into an extensional language—various authors have introduced the idea of possible states—briefly the states of the universe or some (e.g. linguistic) equivalents of them—see [39], [12], p. 9, [13], p. 891, and [13], p. 900.

Possible states could be represented by the admissible models of an extensional axiomatic theory $\mathscr{T}^{(e)}$, with respect to a given universe, as in [13]. In our semantical system for ML^{ν}, however, we simply use a class Γ of elementary possible cases without specifying their logical structure. We assume—in accordance with [12] and [13]—that these cases are exhaustive and mutually incompatible, and that they are elementary in the sense that no proper subcase of any of them exists—cf. N47.

In the case of physical theories the assumption that Γ is formed by the admissible models of $\mathscr{T}^{(e)}$ complies with the view that the physical possibility of a sentence p is its logical compatibility with the physical axioms. However, this view is not satisfying (it looks circular) when one is willing to let (physical) possibility occur within the physical axioms themselves, as happens for instance in Painlevé's axiomatization of classical mechanics—see fn. 52, and N21—and we want our theory of modal logic to fit such axiomatizations. To this purpose, we may assume that any sentence p is physically possible if, up to a suitable large space-time translation (relative to a given inertial space), we are able to realize p in a (suitable ideal) laboratory—cf. fn. 24.

In our semantical theory for ML^{ν} we assume for Γ only that Γ has more than one element. Therefore our theory may comply with any of the preceding views on possibility. In the remainder of the monograph we shall say that γ is a Γ-case when $\gamma \in \Gamma$.

In this section a rough intuitive concept of possible case is being referred to. The following explanations are meant to show a particular approach to the notion of a possible case, though not, however, one that is necessarily presupposed in this monograph.

Let K be a satisfiable class of sentences (or closed matrices) in the extensional theory $\mathscr{T}^{(e)}$. Let $\gamma = \hat{\gamma}(\underline{K})$ be the case where all matrices in K hold (so that their holding is to be considered as [strictly] equivalent to the holding of the case γ). Then we may say that γ is a possible case.

Let $\gamma = \hat{\gamma}(K)$ and $\gamma' = \hat{\gamma}(K')$ be possible cases and let $\hat{\gamma}(K \cup K')$ not be such a case. Then γ and γ' can be said to be incompatible.

The possible cases $\hat{\gamma}(K_i)$ ($i = 1, \ldots, n$) are said to be exhaustive if one of them necessarily holds, i.e. if every sentence in $\mathscr{T}^{(e)}$ is logically implied by K_i for some $i \in \{1, \ldots, n\}$. This notion can easily be extended to (a set of) infinitely many possible cases.

We say that γ' is a proper subcase of γ if (1) K' is a logical consequence of K and (2) the converse is not true. If γ has no proper subcases, we call it elementary possible case.

Generally, when we have n possible cases that are exhaustive and mutually incompatible (with n finite), at least one of them is not elementary.

We can determine a class Γ of exhaustive and mutually incompatible elementary possible cases in the following way. We consider the satisfiable sets of closed sentences in $\mathscr{T}^{(e)}$ that are maximal, i.e. such that, for every sentence p in $\mathscr{T}^{(e)}$, they contain either p or $\sim p$. We divide them into classes of logically equivalent sets. We take one set from each class. Finally we identify Γ with the class of the possible cases characterized by the sets obtained in this way.

N6. Objects for EL^ν and QIs for ML^ν. Models and value assignments for EL^ν and ML^ν. Modal product and sum. Abbreviations.

Requirement (b) in N1 says that in ML^ν every typed expression Δ should be meaningful in every possible case and at every model and value-assignment. Incidentally, this requirement seems to me quite natural except when Δ is a description that does not satisfy its uniqueness condition in a Γ-case; but in connection with these exceptions requirement (b) simplifies a modal calculus (in accordance with Frege's method) even more than an extensional one. (In any case the remainder of N6 does not depend on the validity of requirement (b) [N11].) Let $\bar{\Delta}$ be what Δ denotes in the Γ-case γ.

Following Carnap let us say, as a first intuitive approach, that $\bar{\Delta} = \tilde{\Delta}(\gamma)$ is the extension of the designator Δ in γ; and let us say that the designators Δ and Δ' have the same intension if and only if, in every Γ-case, Δ and Δ' have the same extension. Hence the intension of Δ is determined by the above function $\tilde{\Delta}$ from Γ-cases to objects of the same type as Δ, or by something equivalent to $\tilde{\Delta}$.

We want to express semantical rules for our modal language ML^{ν} in an extensional metalanguage. For this reason we want extensional representatives for the intensions, i.e. quasi intensions or QIs. As a preliminary for constructing our QIs, we introduce a universe for an extensional ν-sorted language EL^{ν}, i.e. a universe based on ν sets $\underline{D}_1, \ldots, \underline{D}_{\nu}$, to be called <u>individual domains.</u>

We assume that each $\underline{D}_i$ has at least two elements for the following reason. A customary axiom on universal quantification—cf. AS12.8—implies the existence of at least one element of $\underline{D}_i$. Furthermore, as will be explained shortly, a particular element of $\underline{D}_i$ is assumed to be the "not existing object." So $\underline{D}_i$ must have at least two elements, one to be an "existing object" and one to be the "not existing object," in order to comply with the customary interpretation of the above implication.

Let us add that $\underline{D}_{\nu}$ will be identified with the class Γ of elementary possible cases for the modal language $ML^{\nu-1}$. Hence $\underline{D}_{\nu}$ must have at least two elements so that the semantics considered for ML^{ν} can be essentially modal—cf. AS12.23.

For $\underline{t} \epsilon \tau^{\nu}$ the sets $\underline{D}_1, \ldots, \underline{D}_{\nu}$ determine the class $\underline{O}^{\nu}_{\underline{t}}$ of the objects (extensions) of type $\underline{t}$ by the following recursive conditions:

(a) $\underline{O}^{\nu}_{\underline{t}} = \underline{D}_{\underline{t}}$ for $\underline{t} = 1, \ldots, \nu$.

(b) $\underline{O}^{\nu}_{(\underline{t}_1, \ldots, \underline{t}_n)}$ is the power set of the Cartesian product $\underline{O}^{\nu}_{\underline{t}_1} \times \ldots \times \underline{O}^{\nu}_{\underline{t}_{\underline{n}}}$, i.e. the class of all subsets of the set of all $\underline{n}$-tuples $(\xi_1, \ldots, \xi_{\underline{n}})$ such that $\xi_1 \epsilon \underline{O}^{\nu}_{\underline{t}_{\underline{n}}}$. (Let us say that these sets are attributes and, in particular, relations of the type $(\underline{t}_1, \ldots, \underline{t}_{\underline{n}})$ for $\underline{n} > 1$ and properties of the type $(\underline{t}_1)$ for $\underline{n} = 1$.)

(c) $\underline{O}^{\nu}_{(\underline{t}_1, \ldots, \underline{t}_{\underline{n}}:\underline{t}_0)}$ is the class of functions from the whole Cartesian product $\underline{O}^{\nu}_{\underline{t}_1} \times \ldots \times \underline{O}^{\nu}_{\underline{t}_{\underline{n}}}$ to $\underline{O}^{\nu}_{\underline{t}_0}$. We may conceive of them as special relations of the type $(\underline{t}_1, \ldots, \underline{t}_{\underline{n}}, \underline{t}_0)$.

Following Frege—and e.g. Carnap and Rosser—we consider a choice function $\underline{a}^{\nu}$ of domain τ^{ν} such that

(5) $\underline{a}^{\nu}_{\underline{t}} \epsilon \underline{O}^{\nu}_{\underline{t}}$ (we write $\underline{a}^{\nu}_{\underline{t}}$ instead of $\underline{a}^{\nu}(\underline{t})$) for $\underline{t} \epsilon \tau^{\nu}$;

and as a further preliminary, we assume that in EL^{ν} every description of type $\underline{t}$ that does not satisfy its existence or uniqueness condition denotes $\underline{a}^{\nu}_{\underline{t}}$. We consider our universe for EL^{ν} to be determined by $\underline{D}_1, \ldots, \underline{D}_{\nu}$ and the function $\underline{a}^{\nu}$. We assume the following conditions on this function:

(A) $\underline{a}^{\nu}_{(\underline{t}_1, \ldots, \underline{t}_{\underline{n}})}$ is an empty class of $\underline{n}$-tuples,

(B) the counterdomain of $\underline{a}^{\nu}_{(\underline{t}_1, \ldots, \underline{t}_{\underline{n}}:\underline{t}_0)}$ is $\{\underline{a}^{\nu}_{\underline{t}_0}\}$.

These conditions are of a natural and rather usual kind—cf. [12], p. 38. However one might object that in some situations it is confusing to identify a "nonexisting object" (to be used in assigning meanings to descriptions) with an empty class,[27] and that it would be better to replace condition (A) with the following:

(A*) $\xi \in \underline{a}^{\nu}_{(\underline{t}_1, \ldots, \underline{t}_n)}$ if and only if $\xi = < \underline{a}^{\nu}_{\underline{t}_1}, \ldots, \underline{a}_{\underline{t}_n} >$.

We prefer (A) to (A*), for on the one hand (A) makes the proof of a basic theorem [Theor. 62.1] considerably shorter; on the other hand the aforementioned inconvenience of (A) can be remedied by introducing an additional description operator, $\imath^*$—cf. Def. 12.1.

In view of the choice of our QIs for ML^{ν}, it is useful to consider the universe for $EL^{\nu+1}$ to be determined by the sets $\underline{D}_1, \ldots, \underline{D}_{\nu+1}$ and the function $\underline{a}^{\nu+1}$, and also to assume

(6) $$\underline{D}_{\nu+1} = \Gamma, \quad \underline{a}^{\nu+1}_{\underline{r}} = \underline{a}^{\nu}_{\underline{r}} \quad (\underline{r} = 1, \ldots, \nu).$$

The following definition is customary in extensional logic:

DEF. 6.1. We say that $\underline{V}$ is a value assignment for EL^{ν} ($\mathfrak{M}$ is a model for EL^{ν}) if $\underline{V}$ ($\mathfrak{M}$) is a function from the variables (constants) of EL^{ν} to objects, and

(7) $$\underline{V}(\underline{v}_{\underline{tn}}) \in \underline{O}^{\nu}_{\underline{t}}, \mathfrak{M}(\underline{c}_{\underline{tn}}) \in \underline{O}^{\nu}_{\underline{t}} \quad \text{for } \underline{t} \in \tau^{\nu} \quad (\underline{n} = 1, 2, \ldots).$$

Now we want to present for ML^{ν} the analogue of $\underline{O}^{\nu}_{\underline{t}}$ and Def. 6.1. Therefore we start introducing our QIs, whose choice will be motivated and justified in the forthcoming section.

We intend to identify the class $\underline{QI}^{\nu}_{\underline{t}}$ of the QIs for ML^{ν} of the type $\underline{t}$ with the class of objects of a suitable type, $\eta(\underline{t})$, to be called the <u>extensional correspondent</u> of the type $\underline{t}$ and to be briefly denoted by $\underline{t}^{\eta}$. We now define $\underline{t}^{\eta}$ recursively by the following four conditions (where $\underline{t}_1, \ldots, \underline{t}_n \in \tau^{\nu}$):

$$0^\eta = (\nu+1), \quad \underline{r}^\eta = (\nu+1:\underline{r}) \quad (\underline{r} = 1, \ldots, \nu),$$

(8) $$(\underline{t}_1, \ldots, \underline{t}_{\underline{n}})^\eta = (\underline{t}_1^\eta, \ldots, \underline{t}_{\underline{n}}^\eta, \nu+1),$$

$$(\underline{t}_1, \ldots, \underline{t}_{\underline{n}}:\underline{t}_0)^\eta = (\underline{t}_1^\eta, \ldots, \underline{t}_{\underline{n}}^\eta:\underline{t}_0^\eta).$$

Then we set

(9) $$\underline{QI}_{\underline{t}}^\nu = \underline{O}_{\underline{t}^\eta}^{\nu+1} \quad \text{for } \underline{t} \epsilon \bar{\tau}^\nu.$$

Following Carnap we assume not only that every intension determines its corresponding QI, but also that every QI corresponds to a suitable intension—cf. [12], p. 181 and [13], p. 895.

The QIs defined by (8) and (9) in compliance with requirement (b) in N1 are relatively simple, I think, and in particular have relatively low levels—cf. (3)—precisely

(10) $$\text{lev}[\underline{t}^\eta] = 1 + \text{lev}[\underline{t}] \quad \text{for } \underline{t} \epsilon \bar{\tau}^\nu.$$

Indeed by $(3)_{1,2}$ and $(8)_{1,2}$ equality $(10)_1$ holds for $\underline{t} = 0, \ldots, \nu$. Let $(10)_1$ hold for $\underline{t} = \underline{t}_0$ and $\underline{t} = \underline{t}_1, \ldots, \underline{t} = \underline{t}_{\underline{n}}$. Then by $(3)_{2,3,4}$ and $(8)_{3,4}$ the equality $(10)_1$ holds for $\underline{t} = (\underline{t}_1, \ldots, \underline{t}_{\underline{n}})$ and for $\underline{t} = (\underline{t}_1, \ldots, \underline{t}_{\underline{n}}:\underline{t}_0)$. Hence it holds for every $\underline{t} \epsilon \bar{\tau}^\nu$.

DEF. 6.2. Let us say that $\mathcal{V}$ is a value assignment for ML^ν ($\mathcal{M}$ is a model for ML^ν) if $\mathcal{V}$ ($\mathcal{M}$) is a function from the variables (constants) of ML^ν to QIs, and—cf. (8), (9)—

(7′) $$\mathcal{V}(\underline{v}_{\underline{tn}}) \epsilon \underline{QI}_{\underline{t}}^\nu, \mathcal{M}(\underline{c}_{\underline{tn}}) \epsilon \underline{QI}_{\underline{t}}^\nu \quad \text{for } \underline{t} \epsilon \tau^\nu \quad (n = 1, 2, \ldots).$$

We now introduce some notations in connection with value assignments, modal products, and modal sums. Let $\underline{x}_1, \ldots, \underline{x}_{\underline{n}}$ be distinct variables of the respective types $\underline{t}_1, \ldots, \underline{t}_{\underline{n}}$, let $\tilde{\underline{x}}_{\underline{i}}$ be an object of type $\underline{t}_{\underline{i}}^\eta$ ($\underline{i} = 1, \ldots, \underline{n}$), and let $\mathcal{V}'$ be the value assignment such that we have $\mathcal{V}'(\underline{y}) = \mathcal{V}(\underline{y})$ for every variable $\underline{y}$ other than $\underline{x}_1, \ldots, \underline{x}_{\underline{n}}$ and $\mathcal{V}'(\underline{x}_{\underline{i}}) = \tilde{\underline{x}}_{\underline{i}}$ ($i = 1, \ldots, \underline{n}$). Then we write

(11) $$\mathcal{V}' = \mathcal{V}\begin{pmatrix} \underline{x}_1 \ldots \underline{x}_n \\ \tilde{\underline{x}}_1 \ldots \tilde{\underline{x}}_n \end{pmatrix}.$$

On the one hand let p be a matrix in ML^ν. Then—cf. [12], p. 182 and [13], III.9.II—Np means "p holds in every Γ-case" and ◇p means "p holds in some Γ-case." On the other hand the intersection sign ∩ for classes is an analogue of ∧, and several authors use, as a generalization of ∩, the sign ⋂ for a manifold product of classes.

In accordance with the interpretation of Np and the uses of ∧, ∩, and ⋂, let us introduce strict identity $=^\frown$, strict implication $\supset^\frown$, strict equivalence $\equiv^\frown$, and the modal product[28] $F^\frown$ of an attribute F as follows:[29]

DEF. 6.3. $a =^\frown b \equiv_D Na = b$.
DEF. 6.4. $p \supset^\frown q \equiv_D N(p \supset q)$.
DEF. 6.5. $p \equiv^\frown q \equiv_D N(p \equiv q)$.
DEF. 6.6. $F^\frown(x_1, \ldots, x_n) \equiv_D NF(x_1, \ldots, x_n)$.

Since ∨, ∪, and ⋃ are the duals of ∧, ∩, and ⋂ respectively, and since ◇ is the dual of N, we introduce e.g. $=^\smile$ and F as respective duals of $=^\frown$ and $F^\frown$ by the following metalinguistic definitions:

DEF. 6.7. $a =^\smile b \equiv_D \Diamond a = b$.
DEF. 6.8. $F^\smile(x_1, \ldots, x_n) \equiv_D \Diamond F(x_1, \ldots, x_n)$.

Here are other notations to be used in the sequel and related to the signs ∧, ∩, ⋂, and their duals:

DEF. 6.9. $\bigwedge_{i=1}^{n} p_i \equiv_D p_1 \wedge p_2 \wedge \ldots \wedge p_n$.

DEF. 6.10. $\bigvee_{i=1}^{n} p_i \equiv_D p_1 \vee p_2 \vee \ldots \vee p_n$.

As further preliminaries we now define in ML^{ν}—following Carnap substantially, see [13], III.9.II—extensional attribute of type $t = (t_1, \ldots, t_n)$ $[Ext_t]$ and extensional function of type $\theta = (t_1, \ldots, t_n : t_0)$ $[Ext_\theta]$:

DEF. 6.11. $Ext_t(F) \equiv_D (\forall x_1, y_1, \ldots, x_n, y_n)\, [F(x_1, \ldots, x_n)$

$$\bigwedge_{i=1}^{n} x_i = y_i \supset F(y_1, \ldots, y_n)],$$

DEF. 6.12. $Ext_\theta(f) \equiv_D (\forall x_1, y_1, \ldots, x_n, y_n)\, [\bigwedge_{i=1}^{n} x_i =$

$$y_i \supset f(x_1, \ldots, x_n) = f(y_1, \ldots, y_n)],$$

where Ext_t and Ext_θ—briefly Ext—are constants of the respective types (t) and (θ) and where f and F are variables of the respective types θ and t while x_i and y_i are the first two variables distinct from $x_1, y_1, \ldots, x_{i-1}, y_{i-1}$ and having the type t_i $(i = 1, \ldots, n)$.

N7. On the QIs for ML^{ν}.

In this section we proceed with rather intuitive considerations aimed at gradually justifying our choice of the QIs for ML^{ν}, in particular—see (8), (9)—our choice of the extensional correspondent t^{η} of t $(t \in \bar{\tau}^{\nu})$.

To know the intension of a sentence p in ML^{ν} (at a model $\mathcal{M}$) means to know in which Γ-cases p holds. Therefore, following Carnap—see [12], pp. 9, 181, and [13], p. 895—we represent this intension by the range of p, i.e. the class $\tilde{p}$ of the Γ-cases [N5] where p holds. Then it is natural to set $0^{\eta} = (\nu{+}1)$ and to let (9) hold for $t = 0$.

In order to provide examples justifying our choice of the individual QIs, we suppose that, speaking intuitively, $\underline{M}_{\underline{r}}$ is the constant $\underline{c}_{\underline{1r}}$ and that it expresses the rocket Mida $\underline{r}$, briefly $\underline{\bar{M}}_{\underline{r}}$ (in ML^{ν}), i.e. that $\underline{M}_{\underline{r}}$ denotes $\bar{M}_{\underline{r}}$ in every Γ-case ($\underline{r} = 1, 2, 3$).

Let $\mathfrak{m}$ express that one, $\bar{\mathfrak{m}}$, of the above rockets which at the terminal instant $\underline{t}_{20}$ of the twentieth century is at the maximum distance from the earth. (If several of these rockets are at this distance, then let $\bar{\mathfrak{m}}$ be the one with the lowest index.) Hence it is reasonable to assert in ML^{ν} that

(12) $\underline{N\underline{M}_{\underline{i}} \neq \underline{M}_{\underline{j}}}$ $(\underline{i} < \underline{j}; \underline{i}, \underline{j} = 1, 2, 3)$, $\diamond \underline{p}_1 \diamond \sim \underline{p}_1$ where $\underline{p}_1 \equiv_D \mathfrak{m} = \underline{M}_1$.

In light of the above considerations it is natural, I think, to represent the intension of $\mathfrak{m}$ by a suitable function $\bar{\mathfrak{m}} = \bar{\mathfrak{m}}(\gamma)$ $(\gamma \in \Gamma)$ such that $\underline{\tilde{m}}(\gamma) \in \{\underline{\bar{M}}_1, \underline{\bar{M}}_2, \underline{\bar{M}}_3\}$ for $\gamma \in \Gamma$.[30] More generally, substantially following Carnap—see [13], III.9.IV—it is natural to take the QIs of type $\underline{r}$ to be the functions from the Γ-cases to objects of type $\underline{r}$, in other words to assume that $\underline{r}^{\eta}$ equals $(\nu+1:\underline{r})$ and that (9) holds for $\underline{t} = \underline{r}$ ($\underline{r} = 1, \ldots, \nu$).

Now, in order to discuss our choice of the QIs for attributes, let us set—see (4), Def. 6.3, and (12)—

$$(13)\quad \begin{aligned} &\underline{F}_1(\underline{x}) \equiv_D \underline{p}_1 \underline{x} = \underline{M}_1 \vee \sim\underline{p}_1(\underline{x} = \underline{M}_2 \vee \underline{x} = \underline{M}_3), \\ &\underline{F}_2(\underline{x}) \equiv_D \underline{p}_1 \underline{x} =^{\cap} \underline{M}_1 \vee \sim\underline{p}_1(\underline{x} =^{\cap} \underline{M}_2 \vee \underline{x} =^{\cap} \underline{M}_3). \end{aligned}$$

We assume that (12) holds so that on the one hand, by $(12)_3$ both $\underline{F}_1(\underline{M}_1)$ and $\underline{F}_1(\mathfrak{m})$ hold in every Γ-case γ where $\underline{p}_1$ holds $(\gamma \in \tilde{\underline{p}}_1)$, whereas on the other hand, $\underline{F}_1(\underline{M}_1)$ does not hold in the other Γ-cases $(\gamma \in \Gamma - \tilde{\underline{p}})$. By $(13)_1$ and Def. 6.11 the property (expressed by) $\underline{F}_1$ is extensional, and assuming $\underline{D}_1 = \{\underline{\bar{M}}_1, \underline{\bar{M}}_2, \underline{\bar{M}}_3\}$, it might be represented by the function $\underline{F}'_1$ of type $(\nu+1:(1))$ such that for $\gamma \in \tilde{\underline{p}}_1$ $\underline{F}'_1(\gamma)$ is $\{\underline{\bar{M}}_1\}$, and for $\gamma \in \Gamma - \tilde{\underline{p}}_1$ $\underline{F}'_1(\gamma)$ is $\{\underline{\bar{M}}_2, \underline{\bar{M}}_3\}$.

Carnap takes functions of this kind to be the QIs for the properties spoken of in his language $\underline{L}_2$—see [13], p. 895. In $\underline{L}_2$ the sign $\underline{N}$ is not admitted within the scope of a lambda or iota operator. More precisely, only extensional properties are expressible in $\underline{L}_2$—cf. [13], III.9.II. Analogously, if we assumed (9) to hold for $\underline{t} = (\theta)$ and if we set $\underline{t}^{\eta} = (\nu+1:(\theta))$, then (taking fn. 30 into account) the only properties of type (θ) to which we would be able to assign QIs would be extensional. Therefore we do otherwise.

Let us now observe that by (12) and $(13)_2$, for $\gamma \in \widetilde{\underline{p}}_1$, in the Γ-case γ both $\underline{F}_2(\underline{M}_1)$ and $\mathfrak{m} = \underline{M}_1$ hold but $\underline{F}_2(\mathfrak{m})$ does not—hence $\underline{F}_2$ is not extensional [Def. 6.11]; furthermore $\underline{F}_2(\underline{M}_1)$ does not hold in γ for $\gamma \in (\Gamma\text{-}\widetilde{p}_1)$. Hence the holding of the matrix $\underline{F}_2(\underline{x})$ in a Γ-case γ depends essentially on the intension of $\underline{x}$ and on γ.

We might take the QI of the property $\underline{F}_2$ to be the function $\underline{F}'_2$ of the type $(\nu+1:(1^{\eta}))$ from Γ-cases to subclasses of QI_1, such that for every $\gamma \in \Gamma$, $\underline{F}'_2(\gamma)$ is $\{\widetilde{\underline{M}}_1\}$ or $\{\widetilde{\underline{M}}_2, \widetilde{\underline{M}}_3\}$ according to whether or not $\gamma \in \widetilde{\underline{p}}_1$ holds. It is simpler (and equivalent) to take the QI of $\underline{F}_2$ to be the binary relation $\widetilde{\underline{F}}_2$ of the type $(1^{\eta}, \nu+1)$, which holds for $\widetilde{\underline{x}}$ and γ if and only if $\gamma \in \Gamma$ and $\widetilde{\underline{x}} \in \underline{F}'_2(\gamma)$.[31] We shall use $\widetilde{\underline{F}}_2$ because it has a lower level than $\underline{F}'_2$.

More generally, and using a framework different from but equivalent to the preceding one, let $\underline{p}$ be a matrix in ML^{ν}, where for the sake of simplicity no quantifier occurs (but $\underline{N}$ possibly does), and let $\underline{x}_1, \ldots, \underline{x}_n$ be the variables occurring in $\underline{p}$, with $\underline{t}_1, \ldots, \underline{t}_n$ being their respective types.

Then for $\gamma \in \Gamma$, the holding of $\underline{p}$ in γ depends on γ and on the intensions $\widetilde{\underline{x}}_1, \ldots, \widetilde{\underline{x}}_n$ assigned to $\underline{x}_1, \ldots, \underline{x}_n$ respectively. Hence we may take the QI of the expression $(\lambda \underline{x}_1, \ldots, \underline{x}_n)\underline{p}$ (of the type $(\underline{t}_1, \ldots, \underline{t}_n)$) to be that (extensional) relation $\widetilde{\underline{F}}$ of type $(\underline{t}_1^{\eta}, \ldots, \underline{t}_n^{\eta}, \nu+1)$ which holds for $\widetilde{\underline{x}}_1, \ldots, \widetilde{\underline{x}}_n, \gamma$ if and only if $\gamma \in \Gamma$, $\widetilde{\underline{x}}_1 \in \underline{QI}_{\underline{t}_1}$,

$\ldots, \underline{\tilde{x}}_n \in QI_{\underline{tn}}$, and the matrix $\underline{p}$ holds in γ for the intensional values of $\underline{x}_1, \ldots, \underline{x}_n$ characterized by the QIs $\underline{\tilde{x}}_1, \ldots, \underline{\tilde{x}}_n$ respectively.

On the basis of the preceding considerations, relation (9) is justified for $\underline{t} = (\underline{t}_1, \ldots, \underline{t}_n)$ with $\underline{t}_1, \ldots, \underline{t}_n \in \tau^\nu$—cf. $(8)_3$. Let $\underline{t} = (\underline{t}_1, \ldots, \underline{t}_n : \underline{t}_0)$ hold. Then it is relatively natural and simple to assume (9) for $\underline{t}^\eta = (\underline{t}_1^\eta, \ldots, \underline{t}_n^\eta : \underline{t}_0^\eta)$.

One may object that such QIs for functions may be superfluous on the grounds that in science one does not use any nonextensional function—see Def. 6.12—i.e. a function $\underline{f}$ such that the object $\underline{f}(\underline{x})$ depends essentially on the intension of $\underline{x}$. We can obtain an example of such a function by assuming (12) and setting (in ML^ν):

$$(14) \quad \underline{f}(\underline{x}) =_D (\imath \underline{y}) [\underline{x} =^\frown \underline{M}_1 \ \underline{y} = \underline{M}_2 \vee \sim \underline{x} =^\frown \underline{M}_1 \ \underline{y} = \underline{M}_3] \text{ [see (4)].}$$

The above choice of $\underline{t}^\eta$ is essential if we are to deal with such a function, and the possibility of dealing with it is implicit in requirement (b) [N1], which asserts that every wff must be meaningful. The choice of $\underline{t}^\eta$ is simple, I think, and conforms through the above requirement, with the general character to be given to ML^ν.

Let us consider a functor $\underline{f}$ in ML^ν of type $(\underline{t}_1, \ldots, \underline{t}_n : \underline{t}_0)$ and suppose $\underline{f}$ to be assigned a QI that is a function $\underline{f}'$ of type $(\underline{t}_1^\eta, \ldots, \underline{t}_n^\eta, \nu+1 : \underline{t}_0^\eta)$ so that $\underline{f}'$ can be thought of as a particular relation $\underline{R}$ of the type $(\underline{t}_1, \ldots, \underline{t}_n, \underline{t}_0)^\eta = (\underline{t}_1^\eta, \ldots, \underline{t}_n^\eta, \underline{t}_0^\eta, \nu+1)$ up to the interchange of the last two arguments of $\underline{R}$. At first sight, to assign $\underline{f}$ such a QI might appear reasonable by analogy with a common practice of extensional logic. However, as a consequence the most natural QI for the designator $\underline{f}(\underline{x}_1, \ldots, \underline{x}_n)$, when each variable $\underline{x}_i$ is assigned an object $\underline{\tilde{x}}_i$ of type $\underline{t}_i^\eta$ as QI ($\underline{i} = 1, \ldots, \underline{n}$), would be the object $\underline{f}'(\underline{\tilde{x}}_1, \ldots, \underline{\tilde{x}}_n, \gamma)$ of type $\underline{t}_0^\eta (\gamma \in \Gamma)$. The dependence of this object on $\gamma(\gamma \in \Gamma)$ would make it unacceptable as a QI for the designator $\underline{f}(\underline{x}_1, \ldots, \underline{x}_n)$. Hence the QI $\widetilde{\underline{f}}$ must be such that we can characterize it by means of a function of the type $(\underline{t}_1^{\eta}, \ldots, \underline{t}_n^{\eta}:\underline{t}_0^{\eta})$. So it is better to choose $\widetilde{\underline{f}}$ among these functions at the outset, as we decided to do by means of (8) and (9).

It is not mandatory to take functions as primitive. At the end of N14 is shown how to construct an interpreted language equivalent to ML^{ν}, where functions are defined (in terms of modally constant attributes [Def. 13.2]).

N8. Equivalent QIs in a Γ-case. Our designation rules, excluding the rule for descriptions.

Let the designators Δ and Δ' have the type $\underline{t}(\underline{t} \in \bar{\tau}^{\nu})$. Then, as Carnap suggests—cf. [12], § 5—we say that in the Γ-case γ they have equivalent intensions (or the same extension) if either $\underline{t} = 0$ and the matrix $\Delta \equiv \Delta'$ holds in γ, or $\underline{t} \in \tau^{\nu}$ and the matrix $\Delta = \Delta'$ holds in γ.[32]

Here we define equivalent QIs in γ in such a way that, as one can expect on the basis of our preceding intuitive considerations, two QIs are equivalent in γ if and only if they represent equivalent intensions in γ.

DEF. 8.1. Let $\gamma \in \Gamma$ hold and $\widetilde{\underline{x}}$ and $\widetilde{\underline{y}}$ be QIs of type $\underline{t}(\underline{t} \in \bar{\tau}^{\nu})$. Then we say that $\widetilde{\underline{x}}$ and $\widetilde{\underline{y}}$ are equivalent QIs of type $\underline{t}$ in the case γ and we write

(15) $$\widetilde{\underline{x}} \overset{\underline{t}}{\underset{\gamma}{=}} \widetilde{\underline{y}}$$

if one of the following alternatives holds:

(a) $\underline{t} = \underline{r}$ where $\underline{r} \in \{1, \ldots, \nu\}$ (hence $\widetilde{\underline{x}}$ and $\widetilde{\underline{y}}$ are functions mapping Γ into $\underline{D}_{\underline{r}}$) and $\widetilde{\underline{x}}(\gamma) = \widetilde{\underline{y}}(\gamma)$.

(b) $\underline{t} = (\underline{t}_1, \ldots, \underline{t}_{\underline{n}})$ (hence $\widetilde{\underline{x}}$ and $\widetilde{\underline{y}}$ are subsets of $\underline{QI}^{\nu}_{\underline{t}_1} \times \ldots \times \underline{QI}^{\nu}_{\underline{t}_{\underline{n}}} \times \Gamma$) and the intersection of $\widetilde{\underline{x}}$ and $\widetilde{\underline{y}}$ with $\underline{QI}^{\nu}_{\underline{t}_1} \times \ldots \times \underline{QI}^{\nu}_{\underline{t}_{\underline{n}}} \times \{\gamma\}$ coincide, i.e. $(\lambda \xi_1, \ldots, \xi_{\underline{n}}) \widetilde{\underline{x}} (\xi_1, \ldots, \xi_{\underline{n}}, \gamma) = (\lambda \xi_1, \ldots, \xi_{\underline{n}}) \widetilde{\underline{y}} (\xi_1, \ldots, \xi_{\underline{n}}, \gamma)$.

(c) $\underline{t} = (\underline{t}_1, \ldots, \underline{t}_{\underline{n}} : \underline{t}_0)$ (hence $\widetilde{\underline{x}}$ and $\widetilde{\underline{y}}$ are functions from $\underline{QI}^{\nu}_{\underline{t}_1} \times \ldots \times \underline{QI}^{\nu}_{\underline{t}_{\underline{n}}}$ to $\underline{QI}^{\nu}_{\underline{t}_0}$) and for every $\underline{n}$-tuple $\langle \xi_1, \ldots, \xi_{\underline{n}} \rangle$ with $\xi_1 \in \underline{QI}^{\nu}_{\underline{t}_1}, \ldots, \xi_{\underline{n}} \in \underline{QI}^{\nu}_{\underline{t}_{\underline{n}}}$ the values $\widetilde{\underline{x}}(\xi_1, \ldots, \xi_{\underline{n}})$ and $\widetilde{\underline{y}}(\xi_1, \ldots, \xi_{\underline{n}})$ taken by the functions $\widetilde{\underline{x}}$ and $\widetilde{\underline{y}}$ for this $\underline{n}$-tuple are equivalent QIs of type $\underline{t}_0$ in the case γ.

(d) $\underline{t} = 0$ (hence $\widetilde{\underline{x}} \subseteq \Gamma, \widetilde{\underline{y}} \subseteq \Gamma$) and γ belongs either to both ranges $\widetilde{\underline{x}}$ and $\widetilde{\underline{y}}$ or to neither of them (i.e. $\{\gamma\} \cap \widetilde{\underline{x}} = \{\gamma\} \cap \widetilde{\underline{y}}$).

Listed below are the quasi-intensional designation rules (δ_1), $\ldots, (\delta_8)$ for ML^{ν}. For every model $\mathcal{M}$ and value assignment $\mathcal{V}$ these rules, together with rule (δ_9) for descriptions [N11], associate with every expression Δ of type $\underline{t}$ in ML^{ν} a <u>quasi-intensional designatum</u> $\widetilde{\Delta}$ with $\widetilde{\Delta} \in \underline{QI}^{\nu}_{\underline{t}}$ (see (8), (9)). This designatum is unique, as will appear from the nature of the rules. Therefore we write $\widetilde{\Delta} = \underline{\widetilde{des}}^{\nu}_{\mathcal{M}\mathcal{V}}(\Delta)$, or $\widetilde{\Delta} = \underline{\widetilde{des}}_{\mathcal{M}\mathcal{V}}(\Delta)$.

For $\underline{r} = 2, \ldots, 8$ the antecedent of rule $(\delta_{\underline{r}})$ is the antecedent of the formation rule $(\underline{f}_{\underline{r}})$ considered in Def. 3.1. Therefore in the middle column of the table following, only the main part of the antecedent of $(\delta_{\underline{r}})$ is written, the remaining part being understood. For the sake of simplicity the equalities

(16) $\widetilde{\Delta}_{\underline{i}} = \underline{\widetilde{des}}_{\mathcal{M}\mathcal{V}}(\Delta_{\underline{i}})$ $(\underline{i} = 1, \ldots, \underline{n})$, $\widetilde{\underline{R}} = \underline{\widetilde{des}}_{\mathcal{M}\mathcal{V}}(R)$, $\widetilde{\phi} = \underline{\widetilde{des}}_{\mathcal{M}\mathcal{V}}(\phi)$

are assumed in the following table:

Rule	If Δ is	then $\widetilde{\Delta} = \widetilde{\text{des}}_{\mathscr{M}\mathscr{V}}(\Delta)$ is
(δ_1)	$\underline{v}_{\underline{tn}}$ or $\underline{c}_{\underline{tn}}$,	$\mathscr{V}(\underline{v}_{\underline{tn}})$ or $\mathscr{M}(\underline{c}_{\underline{tn}})$ respectively.
(δ_2)	$\Delta_1 = \Delta_2$ where Δ_1, $\Delta_2 \in \mathscr{E}^{\nu}_{\underline{t}_1}$ [Def. 3.1],	the class of the Γ-cases γ such that $\widetilde{\Delta}_1 \overset{\gamma}{=}_{\underline{t}_1} \widetilde{\Delta}_2$ [Def. 8.1 and $(16)_1$].
(δ_3)	the matrix $\underline{R}(\Delta_1, \ldots, \Delta_{\underline{n}})$,	the class of the Γ-cases γ such that the $(\underline{n}+1)$-ary relation $\widetilde{\underline{R}}$ holds for $\widetilde{\Delta}_1, \ldots, \widetilde{\Delta}_{\underline{n}}$ and γ [see $(16)_{1,2}$].
(δ_4)	the term $\phi(\Delta_1, \ldots, \Delta_{\underline{n}})$,	$\widetilde{\phi}(\widetilde{\Delta}_1, \ldots, \widetilde{\Delta}_{\underline{n}})$ [cf. $(16)_{1,3}$].
(δ_5)	$\sim\Delta_1$,	$\Gamma - \widetilde{\Delta}_1$.
(δ_6)	$\Delta_1 \wedge \Delta_2$,	$\widetilde{\Delta}_1 \cap \widetilde{\Delta}_2$.
(δ_7)	$(\forall\underline{x})\Delta_1$,	the class of the Γ-cases γ such that for every quasi intension $\widetilde{\underline{x}}$ of the same type as the variable $\underline{x}$, $\gamma \in \widetilde{\text{des}}_{\mathscr{M}\mathscr{V}'}(\Delta_1)$ where $\mathscr{V}' = \mathscr{V}(\frac{\underline{x}}{\widetilde{\underline{x}}})$ [see (11)].
(δ_8)	$N\Delta_1$,	Γ or the empty set Λ according to whether or not the range $\widetilde{\Delta}_1$ of Δ_1 is Γ.

If $\underline{p}$ is a matrix, $\widetilde{\text{des}}_{\mathscr{M}\mathscr{V}}(\underline{p})$ will be called the <u>range</u> of $\underline{p}$ at $\mathscr{M}$ and $\mathscr{V}$. Note that the rules for ranges (δ_5), (δ_6), and (δ_8) are substantially the same as Carnap's rules (c), (e), and (g) in [12], p. 184, and rules implicit in [9], III.9. The rest of the rules above are similar to the corresponding rules by Carnap but somewhat more complex because in ML^{ν} the sign N is admitted within the scope of the lambda operator, and, simply, because in ML^{ν} equivalences such as (13) are admitted, where F_1 and F_2 may be understood to be property variables.

N9. On some semantical concepts. Some fundamental formulas, considered by Carnap, which combine modalities with quantification.

The terminology and symbolism for ML^{ν} introduced by the following definitions have rather common analogues for extensional languages.

DEF. 9.1. Let $\mathscr{A}$ be a set of matrices in ML^{ν}, $\mathscr{M}$ a model, and $\mathscr{V}$ a value assignment, and let $\tilde{\mathscr{A}}$ be the class of the Γ-cases that the ranges of the matrices belonging to $\mathscr{A}$ have in common. Then we say that $\tilde{\mathscr{A}}$ is the range (or the quasi-intensional designatum) of $\mathscr{A}$ at $\mathscr{M}$ and $\mathscr{V}$, and we write $\tilde{\mathscr{A}} = \widetilde{\mathrm{Des}}_{\mathscr{M}\mathscr{V}}(\mathscr{A})$.

DEF. 9.2. The model $\mathscr{M}$ is said to satisfy the set $\mathscr{A}$ of matrices if, for every value assignment $\mathscr{V}$, $\widetilde{\mathrm{Des}}_{\mathscr{M}\mathscr{V}}(\mathscr{A}) = \Gamma$.

DEF. 9.3. $\mathscr{M}$ is called an admissible model (with respect to the set $\mathscr{A}$ of axioms) if $\mathscr{M}$ satisfies the set $\mathscr{A}$ of matrices and the definitions (laid down in ML^{ν}).

DEF. 9.4.

(a) Let B be a set of matrices and p a matrix. Then we say that B $\mathscr{A}$-implies p—or that p is an admissible consequence of B—and we write B $\Vdash$ p if for every admissible model $\mathscr{M}$ (with respect to the set $\mathscr{A}$ of axioms) and for every value assignment $\mathscr{V}$ we have $\widetilde{\mathrm{Des}}_{\mathscr{M}\mathscr{V}}(\mathrm{B}) \subseteq \widetilde{\mathrm{des}}_{\mathscr{M}\mathscr{V}}(\mathrm{p})$.

(b) For $\underline{B} = \{\underline{p}_1, \ldots, \underline{p}_n\}$ we say that $\underline{p}_1, \ldots, \underline{p}_n$ $\mathscr{A}$-imply $\underline{p}$ and we write $\underline{p}_1, \ldots, \underline{p}_n \Vdash \underline{p}$.

(c) If the above set $\mathscr{A}$ of axioms is empty ($\mathscr{A} = \Lambda$), then we say that $\underline{p}$ is $\underline{L}$-implied by $\underline{B}$ and, in case $\underline{B} = \{\underline{p}_1, \ldots, \underline{p}_n\}$, also by $\underline{p}_1, \ldots, \underline{p}_n$.

DEF. 9.5.

(a) Let $\Lambda \Vdash \underline{p}$ [Def. 9.4a]. Then we say that $\underline{p}$ is $\mathscr{A}$-true—cf. [13], III.10.I—(in ML^ν) and we write $\Vdash \underline{p}$.

(b) Let $\mathscr{A} = \Lambda$. Then we say that $\underline{p}$ is logically true, briefly $\underline{L}$-true (in ML^ν).

THEOR. 9.1. If $\underline{p} \Vdash \underline{q}$ and $\underline{q} \Vdash \underline{r}$, then $\underline{p} \Vdash \underline{r}$ [Def. 9.4].

THEOR. 9.2. $\underline{p}_1, \ldots, \underline{p}_n, \underline{p} \Vdash \underline{q}$ if and only if $\underline{p}_1, \ldots, \underline{p}_n \Vdash \underline{p} \supset \underline{q}$ ($\underline{n} = 0, 1, \ldots$) [Def. 9.4 and rules (δ_5), (δ_6)].

THEOR. 9.3. $\Vdash \underline{p} \supset \underline{q}$ and $\Vdash \underline{p}$ imply $\Vdash \underline{q}$.

THEOR. 9.4. $\Vdash \underline{p}$ implies $\Vdash (\forall \underline{x})\underline{p}$ and $\Vdash \underline{Np}$.

DEF. 9.6.

(a) Let γ belong to the range of the matrix $\underline{p}$ at the model $\mathscr{M}$ and the value assignment $\mathscr{V}$ $[\gamma \epsilon \widetilde{\text{des}}_{\mathscr{M}\mathscr{V}}(\underline{p})]$. Then we say that $\underline{p}$ holds, or is true, in γ at $\mathscr{M}$ and $\mathscr{V}$.

(b) If $\underline{p} \supset \underline{q}$ holds (or is true) in γ at $\mathscr{M}$ and $\mathscr{V}$, then we say that $\underline{p}$ implies $\underline{q}$ in γ at $\mathscr{M}$ and $\mathscr{V}$.

(c) If $\widetilde{\text{des}}_{\mathscr{M}\mathscr{V}}(\underline{p}) = \Gamma$, we say that $\underline{p}$ is Γ-true at $\mathscr{M}$ and $\mathscr{V}$, and in case $\underline{p}$ is $\underline{p}_1 \supset \underline{q}$ we also say that $\underline{p}_1$ Γ-implies $\underline{q}$.

THEOR. 9.5. The matrix $\underline{p}$ is $\mathscr{A}$-true in ML^ν (hence $\underline{L}$-true for $\mathscr{A} = \Lambda$) if and only if $\underline{p}$ is Γ-true at every admissible model and every value assignment.

In N5 we considered the alternative of characterizing the Γ-cases by the admissible models of a theory $\mathcal{T}^{(e)}$ based on an extensional language, say EL^{ν}. A sentence of EL^{ν} holding for every admissible model of $\mathcal{T}^{(e)}$ is said to be $\mathcal{A}$-true in EL^{ν}—cf. [13], III.10.I; it is also said to be $\underline{L}$-true in EL^{ν} when $\mathcal{T}^{(e)}$ has no nonlogical axioms.

The above considerations (on models) somehow relate Γ-truth in ML^{ν} [Def. 9.6(c)] to $\mathcal{A}$-truth or $\underline{L}$-truth in EL^{ν}. But let us remark that in connection with EL^{ν} the concept of $\mathcal{A}$- or $\underline{L}$-truth in EL^{ν} and the concept of $\mathcal{A}$- or $\underline{L}$-provability have similar roles in semantics and syntax respectively.[33] They are <u>corresponding</u> concepts, and the same can be said of the semantical concept of $\mathcal{A}$- or $\underline{L}$-implication and the syntactical concept of $\mathcal{A}$- or $\underline{L}$-deducibility (or inference).

From the point of view of this correspondence between semantical and syntactical concepts, as analogues for ML^{ν} of $\mathcal{A}$- and $\underline{L}$-truth and $\mathcal{A}$- and $\underline{L}$-implication in EL^{ν} we can choose $\mathcal{A}$- and $\underline{L}$-truth in ML^{ν} and $\mathcal{A}$- and $\underline{L}$-implication in ML^{ν} [Def. 9.4] respectively in that Theor. 9.3 assures the validity of modus ponens in ML^{ν} in connection with $\underline{L}$-implication and the analogue holds for EL^{ν}.

The following theorem assures the validity of modus ponens in ML^{ν} in a somewhat stronger way.

THEOR. 9.6. If $\underline{p}$ and $\underline{p} \supset \underline{q}$ hold in γ at $\mathcal{M}$ and $\mathcal{V}$, then so does $\underline{q}$.

As might be expected from the similarity between our semantical rules $(\delta_1), \ldots, (\delta_8)$ and Carnap's, the formulas combining modalities and quantification, which hold for Carnap's modal languages and are considered in [12], Theor. 41.5, p. 185, also hold in ML^{ν}. This is apparent from the next theorem.

THEOR. 9.7. Let p and q be matrices [not containing descriptions].[34] Then the following matrices are L-true [Def. 9.5]:[35]

(17) $\underline{Np} \supset^{\frown} \underline{p}$, $\underline{p} \supset^{\frown} \diamond \underline{p}$, $(\underline{p} \supset^{\frown} \underline{q}) \supset^{\frown} (\underline{Np} \supset^{\frown} \underline{Nq})$.

(18) $\underline{N(pq)} \equiv \underline{NpNq}$, $\diamond(\underline{p} \vee \underline{q}) \equiv \diamond\underline{p} \vee \diamond\underline{q}$, $\underline{NNp} \equiv \underline{Np}$, $\underline{N} \sim \underline{Np} \equiv \sim\underline{Np}$.

(19) $\diamond\diamond\underline{p} \equiv \diamond\underline{p}$, $\diamond\underline{Np} \equiv \underline{Np}$, $\underline{N}\diamond\underline{p} \equiv \diamond\underline{p}$.

(20) $(\underline{x})\underline{Np} \equiv \underline{N}(\underline{x})\underline{p}$, $(\exists\underline{x})\underline{Np} \supset^{\frown} \underline{N}(\exists\underline{x})\underline{p}$, $(\exists\underline{x})\diamond\underline{p} \equiv \diamond(\exists\underline{x})\underline{p}$, $\diamond(\underline{x})\underline{p} \supset^{\frown} (\underline{x})\diamond\underline{p}$.

(21) $\diamond(\underline{p} \supset \underline{q}) \supset^{\frown} (\diamond\underline{p} \supset \diamond\underline{q})$, $(\underline{p} \supset^{\frown} \underline{Nq}) \supset (\diamond\underline{p} \supset \underline{Nq})$.

N10. Useful theorems on equivalent QIs.

The following three theorems will be essential for laying down our semantical rule for descriptions.

THEOR. 10.1. Let $\underline{t} \in \bar{\tau}^{\nu}$ and $\xi, \eta, \zeta \in \underline{QI}^{\nu}_{\underline{t}}$ [see (8) and (9)], and $\gamma \in \Gamma$. Then [Def. 8.1]:

(a) $\xi \overset{t}{\underset{\gamma}{=}} \xi$,

(b) $\xi \overset{t}{\underset{\gamma}{=}} \eta$ implies $\eta \overset{t}{\underset{\gamma}{=}} \xi$,

(c) $\xi \overset{t}{\underset{\gamma}{=}} \eta$ and $\eta \overset{t}{\underset{\gamma}{=}} \zeta$ imply $\xi \overset{t}{\underset{\gamma}{=}} \zeta$.

THEOR. 10.2. Let $\underline{t} \in \bar{\tau}^{\nu}$ and $\xi, \eta \in \underline{QI}^{\nu}_{\underline{t}}$ [see (8), (9)]. Then $\xi \overset{t}{\underset{\gamma}{=}} \eta$ [Def. 8.1] holds for every Γ-case γ if and only if ξ and η coincide ($\xi = \eta$).

Proof. By Def. 8.1(a), (b), and (d) our theorem obviously holds in the following three cases: $\underline{t} = \underline{r}$ ($\underline{r} = 1, \ldots, \nu$), $\underline{t} = (\underline{t}_1, \ldots, \underline{t}_{\underline{n}})$, and $\underline{t} = 0$, hence for $\underline{\text{lev}}[\underline{t}] = 0$.

Case 4. $\underline{t} = (\underline{t}_1, \ldots, \underline{t}_{\underline{n}} : \underline{t}_0)$. Let m be any positive integer and, as inductive hypothesis, let Theor. 10.2 hold for every t with

$\underline{\text{lev}}[\underline{t}] < \underline{m}$. Suppose $\underline{\text{lev}}[\underline{t}] = \underline{m}$. Then, since by $(3)_4$ $\underline{\text{lev}}[\underline{t}_0] < \underline{m}$, Theor. 10.2 holds for $\underline{t} = \underline{t}_0$.

In case 4 the validity of the condition $\xi \overset{t}{\underset{\gamma}{=}} \eta$ for every $\gamma \in \Gamma$ implies, by Def. 8.1(c), that for every $\underline{n}$-tuple $\langle \zeta_1, \ldots, \zeta_{\underline{n}} \rangle$ with $\zeta_1 \in QI_{\underline{t}_1}, \ldots, \zeta_{\underline{n}} \in QI_{\underline{t}_{\underline{n}}}$, the condition $\xi(\zeta_1, \ldots, \zeta_{\underline{n}}) \overset{t_0}{\underset{\gamma}{=}} \eta(\zeta_1, \ldots, \zeta_{\underline{n}})$ holds for every $\gamma \in \Gamma$. Also $\xi(\zeta_1, \ldots, \zeta_{\underline{n}})$ and $\eta(\zeta_1, \ldots, \zeta_{\underline{n}})$ are QIs of type $\underline{t}_0$ and Theor. 10.2 holds for $\underline{t} = \underline{t}_0$. Hence $\xi(\zeta_1, \ldots, \zeta_{\underline{n}}) = \eta(\zeta_1, \ldots, \zeta_{\underline{n}})$. By the arbitrariness of $\zeta_1, \ldots, \zeta_{\underline{n}}$ we conclude $\xi = \eta$.

Conversely $\xi = \eta$ obviously implies that $\xi \overset{t}{\underset{\gamma}{=}} \eta$ holds for every $\gamma \in \Gamma$. Hence Theor. 10.2 holds for $\underline{\text{lev}}[\underline{t}] = \underline{m}$ in case 4.

By the principle of induction our theorem also holds in case 4.

QED

THEOR. 10.3. Let $\underline{t} \in \bar{\tau}^\nu$ and $\{\xi_\gamma\}_{\gamma \in \Gamma}$ be a family of QIs of type $\underline{t}$; i.e. $\xi_\gamma \in QI^\nu_{\underline{t}}$ for every $\gamma \in \Gamma$. Then there is exactly one quasi intension ξ' of type $\underline{t}$ such that for every $\gamma \in \Gamma$ the quasi intensions ξ' and ξ_γ are equivalent in γ, i.e. $\xi' \overset{t}{\underset{\gamma}{=}} \xi_\gamma$ [Def. 8.1].

Proof. We need prove only the existence of the above ξ', its uniqueness being an obvious consequence of Theors. 10.1 and 10.2.

Case 1. $\underline{t} = 0$. Theor. 10.3 obviously holds. [Let ξ' be the range containing the Γ-cases γ with $\gamma \in \xi_\gamma$, briefly $\xi' = (\lambda\gamma)\gamma \in \xi$.]

Case 2. $\underline{t} = \underline{r}(1 \leq \underline{r} \leq \nu)$. By (8) and (9), Theor. 10.3 obviously holds. [Let ξ' be that function mapping Γ into $\underline{D}_{\underline{r}}$, which for every Γ-case γ takes $\xi_\gamma(\gamma)$ as value, briefly $\xi' = (\lambda\gamma)\xi_\gamma(\gamma)$.]

Case 3. $\underline{t} = (\underline{t}_1, \ldots, \underline{t}_{\underline{n}})$. Remembering that by (8) and (9) $\xi_\gamma \subset QI^\nu_{\underline{t}_1} \times \ldots \times QI^\nu_{\underline{t}_{\underline{n}}} \times \Gamma$ for every $\gamma \in \Gamma$, let ξ' be the class of the $(\underline{n}+1)$-tuples $\langle \zeta_1, \ldots, \zeta_{\underline{n}}, \gamma \rangle$ belonging to ξ_γ as elements $(\gamma \in \Gamma)$, briefly $\xi' = (\lambda\zeta_1, \ldots, \zeta_{\underline{n}}, \gamma)\, (\langle \zeta_1, \ldots, \zeta_{\underline{n}}, \gamma \rangle \in \xi_\gamma$ and $\gamma \in \Gamma)$.

Then by Def. 8.1(b), for every $\gamma \in \Gamma$, $\xi' \overset{t}{\underset{\gamma}{=}} \xi_\gamma$. Since by Theor. 10.2 ξ' is unique, Theor. 10.3 holds in case 3.

Case 4. $\underline{t} = (\underline{t}_1, \ldots, \underline{t}_n : \underline{t}_0)$. Theor. 10.3 holds under the hypothesis $\underline{lev}[\underline{t}] = 0$. Let $\underline{m}$ be any positive integer and as inductive hypothesis let Theor. 10.3 hold for every $\underline{t}$ with $\underline{lev}[\underline{t}] < \underline{m}$. Suppose $\underline{lev}[\underline{t}] = \underline{m}$; hence since $\underline{lev}[\underline{t}_0] < \underline{t}$, Theor. 10.3 holds for $\underline{t} = \underline{t}_0$.

Let $\zeta_1 \in \underline{QI}^\nu_{\underline{t}_1}, \ldots, \zeta_{\underline{n}} \in \underline{QI}^\nu_{\underline{t}_n}$ and set $\eta_\gamma = \xi_\gamma(\zeta_1, \ldots, \zeta_{\underline{n}})$ for $\gamma \in \Gamma$. Then $\{\eta_\gamma\}_{\gamma \in \Gamma}$ is a family of QIs of type $\underline{t}_0$. Then, since Theor. 10.3 holds for $\underline{t} = \underline{t}_0$, there is exactly one QI η' of type $\underline{t}_0$ such that $\eta' \overset{t}{\underset{\gamma}{=}} \eta_\gamma$ for every $\gamma \in \Gamma$.

We can write $\eta' = \xi'(\zeta_1, \ldots, \zeta_{\underline{n}})$ and consider this equality as a functional equation in the unknown function $\xi'(\xi' \in \underline{QI}^\nu_{\underline{t}})$. By the arbitrariness of $\zeta_1, \ldots, \zeta_{\underline{n}}$ this equation is satisfied by exactly one element ξ' of $\underline{QI}^\nu_{\underline{t}}$.

We conclude that Theor. 10.3 also holds for $\underline{lev}[\underline{t}] = \underline{m}$. By the principle of induction (in the strong sense), Theor. 10.3 also holds in case 4.

QED

N11. The semantical rule for descriptions in ML^ν.

After some preliminaries (including a suitable uniqueness definition in ML^ν) we lay down and discuss some requirements for descriptions. Among them there are the requirements (b) and (c)—see p. 5—which will suggest our rule (δ_9) for descriptions in ML^ν. Theor. 11.1 will assure the logical correctness of rule (δ_9) and Theor. 11.3 will assure, I believe, that rule (δ_9) is acceptable especially since ML^ν is to be applied to sciences such as physics where every two possible cases are, a priori, on a par (see below).

As preliminaries we first assume the following hypothesis:

HYP. 11.1. $\phi(\underline{x})$ is a matrix in ML^{ν}, $\underline{y}$ is the first variable of the same type as $\underline{x}$ that does not occur in $\phi(\underline{x})$, and $\phi(\underline{y})$ is the matrix obtained from $\phi(\underline{x})$ by replacing the free occurrences of the variable $\underline{x}$ in $\phi(\underline{x})$ by occurences of $\underline{y}$.

We associate with every matrix $\underline{p}$ its <u>extensionalization</u> $\underline{p}^{(ex)}$ <u>with respect to</u> $\underline{x}$ <u>as follows</u>:

DEF. 11.1. $\phi(\underline{x})^{(ex)} \equiv_D (\exists \underline{y})\,[\phi(\underline{y}) \wedge \underline{y} = \underline{x}]$ where Hyp. 11.1 holds.[36]

Now let us remark that the problem of assigning a meaning to a description $(\imath \underline{x})\,\phi(\underline{x})$ in ML^{ν} looks rather complex if the matrix is arbitrary, but if in some Γ-case γ $\phi(\underline{x})$ is <u>extensional with respect to</u> $\underline{x}$, i.e. if $\phi(\underline{x}) \equiv \phi(\underline{x})^{(ex)}$ holds in γ, then in connection with γ our problem is almost as simple as in extensional logic.

For $\nu = 1, 2, \ldots$ and $\underline{t} \epsilon \tau^{\nu}$, as preliminary metalinguistic definitions in ML^{ν} (and EL^{ν}), we first set

DEF. 11.2. $\underline{a}^{*}_{\underline{t}} =_D (\imath \underline{v}_{\underline{t}1})\,(\underline{v}_{\underline{t}1} \neq \underline{v}_{\underline{t}1})$ for $\underline{t} \epsilon \tau^{\nu}$

so that $\underline{a}^{*}_{\underline{t}}$ is not a constant but a closed wff; moreover, we define "<u>there is exactly one</u> $\underline{x}$ <u>such that</u> $\phi(\underline{x})$ <u>holds</u>" by

DEF. 11.3. $(\exists_1 \underline{x})\,\phi(\underline{x}) \equiv_D (\exists \underline{x})\,\{\phi(\underline{x})\,(\underline{y})\,[\phi(\underline{y}) \supset \underline{y} = \underline{x}]\}$ where Hyp. 11.1 holds.

Note that within ML^{ν}, the definiens in Def. 11.3 cannot be replaced by $(\exists \underline{x})\,(\underline{y})\,[\phi(\underline{y}) \equiv \underline{x} = \underline{y}]$. The resulting definition can be accepted within a theory of extensional logic—see e.g. [32], Theor. VII.2.1, p. 167—but within ML^{ν}, after the above replacement, $(\exists_1 \underline{x})\,\phi(\underline{x})$ would $\underline{L}$-imply that $\phi(\underline{x})$ should be extensional with respect to $\underline{x}$.[37]

We shall follow Frege's method for descriptions for the sake of simplicity. Then it is natural to demand that our next rule (δ_9) for descriptions fulfill the following requirements for every matrix $\phi(\underline{x})$.

(a) $[\phi(\underline{x}) \equiv \phi(\underline{x})^{(ex)}]\ (\exists_1 \underline{x})\phi(\underline{x}) \Vdash \phi[(\imath\underline{x})\phi(\underline{x})]$ [Defs. 9.4, 11.1, 11.3].

(b) $\sim(\exists_1 \underline{x})\phi(\underline{x}) \Vdash (\imath\underline{x})\phi(\underline{x}) = \underline{a}_{\underline{t}}^*$ [Def. 11.2].

Let us remark that we cannot omit $\phi(\underline{x}) \equiv \phi(\underline{x})^{(ex)}$ in requirement (a). Indeed let

(a′) $(\exists_1 \underline{x})\phi(\underline{x}) \Vdash \phi[(\imath\underline{x})\phi(\underline{x})]$

hold for every matrix $\phi(\underline{x})$ as a hypothesis for reductio ad absurdum. Let us assume (12) as our set of axioms and let us set

(22) $\phi_3(\underline{x}) \equiv_D \underline{p}_1(\underline{x} =^\cap \underline{M}_1) \vee \sim\underline{p}_1(\underline{x} =^\cap \underline{M}_2),\quad \xi =_D (\imath\underline{x})\phi_3(\underline{x}).$

Then by Def. 11.3 $\Vdash(\exists_1\underline{x})\phi_3(\underline{x})$, hence by (a′) and $(22)_2$:

(23) $\Vdash \underline{p}_1 \supset^\cap \xi =^\cap \underline{M}_1, \qquad \Vdash \sim\underline{p}_1 \supset^\cap \xi =^\cap \underline{M}_2.$

From (23) and the $\underline{L}$-truth of $(21)_2$ we deduce [Theor. 9.3] $\Vdash \Diamond\underline{p}_1 \supset \xi =^\cap \underline{M}_1$ and $\Vdash \Diamond\sim\underline{p}_1 \supset \xi =^\cap \underline{M}_2$, whence by $(12)_2$ we obtain $\Vdash \xi =^\cap \underline{M}_{\underline{i}}$ ($\underline{i} = 1, 2$). Then $\Vdash \underline{M}_1 =^\cap \underline{M}_2$ in contrast to $(12)_1$. Hence <u>requirement</u> ($\underline{a}$′) <u>cannot be accepted.</u>

We note that in the above considerations it is essential that requirement ($\underline{a}$′) should behave in the same way with respect to all Γ-cases and in particular with respect to at least two of them, in which $\underline{p}_1$ and $\sim\underline{p}_1$ hold respectively. Therefore we need not deny that $(\exists_1\underline{x})\phi(\underline{x}) \supset \phi[(\imath\underline{x})\phi(\underline{x})]$ might hold for every matrix $\phi(\underline{x})$ in one privileged Γ-case, for instance the real one $\gamma_{\mathfrak{R}}$.

In sciences such as astronomy, geology, and geography, $\gamma_{\mathfrak{R}}$ has a privileged role, but for mechanics, physics, chemistry, and biology, all possible cases must be regarded, I think, in the same way;[38] this holds in particular for foundations of mechanics according to Painlevé's ideas. We intend to apply our semantical theory for ML^{ν} to the problem of presenting such foundations rigorously; hence we consider all Γ-cases in the same way and consequently decide to reject requirement (a′).

On the basis of the observations regarding (a′) and the uniqueness definition [Def. 11.3], it is natural, I think, to strengthen (a) into the following requirement, which is equivalent to (a) together with the condition $\Vdash (\iota \underline{x})\phi(\underline{x}) = (\iota \underline{x})\phi(\underline{x})^{(\underline{ex})}$ [Def. 11.1]:

(c) $(\exists_1 \underline{y})\phi(\underline{y}) \Vdash (\underline{y})[\phi(\underline{y}) \supset \underline{y} = (\iota \underline{x})\phi(\underline{x})]$ under Hyp. 11.1.

The following rule for descriptions, suggested by requirements (b) and (c) above, will be justified by the two forthcoming theorems.

RULE (δ_9). Assume that Δ is $(\iota \underline{x})\phi(\underline{x})$ where $\phi(\underline{x})$ is a matrix (in ML^{ν}) and $\underline{x}$ is a variable of type $\underline{t}$ $(\underline{t} \in \tau^{\nu})$ not necessarily occurring free in $\phi(\underline{x})$; furthermore let $\mathcal{M}$ be a model and $\mathcal{V}$ a value assignment.

Then the quasi-intensional designatum $\widetilde{des}_{\mathcal{M}\mathcal{V}}(\Delta)$ of Δ at $\mathcal{M}$ and $\mathcal{V}$ is the quasi intension $\widetilde{\Delta}$ of type $\underline{t}$ fulfilling the following conditions (b′) and (c′)—cf. (b) and (c)—(the existence and uniqueness of this $\widetilde{\Delta}$ is assured by the next theorem):

(b′) In every Γ-case γ where $(\exists_1 \underline{x})\phi(\underline{x})$ [Def. 11.3] does not hold at $\mathcal{M}$ and $\mathcal{V}$, $\widetilde{\Delta}$ is equivalent to the improper QI of type $\underline{t}$—i.e. to $a^{\nu+1}_{\underline{t}\eta}$, see (5), (8), (9)—in other words $\widetilde{\Delta} = {}^{t}_{\gamma} a^{\nu}_{\underline{t}\eta}$ [Def. 8.1].

(c′) In every Γ-case γ where $(\exists_1 \underline{x})\phi(\underline{x})$ holds at $\mathcal{M}$ and $\mathcal{V}$, $\widetilde{\Delta}$ is equivalent to every quasi intension $\widetilde{\underline{y}}$ of type $\underline{t}$ such that the matrix $\phi(\underline{x})$ holds in γ at the model $\mathcal{M}$ and the value assignment $\mathcal{V}\binom{\underline{x}}{\widetilde{\underline{y}}}$—see (11).

THEOR. 11.1. Under the assumptions about Δ, $\phi(\underline{x})$, $\underline{x}$, $\underline{t}$, $\mathcal{M}$, and $\mathcal{V}$ made in rule (δ_9), there is exactly one quasi intension $\widetilde{\Delta}$ fulfilling conditions (b′) and (c′).

Proof. Let Γ_1 be the class of the Γ-cases in which $(\exists_1 \underline{x})\phi(\underline{x})$ holds at $\mathcal{M}$ and $\mathcal{V}$. Moreover let $\gamma \in \Gamma_1$. Then by Def. 11.3 (and the designation rules (δ_1) to (δ_8) in N8) there is a quasi intension $\widetilde{\underline{x}}$ of type $\underline{t}$ such that $\phi(\underline{x})$ holds in γ at $\mathcal{M}$ and $\mathcal{V}\binom{\underline{x}}{\widetilde{\underline{x}}}$—see (11). Hence by the axiom of choice and the above arbitrariness of γ, there is a family $\{\xi_\gamma\}_{\gamma\in\Gamma}$ of QIs of type $\underline{t}$ such that, for every $\gamma \in \Gamma$, either $\gamma \in \Gamma_1$ and $\phi(\underline{x})$ holds in γ at $\mathcal{M}$ and $\mathcal{V}\binom{\underline{x}}{\xi_\gamma}$—see (11)—or $\gamma \in (\Gamma - \Gamma_1)$ and ξ_γ is the improper quasi intension $\underline{a}_{t\eta}^{\nu+1}$ of type $\underline{t}$. Then by Theor. 10.3 there is exactly one quasi intension $\widetilde{\Delta}$ $(\widetilde{\Delta} = \xi')$ of type $\underline{t}$ such that $\widetilde{\Delta} \overset{t}{\underset{\gamma}{=}} \xi_\gamma$ for every $\gamma \in \Gamma$.

We remark that $\widetilde{\Delta}$ obviously fulfills condition (b′). (Indeed for $\gamma \in (\Gamma - \Gamma_1)$, ξ_γ is $\underline{a}_{t\eta}^{\nu+1}$ and $\widetilde{\Delta} \overset{t}{\underset{\gamma}{=}} \xi_\gamma$.)

As to condition (c′) we now assume that γ is an arbitrary element of Γ_1, so that $(\exists_1 \underline{x})\phi(\underline{x})$ holds in γ at $\mathcal{M}$ and $\mathcal{V}$. Then by Def. 11.3 there is a quasi intension $\widetilde{\underline{z}}$ of type $\underline{t}$ such that (1) $\phi(\underline{x})$ holds in γ at $\mathcal{M}$ and $\mathcal{V}\binom{\underline{x}}{\widetilde{\underline{z}}}$, and (2) if $\phi(\underline{x})$ holds in γ at $\mathcal{M}$ and $\mathcal{V}\binom{\underline{x}}{\widetilde{\underline{y}}}$, then $\widetilde{\underline{z}} \overset{t}{\underset{\gamma}{=}} \widetilde{\underline{y}}$. So in particular, $\widetilde{\underline{z}} \overset{t}{\underset{\gamma}{=}} \xi_\gamma$ holds.

Since we also know $\widetilde{\Delta} \overset{t}{\underset{\gamma}{=}} \xi_\gamma$, by Theor. 10.1 we deduce $\widetilde{\Delta} \overset{t}{\underset{\gamma}{=}} \widetilde{\underline{y}}$, and by the arbitrariness of γ $(\gamma \in \Gamma_1)$ we conclude that $\widetilde{\Delta}$ also fulfills condition (c′). We conclude that a quasi intension $\widetilde{\Delta}$ of type $\underline{t}$ satisfying (b′) and (c′) exists.

Now we suppose that conditions (b′) and (c′) hold both for $\widetilde{\Delta} = \eta$ and for $\widetilde{\Delta} = \zeta$ $(\eta, \zeta \in \underline{QI}_t)$. Then the condition $\eta \overset{t}{\underset{\gamma}{=}} \zeta$ holds for $\gamma \in \Gamma_1$ by requirement (c′), and also for $\gamma \in \Gamma - \Gamma_1$ by requirement (b′). Then it holds for every $\gamma \in \Gamma$ so that by Theor. 10.2 $\eta = \zeta$.

QED

The preceding theorem assures that rule (δ_9) is acceptable from the logical point of view. Theor. 11.2 is a preliminary concerning $\underline{a}_{\underline{t}}^*$ [Def. 11.2] and Theor. 11.3 shows that (δ_9) satisfies our purposes.

THEOR. 11.2. Rule (δ_9) implies that the improper quasi intension $\underline{a}_{\underline{t}\eta}^{\nu}$ of type $\underline{t}$ is $\widetilde{\mathrm{des}}_{\mathcal{M}\mathcal{V}}(\underline{a}_{\underline{t}}^*)$ [Def. 11.2] for every model $\mathcal{M}$ and value assignment $\mathcal{V}$.

Proof. Let $\phi(\underline{x})$ be the matrix $\underline{x} \neq \underline{x}$. Hence $\sim(\exists_1 \underline{x})\phi(\underline{x})$ holds in every Γ-case. Then by rule (δ_9)—see (b′)—$\widetilde{\mathrm{des}}_{\mathcal{M}\mathcal{V}}[(\imath \underline{x})\phi(\underline{x})] \overset{t}{\underset{\gamma}{=}} \underline{a}_{\underline{t}\eta}^{\nu+1}$ holds for every $\gamma \in \Gamma$. Furthermore, by Def. 11.2, $\underline{a}_{\underline{t}}^*$ is $(\imath \underline{x})\phi(\underline{x})$. Then by Theor. 10.2, we have $\widetilde{\mathrm{des}}_{\mathcal{M}\mathcal{V}}(\underline{a}_{\underline{t}}^*) = \underline{a}_{\underline{t}\eta}^{\nu+1}$.

QED

THEOR. 11.3. If, besides rules (δ_1) to (δ_8), rule (δ_9) is also fulfilled, then requirements (b) and (c) on descriptions hold for every matrix $\phi(\underline{x})$ in ML^{ν}.

Proof. Assume rule (δ_9). Let $\sim(\exists_1 \underline{x})\phi(\underline{x})$ hold in γ (at $\mathcal{M}$ and $\mathcal{V}$)—briefly $\gamma \in \Gamma - \Gamma_1$—moreover let Δ be $(\imath \underline{x})\phi(\underline{x})$ and $\widetilde{\Delta} = \widetilde{\mathrm{des}}_{\mathcal{M}\mathcal{V}}(\Delta)$. Then by rule (δ_9)—see (b′)—$\widetilde{\Delta} \overset{t}{\underset{\gamma}{=}} \underline{a}_{\underline{t}\eta}^{\nu+1}$. By Theor. 11.2, $\widetilde{\mathrm{des}}_{\mathcal{M}\mathcal{V}}(\underline{a}_{\underline{t}}^*) = \underline{a}_{\underline{t}\eta}^{\nu+1}$. Hence by rule (δ_2) $(\imath \underline{x})\phi(\underline{x}) = \underline{a}_{\underline{t}}^*$ holds in γ. We conclude that condition (b) is satisfied.

Now we assume that $(\exists_1 \underline{x})\phi(\underline{x})$ holds in γ at $\mathcal{M}$ and $\mathcal{V}$. Then by rule (δ_9)—see (c′)—$\widetilde{\Delta}$ is equivalent [Def. 8.1] to every quasi intension $\widetilde{\underline{y}}$ such that the matrix $\phi(\underline{x})$ holds in γ at $\mathcal{M}$ and $\mathcal{V}\binom{\underline{x}}{\widetilde{\underline{y}}}$—see (11). Hence under Hyp. 11.1 the matrix $(\underline{y})\,[\phi(\underline{y}) \supset \underline{y} = (\imath \underline{x})\phi(\underline{x})]$ holds in γ at $\mathcal{M}$ and $\mathcal{V}$. We conclude that condition (c) is also fulfilled.

QED

N12. The L-truth in ML^ν of some logical axioms proposed for a ν-sorted modal calculus MC^ν.

Here we propose some axioms of a ν-sorted modal calculus MC^ν. Their L-truth in ML^ν [Def. 9.5] can be proved essentially by means of rules (δ_1) to (δ_9) [Ns8, 11]. Among these axioms are the analogues for ML^ν of the axioms considered in Rosser's theory of extensional logic—see axioms 1 to 12 in [32], p. 212—for the propositional calculus, the lower predicate calculus, and identity and classes, or properties. The above analogy is not always strong; the analogue for ML^ν of certain axioms in [32] is not very similar to the latter—for example in modal logic identity cannot be defined by means of classes as it is in extensional logic, as certain remarks by Carnap which we will note later indicate. We also prove the L-truth in ML^ν of an admissible axiom for attributes [AS12.19] which has no extensional analogue.

As in [11], by (. . .) we shall mean any (possibly empty) string of universal quantifiers $(\forall \underline{x}_1) \ldots (\forall \underline{x}_n)$; furthermore by $(\underline{N})$ we shall mean any string $\underline{q}_1 \underline{q}_2 \ldots \underline{q}_n$ where $\underline{q}_i$ $(\underline{i} = 1, \ldots, \underline{n})$ is either a universal quantifier or the modal one, i.e. $\underline{N}$, and where for $1 \leq \underline{i} < \underline{j} \leq \underline{n}$, $\underline{q}_i$ and $\underline{q}_j$ may not be distinct.

As for dropping parentheses we decide that the string signs (. . .) and $(\underline{N})$ have the smallest cohesive power, i.e.

$$(24) \qquad \Vdash (\ldots)\underline{p} \equiv (\ldots)(\underline{p}), \qquad \Vdash (\underline{N})\underline{p} \equiv (\underline{N})(\underline{p}).$$

Let $\underline{p}$, $\underline{q}$, and $\underline{r}$ be matrices and $\underline{x}$ a variable (of any ν-type) in ML^ν. Then the following matrices—to be called axioms because they are proposed as axioms for MC^ν—are L-true in ML^ν.

AS12.1. $(\underline{N})\underline{p} \supset \underline{pp}$.

AS12.2. $(\underline{N})\underline{pq} \supset \underline{p}$.

AS12.3. $(\underline{N})(\underline{p} \supset \underline{q}) \supset [\sim(\underline{qr}) \supset \sim(\underline{rp})]$.

AS12.4. $(\underline{N})(\underline{x})(\underline{p} \supset \underline{q}) \supset [(\underline{x})\underline{p} \supset (\underline{x})\underline{q}]$.

AS12.5. $(\underline{N})\underline{N}(\underline{p} \supset \underline{q}) \supset (\underline{Np} \supset \underline{Nq})$.

We now make some remarks concerning all of our axioms. Following Rosser—cf. [32]—we say, for instance, that the class of the matrices in ML^ν (EL^ν) which have the form $(\underline{N})\underline{p} \supset \underline{pp}$ is *axiom scheme* AS12.1—or briefly AS12.1—in MC^ν (EC^ν), and that these matrices are the instances of AS12.1. Furthermore the instances of the axiom schemes in ML^ν (EL^ν) will be called the axioms of MC^ν (EC^ν). For the sake of brevity we shall often say e.g. axiom AS12.1 instead of axiom scheme AS12.1.

The axioms in this section, AS12.$\underline{r}$ (in ML^ν) for $\underline{r} = 1$ to 23, a forthcoming strong axiom, AS25.1, and a modal axiom on natural numbers—cf. $(78)_1$—determine the modal calculus MC^ν.

It will turn out that for $\underline{r} = 5, 7, 9, 19, 23$ the axiom scheme AS12.$\underline{r}$ in EL^ν is empty and AS12.$\underline{r}$ in ML^ν has no extensional analogue, while AS12.13 in EL^ν is empty but AS12.13 has an extensional analogue, which is AS12.13' below. The axiom schemes AS12.$\underline{r}$ in EL^ν ($\underline{r} = 1, \ldots, 22$), AS12.13', and a forthcoming axiom on natural numbers—cf. $(78)_2$—determine an extensional calculus that is based on EL^ν and which will be denoted by EC^ν.

Let us remark that ASs12.1–3 are the analogues for MC^ν of Rosser's axioms for the propositional calculus—cf. axioms 1–3 in [32], p. 212. AS12.4 (in ML^ν) can be called the ordinary analogue for MC^ν of axiom 4 in [32], p. 212, or else the ordinary analogue for MC^ν of AS12.4 in EC^ν.

By the known analogy between the universal quantifier and the necessity sign N—see [12], p. 186, and [30], p. 188—AS12.5 may be called the modal analogue of AS12.4 (or of AS12.4 in EC^{ν} or else of axiom 4 in [32]).

Rules (δ_5) to (δ_8) imply the L-truth of these matrices:

AS12.6. (N) $p \supset (x)p$ where p is closed with respect to x.

AS12.7. (N) $p \supset Np$ where p is modally closed [Def. 4.3].

We regard AS12.7 to be the modal analogue of AS12.6 and of axiom 5 in [32], p. 213.

Now let Δ be a term free for x in the matrix $\phi(x)$, i.e. let no free occurrence of x in $\phi(x)$ belong to the scope of any operator $(\forall z)$ or (ιz) where z is a variable free in Δ. Moreover let $\phi(\Delta)$ result from the substitution of Δ for the free occurrences of x in $\phi(x)$. Then the following matrices are L-true in ML^{ν}:

AS12.8. (N) $(x)\phi(x) \supset \phi(\Delta)$.

AS12.9. (N) $Np \supset p$.

AS12.8 is the analogue for ML^{ν} of the conjunction of axioms 6 and 8 in [32], p. 213. The particular instance $(x)\phi(x) \supset \phi(x)$ of AS12.8 belongs to the lower predicate calculus and is the analogue for ML^{ν} of axiom 6 in [32]. Note that AS12.9 is to be taken as the precise modal analogue of either this instance of AS12.8 or axiom 6 in [32].

Let x, y, z be distinct variables. Then the matrices

AS12.10. (N) $x = x$.

AS12.11. (N) $x = y \supset y = x$.

AS12.12. (N) $(x = y)(y = z) \supset x = z$.

are L-true [Theor. 10.1]. Now let $\phi(y)$ result from $\phi(x)$ by substituting y for the free occurrences of x, and let none of these occurrences belong to the scope of the operator $(\forall \underline{y})$ or $(\imath \underline{y})$. Then the following matrix is $\underline{L}$-true:[39]

AS12.13. $(\underline{N})\underline{x} =^{\frown} \underline{y} \supset [\phi(\underline{x}) \equiv \phi(\underline{y})]$.

Let us call AS12.13 the principle of strict identity. It is similar to the principle of identity

AS12.13'. $(\ldots)\underline{x} = \underline{y} \supset [\phi(\underline{x}) \equiv \phi(\underline{y})]$

of extensional logic, precisely P8 in [11], p. 86, and axiom scheme 7A in [32], p. 163, but the above principle of identity is false in ML^{ν}. For instance this can be shown by choosing $\phi(\underline{x})$ to be $\underline{x} =^{\frown} \underline{y}$.

In connection with this principle and its converse, note that in extensional logic we can define the identity $\underline{x} = \underline{y}$ (also when $\underline{x}$ and $\underline{y}$ are individuals, cf. fn. 41 below) by means of the universal equivalence $(\underline{F})[\underline{F}(\underline{x}) \equiv \underline{F}(\underline{y})]$, and we can prove that identity is an equivalence relation—i.e. we can prove ASs12.10–12.

As Carnap remarks in [13], III.9.III, on the basis of the semantical theory (of extension and intension), it is quite obvious that, as was already known, in modal logic the universal equivalence $(\underline{F})[\underline{F}(\underline{x}) \equiv \underline{F}(\underline{y})]$ defines—i.e. is equivalent to—the strict identity $\underline{x} =^{\frown} \underline{y}$;[40] moreover if we take the concept of extensional property ($\underline{\text{Ext}}$) as undefined, then we can define simple identity.[41]

Now we assume that $\underline{t}_0 \in \tau^{\nu}, \ldots, \underline{t}_n \in \tau^{\nu}, \underline{t} = (\underline{t}_1, \ldots, \underline{t}_n)$, and $\theta = (\underline{t}_1, \ldots, \underline{t}_n : \underline{t}_0)$ hold and that $\underline{x}_1, \ldots, \underline{x}_n$, $\underline{F}$, $\underline{G}$, $\underline{f}$, and $\underline{g}$ are distinct variables of the respective types $\underline{t}_1, \ldots, \underline{t}_n, \underline{t}, \underline{t}, \theta$, and θ. Then the following matrices are $\underline{L}$-true:[42]

AS12.14. $(\underline{N})\underline{F} = \underline{G} \equiv (\forall \underline{x}_1, \ldots, \underline{x}_n)[\underline{F}(\underline{x}_1, \ldots, \underline{x}_n) \equiv \underline{G}(\underline{x}_1, \ldots, \underline{x}_n)]$.

AS12.15. $(\underline{N})\underline{f} = \underline{g} \equiv (\forall \underline{x}_1, \ldots, \underline{x}_n)[\underline{f}(\underline{x}_1, \ldots, \underline{x}_n) = \underline{g}(\underline{x}_1, \ldots, \underline{x}_n)]$.

Let us first remark that AS12.14 looks just like the primitive sentence P9 considered by Carnap in [11], p. 86, for a certain extensional language $\underline{B}$. P9 is called the extensionality principle because it says (in $\underline{B}$) that $\underline{F} = \underline{G}$ holds if and only if $\underline{F}$ and $\underline{G}$ hold for the same extensions—see [11], p. 113. On the other hand AS12.14 says (in ML^{ν}) that $\underline{F} = \underline{G}$ holds (in the actual Γ-case γ) if and only if $\underline{F}$ and $\underline{G}$ hold (in γ) for the same intensions, and AS12.15 is the analogue for functions of AS12.14 (concerning attributes). Therefore we may call ASs12.14 and 15 the intensionality principles.

Now we suppose, in addition to the assumptions about $\underline{x}_1, \ldots, \underline{x}_n$, $\underline{F}$, and $\underline{f}$ mentioned above, that $\underline{F}$ and $\underline{f}$ do not occur free either in the matrix $\underline{p}$ or in the term Δ. Then the following matrices are $\underline{L}$-true—see fn. 44 below—

AS12.16. $(\underline{N})(\exists \underline{F})(\forall \underline{x}_1, \ldots, \underline{x}_n)[\underline{F}(\underline{x}_1, \ldots, \underline{x}_n) \equiv \underline{p}]$. [Cf. axiom 12 in [32], p. 213.]

AS12.17. $(\underline{N})(\exists \underline{f})(\forall \underline{x}_1, \ldots, \underline{x}_n)[\underline{f}(\underline{x}_1, \ldots, \underline{x}_n) = \Delta]$.

By Theor. 11.3, requirements (b) and (c) on descriptions, considered in N11, hold for every matrix $\phi(\underline{x})$. Hence the following matrices[43] are $\underline{L}$-true under Hyp. 11.1:[44]

AS12.18. (I) $(\underline{N})(\exists_1 \underline{x})\phi(\underline{x})\phi(\underline{y}) \supset \underline{y} = (\iota \underline{x})\phi(\underline{x})$,
(II) $(\underline{N}) \sim(\exists_1 \underline{x})\phi(\underline{x}) \supset (\iota \underline{x})\phi(\underline{x}) = \underline{a}^*$ [Defs. 11.2, 3].

Now we also assume that $\mathcal{M}$ is a model, $\mathcal{V}$ is a value assignment, and γ is a Γ-case. Let ξ be the class of the $(\underline{n}+1)$-tuples $\langle \zeta_1, \ldots, \zeta_n, \gamma' \rangle$ such that in the first place $\zeta_1 \in \underline{QI}_{\underline{t}_1}, \ldots, \zeta_n \in \underline{QI}_{\underline{t}_n}$ and $\gamma' \in \Gamma$—hence $\xi \in \underline{QI}_{\underline{t}}$ for $\underline{t} = (\underline{t}_1, \ldots, \underline{t}_n)$—and in the second place the matrix $\underline{p}$ holds in γ (but not necessarily in γ') at $\mathcal{M}$ and $\mathcal{V}\begin{pmatrix} \underline{x}_1 \ldots \underline{x}_n \\ \zeta_{\underline{n}} \ldots \zeta_{\underline{n}} \end{pmatrix}$—see (11).

Now we remark that, by our definition of ξ, the matrix $(\forall \underline{x}_1, \ldots, \underline{x}_n)[\underline{F}(\underline{x}_1, \ldots, \underline{x}_n) \equiv p]$ holds in γ at $\mathcal{M}$ and $\mathcal{V}\binom{F}{\xi}$; moreover the matrix $(\forall \underline{x}_1, \ldots, \underline{x}_n)[\diamond \underline{F}(\underline{x}_1, \ldots, \underline{x}_n) \equiv \underline{NF}(\underline{x}_1, \ldots, \underline{x}_n)]$ holds at $\mathcal{M}$ and $\mathcal{V}\binom{F}{\xi}$ in every Γ-case γ'. Therefore we may assert the $\underline{L}$-truth of the following matrix:

AS12.19. $(\underline{N})(\exists \underline{F})(\forall \underline{x}_1, \ldots, \underline{x}_n)\{[\diamond \underline{F}(\underline{x}_1, \ldots, \underline{x}_n) \equiv \underline{NF}(\underline{x}_1, \ldots, \underline{x}_n)][\underline{F}(\underline{x}_1, \ldots, \underline{x}_n) \equiv \underline{p}]\}$ where $\underline{x}_1, \ldots, \underline{x}_n$ and $\underline{F}$ are distinct variables of the respective types $\underline{t}_1, \ldots \underline{t}_n$, and $(\underline{t}_1, \ldots, \underline{t}_n)$, and $\underline{p}$ is any matrix with no free occurrences of $\underline{F}$.

This axiom, proposed for MC^{ν}, has no extensional analogue, strictly speaking; however it can be connected to AS12.16 and hence to axiom 12 in [32], p. 213.

Zermelo's principle (the axiom of choice) can be expressed in ML^{ν} in the same way as in the extensional languages, i.e. as follows—cf. Def. 4.6—

AS12.20. $(\forall \underline{F}, \underline{G})[(\underline{F}, \underline{G} \in \Psi) \supset \sim(\exists \underline{x})(\underline{x} \in \underline{F} \wedge \underline{x} \in \underline{G})(\exists \underline{x})\underline{x} \in \underline{F}] \supset (\exists \underline{G})(\underline{F})[\underline{F} \in \Psi \supset (\exists_1 \underline{x})(\underline{x} \in \underline{F} \wedge \underline{x} \in \underline{G})]$.

The following two axioms are surely nonessential in modal or extensional logic, in that they mirror conditions (A) and (B) [N6] respectively and these conditions—on the nonexisting object $\underline{a}_{\underline{t}}^{\nu}$ $(\underline{t} \in \tau^{\nu})$—have a conventional character:

AS12.21. $(\underline{N}) \sim \underline{a}_{\underline{t}}^{*}(\underline{x}_1, \ldots, \underline{x}_n)$ for $\underline{t} = (\underline{t}_1, \ldots, \underline{t}_n)$,

AS12.22. $(\underline{N})\, \underline{a}^{*}_{(\underline{t}_1, \ldots, \underline{t}_n : \underline{t}_0)}(\underline{x}_1, \ldots, \underline{x}_n) = \underline{a}^{*}_{\underline{t}_0}$ $(\underline{t}_0, \ldots, \underline{t}_n \in \tau^{\nu})$, where $\underline{x}_1, \ldots, \underline{x}_n$ are $\underline{n}$ distinct variables of the respective types $\underline{t}_1, \ldots, \underline{t}_n$.

The instances of ASs12.21 and 12.22 that belong to EL^{ν} are axioms of EC^{ν}.

In accordance with some assertions made in N6 one might object that AS12.21 is confusing and that it would be better to replace it with the following axiom, which mirrors condition (A*) on the function $\underline{a}^{\nu}$ [N6]:

AS12.21*. $(\underline{N})\ \underline{a}_{\underline{t}}^{*}(\underline{x}_1, \ldots, \underline{x}_{\underline{n}}) \equiv \bigwedge_{\underline{i}=1}^{\underline{n}} \underline{x}_{\underline{i}} = \underline{a}_{\underline{t}_{\underline{i}}}^{*}$ for $\underline{t} = (\underline{t}_1, \ldots, \underline{t}_{\underline{n}})$

where $\underline{x}_1, \ldots, \underline{x}_{\underline{n}}$ are as in AS12.21.

It is preferable to include AS12.21 instead of AS12.21* into MC^{ν} or EC^{ν}, because on the one hand AS12.21 is more useful in the proof of Theor. 62.1, which is basic for showing the relative completeness of MC^{ν} [Theor. 63.1]; and on the other hand we can remedy the confusion caused by AS12.21 by introducing in ML^{ν} or EL^{ν} an additional description operator, $\imath^*$, by means of the following abbreviating recursive definition:

DEF. 12.1(a). $(\imath^* \underline{v}_{\underline{tr}})\underline{p} =_D (\imath \underline{v}_{\underline{tr}})\underline{p}$ for $\underline{t} \in \{1, \ldots, \nu\}$ or $\underline{t} = (\underline{t}_1, \ldots, \underline{t}_{\underline{n}} : \underline{t}_0)$,

(b). $(\imath^* \underline{v}_{\underline{tr}})\underline{p} =_D (\imath \underline{v}_{\underline{tr}}) [(\exists_1 \underline{x})\underline{p}\ \underline{v}_{\underline{tr}} = (\imath \underline{v}_{\underline{tr}})\underline{p} \vee \sim(\exists_1 \underline{x})\underline{p}\ \underline{v}_{\underline{tr}} = (\lambda \underline{x}_1, \ldots, \underline{x}_{\underline{n}}) \bigwedge_{\underline{i}=1}^{\underline{n}} \underline{x}_{\underline{i}} = \underline{a}_{\underline{t}_{\underline{i}}}^{*}]$ for $\underline{t} = (\underline{t}_1, \ldots, \underline{t}_{\underline{n}})$

where, for $\underline{i} = 1, \ldots, \underline{n}$, $\underline{x}_{\underline{i}}$ is the variable of type $\underline{t}_{\underline{i}}$ which is distinct from $\underline{x}_1$ to $\underline{x}_{\underline{i}-1}$ and has the smallest index.

Let us consider the case where condition (A*) in N6 is used instead of (A), and $\underline{p}$ is a matrix in ML^{ν} (EL^{ν}). Then we can prove that $(\imath^* \underline{x})\underline{p}$ is strictly equal (is equal) to $(\imath \underline{x})\underline{p}$ in MC^{ν} (EC^{ν}); furthermore in the same case the intensional designatum (the designatum) of $(\imath \underline{x})\underline{p}$ is the one of $(\imath^* \underline{x})\underline{p}$ in correspondence with condition (A)—i.e. the one of $(\imath^*\underline{x})\underline{p}$ in ML^ν (EL^ν). Hence for $(\imath^*\underline{x})\underline{p}$ the analogue of AS12.21* holds in MC^ν (EC^ν).

Thus those who find it confusing to use condition (A) in N6 and the corresponding axiom AS12.21 (in connection with descriptions) may be satisfied by using $\imath^*$ instead of $\imath$.

Since we assume that Γ has at least two elements, the following matrix, which characterizes a modal calculus in that it asserts the existence of a contingent proposition, is $\underline{L}$-true in ML^ν:

AS12.23. $(\underline{N})(\exists \underline{v}_{(1)1}, \underline{v}_{11})(\underline{v}_{11} \in^{\cup} \underline{v}_{(1)1} \wedge \diamond \sim \underline{v}_{11} \in \underline{v}_{(1)1})$.

N13. On extensions of designators in ML^ν. Comparison with those for some languages by Carnap.

Following Carnap, we say that in the Γ-case γ two designators in ML^ν have the same extension if they have equivalent intensions—see N8—hence equivalent QIs [Def. 8.1]. So one could represent the extensions in γ by the equivalence classes of the relation of being equivalent QIs in γ. As a consequence the representations of any two extensions $\underline{E}$ and $\underline{E}'$ connected to distinct Γ-cases would necessarily be distinct. Yet $\underline{E}$ and $\underline{E}'$ can be chosen so that one can consider them to be coinciding.[45] Such a discrepancy disappears if, in accordance with [12], §23, and on the basis of the next theorem, we represent the extensions for ML^ν by the $\underline{L}$-determinate intensions defined as follows:

DEF. 13.1. For $\underline{t} \in \bar{\tau}^\nu$ we say that ξ is an $\underline{L}$-determinate QI of type $\underline{t}$ (for ML^ν) if $\xi \in QI^\nu_{\underline{t}}$ and one of the following conditions holds:

(a) For some $\underline{r} \in \{\bar{1}, \ldots, \nu\}$ $\underline{t} = \underline{r}$ and ξ is a constant function of type $(\nu+1:\underline{r})$ (so that the counterdomain of ξ has exactly one element)—see the recursive condition (c) in N6.

(b) $\underline{t} = (\underline{t}_1, \ldots, \underline{t}_{\underline{n}})$ and $\xi \in \underline{QI}^{\nu}_{\underline{t}}$—see (8), (9)—[so that ξ is an $(\underline{n}+1)$-ary relation $\widetilde{R}$ of type $(\underline{t}^{\eta}_1, \ldots, \underline{t}^{\eta}_{\underline{n}}, \nu+1)$] and for every $\underline{n}$-tuple of objects $\widetilde{x}_1, \ldots, \widetilde{x}_{\underline{n}}$ and every pair of Γ-cases γ and γ', the relation $\widetilde{R}$ holds for $\widetilde{x}_1, \ldots, \widetilde{x}_{\underline{n}}$ and γ if and only if $\widetilde{R}$ holds for $\widetilde{x}_1, \ldots, \widetilde{x}_{\underline{n}}$ and γ' (briefly the $\underline{n}$-ary relation $(\lambda\widetilde{x}_1, \ldots, \widetilde{x}_{\underline{n}})\widetilde{R}(\widetilde{x}_1, \ldots, \widetilde{x}_{\underline{n}}, \gamma)$ is independent of γ).

(c) $\underline{t} = (\underline{t}_1, \ldots, \underline{t}_{\underline{n}}:\underline{t}_0)$ and ξ is a function of type $(\underline{t}^{\eta}_1, \ldots, \underline{t}^{\eta}_{\underline{n}}:\underline{t}^{\eta}_0)$—see (8) and condition (c) on p. 6—taking only $\underline{L}$-determinate QIs of type $\underline{t}_0$ as values.

(d) $\underline{t} = 0$ and either ξ is Γ, or ξ is the empty class.

Note that only the condition (c) involves a recursion, whereas (b) complies with the assertion that any $\underline{L}$-determinate $\underline{n}$-ary relation holds, in all Γ-cases, for the same $\underline{n}$-tuples of intensions. More precisely the $\underline{L}$-determinate QIs of type $\underline{t} = (\underline{t}_1, \ldots, \underline{t}_{\underline{n}})$ are the designata for the modally constant attributes of type $\underline{t}$ ($\underline{\text{Mconst}}_{\underline{t}}$) according to the following definition:

DEF. 13.2. $\underline{\text{Mconst}}_{\underline{t}}(\underline{F}) \equiv_D (\forall \underline{x}_1, \ldots, \underline{x}_{\underline{n}}) [\Diamond \underline{F}(\underline{x}_1, \ldots, \underline{x}_{\underline{n}}) \equiv \underline{NF}(\underline{x}_1, \ldots, \underline{x}_{\underline{n}})]$

where $\underline{t} = (\underline{t}_1, \ldots, \underline{t}_{\underline{n}})$, $\underline{F}$ is a variable of type $\underline{t}$, and $\underline{x}_{\underline{i}}$ is the first variable of type $\underline{t}_{\underline{i}}$ distinct from $\underline{x}_1, \ldots, \underline{x}_{\underline{i}-1}$ $(\underline{i} = 1, \ldots, \underline{n})$.

THEOR. 13.1. Let $\underline{t} \in \bar{\tau}^{\nu}$, $\xi \in \underline{QI}^{\nu}_{\underline{t}}$, and $\gamma \in \Gamma$. Then there is exactly one $\underline{L}$-determinate intension ζ equivalent to ξ in γ.

Proof. In addition to the hypotheses of the theorem let $\bar{\zeta}$ be the class of the $\underline{L}$-determinate intensions equivalent to ξ in γ.

Then for $\underline{t} = 0$, by Defs. 8.1(d) and 13.1(d), $\zeta \in \bar{\zeta}$ if and only if ζ is Γ or Λ according to whether $\gamma \in \xi$ holds or not; hence Theor. 13.1 holds for $\underline{t} = 0$.

Let $\underline{r} \in \{1, \ldots, \nu\}$ hold. Then, by Defs. 8.1(a) and 13.1(a), we have $\zeta \in \bar{\xi}$ for $\underline{t} = \underline{r}$ if and only if ζ is the function of type $(\nu+1:\underline{r})$ defined as follows: $\zeta(\gamma') = \xi(\gamma)$ for every $\gamma' \in \Gamma$; hence Theor. 13.1 holds for $\underline{t} = \underline{r}$.

For $\underline{t} = (\underline{t}_1, \ldots, \underline{t}_n)$, by Defs. 8.1(b) and 13.1(b), $\zeta \in \bar{\xi}$ if and only if ζ is the set of the $\underline{n}$-tuples $\langle \tilde{\underline{x}}_1, \ldots, \tilde{\underline{x}}_n, \gamma' \rangle$ such that $\langle \tilde{\underline{x}}_1, \ldots, \tilde{\underline{x}}_n, \gamma \rangle \in \xi$ and $\gamma' \in \Gamma$; hence Theor. 13.1 holds for $\underline{t}$.

Now let $\underline{m}$ be any positive integer and, as an inductive hypothesis, let our theorem hold for every type $\underline{t}$ with $\underline{\text{lev}[t]} < \underline{m}$—see (3).

Suppose $\underline{\text{lev}[t]} = \underline{m}$. We already know that our theorem holds for such $\underline{t}$, except in the case $\underline{t} = (\underline{t}_1, \ldots, \underline{t}_n : \underline{t}_0)$. Hence let us consider this case: Let $\tilde{\underline{x}}_1 \in \underline{\text{QI}}^{\nu}_{\underline{t}_1}, \ldots, \tilde{\underline{x}}_n \in \underline{\text{QI}}^{\nu}_{\underline{t}_n}$. Then, since $\xi \in \underline{\text{QI}}^{\nu}_{\underline{t}}$, $\xi(\tilde{\underline{x}}_1, \ldots, \tilde{\underline{x}}_n) \in \underline{\text{QI}}^{\nu}_{\underline{t}_0}$. Furthermore by (3) $\underline{\text{lev}[t_0]} < \underline{\text{lev}[t]} = \underline{m}$. Hence by our inductive hypothesis, there is exactly one $\underline{L}$-determinate quasi intension $\tilde{\underline{y}}$ of type $\underline{t}_0$ equivalent to $\xi(\tilde{\underline{x}}_1, \ldots, \tilde{\underline{x}}_n)$ in γ. Let us set $\tilde{\underline{y}} = \zeta(\tilde{\underline{x}}_1, \ldots, \tilde{\underline{x}}_n)$. Hence by the above arbitrariness of $\tilde{\underline{x}}_1, \ldots, \tilde{\underline{x}}_n$ the preceding equality defines (exactly) one quasi intension ζ of type $\underline{t}$. By Def. 13.1(c) this ζ is an $\underline{L}$-determinate QI. Moreover by Def. 8.1(c) ζ is equivalent to ξ in γ. Hence $\zeta \in \bar{\xi}$.

We now add the assumption $\zeta' \in \bar{\xi}$. Then by the above uniqueness property of $\tilde{\underline{y}}$, $\zeta'(\tilde{\underline{x}}_1, \ldots, \tilde{\underline{x}}_n) = \tilde{\underline{y}} = \zeta(\tilde{\underline{x}}_1, \ldots, \tilde{\underline{x}}_n)$ for every quasi intension $\tilde{\underline{x}}_1, \ldots, \tilde{\underline{x}}_n$ of the respective types $\underline{t}_1, \ldots, \underline{t}_n$. Hence $\zeta' = \zeta$. We conclude that our theorem holds for $\underline{\text{lev}[t]} = \underline{m}$ also in the case $\underline{t} = (\underline{t}_1, \ldots, \underline{t}_n : \underline{t}_0)$.

Hence by the principle of induction our theorem holds for every $\underline{t} \in \bar{\tau}^{\nu}$.

QED

Let us observe that, by Def. 13.1(a), (d) for $\underline{t} = 0, \ldots, \nu$ our $\underline{L}$-determinate QIs substantially coincide with those considered by

Carnap in [13] for his language $\underline{L}_2$. But by Def. 13.1(b), for $\underline{t}$ = $(\underline{t}_1, \ldots, \underline{t}_n)$ our $\underline{L}$-determinate QIs are in a natural one-to-one correspondence with the relations forming the class $\underline{O}^{\nu}_{\underline{t}^e}$ where $\underline{t}^{\underline{e}} = (\underline{t}_1^{\eta}, \ldots, \underline{t}_n^{\eta})$—see (8).[46] On the other hand the extensions of type $\underline{t}$ for Carnap's language $\underline{L}_2$ may be represented simply by the elements of $\underline{O}^{\nu}_{\underline{t}}$ [N6]. We mean that, roughly speaking, the extensions for relations in Carnap's modal language $\underline{L}_2$ are classes of $\underline{n}$-tuples of extensions (for an extensional language), whereas the extensions for relations in our modal language ML^{ν} are classes of $\underline{n}$-tuples of intensions.

Our extensions are more complex, but this is essentially connected with the advantage that in ML^{ν} the sign $\underline{N}$ can be also used within the scopes of the iota and lambda operators [Def. 4.4]. Incidentally in Memoir 3, N56, we shall introduce within ML^{ν} a set of systems $(\underline{v}_1^*, \ldots, \underline{v}_2^*)$ where $\underline{v}_{\underline{r}}^*$ is an analogue of the domain $\underline{D}_{\underline{r}}$ ($\underline{r} = 1, \ldots, \nu$). On the basis of $\underline{v}_1^*, \ldots, \underline{v}_2^*$ and another entity we shall define an analogue of $\underline{QI}_{\underline{t}}$ within ML^{ν} for $\underline{t} \epsilon \bar{\tau}^{\nu}$. Using these tools it is easy to write an analogue of Def. 13.1; thus one introduces within ML^{ν} the $\underline{L}$-determinate intensions based on the system $(\underline{v}_1^*, \ldots \underline{v}_2^*)$ (and another entity of little relevance).

N14. Some L-true matrices in ML^{ν} and some fallacies. The lambda operator in ML^{ν}.

Here are some $\underline{L}$-true matrices in ML^{ν}. Some concern existential quantification possibly combined with modalities; others concern uniqueness properties and descriptions. The $\underline{L}$-truth of these matrices can be proved directly by using the preceding designation rules, and also indirectly by deducing them from AS12.1–18 by means of valid rules of inference.

These $\underline{L}$-true matrices are useful, I think, for demonstrating the adequacy of the preceding designation rules. To the same purpose we shall prove that certain material implications in ML^{ν} are not $\underline{L}$-true.

We assume Hyp. 11.1 and define "there is a strictly unique $\underline{x}$ such that," briefly $(\exists_1^{\frown}\underline{x})$, as follows:

DEF. 14.1. $(\exists_1^{\frown}\underline{x})\phi(\underline{x}) \equiv_D (\exists\underline{x})\{\phi(\underline{x})(\underline{y})[\phi(\underline{y}) \supset \underline{y} = {}^{\frown}\underline{x}]\}$.

One can prove that:[47]

(25) $\Vdash (\exists_1^{\frown}\underline{x})\underline{p} \supset (\exists_1\underline{x})\underline{p}$, $\Vdash \Diamond(\exists_1^{\frown}\underline{x})\underline{Np} \supset N(\exists_1^{\frown}\underline{x})\underline{Np}$,

(26) $\Vdash N(\exists_1\underline{x})\underline{p} \equiv (\exists_1^{\frown}\underline{x})\underline{Np}^{(ex)}$ [Def. 11], $\Vdash (\exists_1\underline{x})\underline{p}(\exists\underline{x})\underline{Np} \supset (\exists_1\underline{x})\underline{Np}$,

(27) $\Vdash \Diamond(\exists\underline{x})\underline{Np} \supset \underline{N}(\exists\underline{x})\underline{Np}$, $\Vdash \underline{N}(\exists_1\underline{x})\underline{p}\,\Diamond(\exists\underline{x})\underline{Np} \supset \underline{N}(\exists_1^{\frown}\underline{x})\underline{Np}$.

THEOR. 14.1. The following formulas are not $\underline{L}$-true in ML^{ν} for a suitable choice of the matrix $\underline{p}$:

(28)
$$(\exists_1\underline{x})\underline{Np} \supset (\exists_1^{\frown}\underline{x})\underline{Np}, \qquad (\exists_1\underline{x})\underline{Np} \supset (\exists_1^{\frown}\underline{x})\underline{p},$$
$$(\exists_1^{\frown}\underline{x})\underline{Np} \supset (\exists_1\underline{x})\underline{p}, \qquad (\exists_1^{\frown}\underline{x})\underline{Np} \supset (\exists_1^{\frown}\underline{x})\underline{p}.$$

Proof. Let the matrices (12) and $\underline{p}_1$ hold in the Γ-case γ (at $\mathcal{M}$ and $\mathcal{V}$). As to $(28)_1$ let $\phi_1(\underline{x})$ be $\underline{x} = \underline{M}_1 \vee \sim\underline{p}_1$ ($\underline{x} = \mathfrak{m}$). Then, since $\underline{p}_1$ holds in γ, by (12), $\phi_1(\underline{M}_1)$ and $(\exists_1\underline{x})N\phi_1(\underline{x})$ also hold. But by Def. 14.1 $(\exists_1^{\frown}\underline{x})\underline{N}\phi_1(\underline{x})$ does not (indeed by (12), $N\phi_1(\underline{M}_1)$, $\underline{N}\phi_1(\mathfrak{m})$, and $\sim\underline{M}_1 = {}^{\frown}\mathfrak{m}$ hold in γ). Hence, taking $\underline{p}$ to be $\phi_1(\underline{x})$, $(28)_1$ does not hold in γ.

As an obvious consequence, taking $\underline{p}$ to be $\underline{N}\phi_1(\underline{x})$, $(28)_2$ does not hold in γ.

As to $(28)_3$ let $\phi_2(\underline{x})$ be $\underline{p}_1(\underline{x} = {}^{\frown}\underline{M}_2) \vee \underline{x} = {}^{\frown}\mathfrak{m}$.

Hence, since by (12) $\Diamond\sim\underline{p}_1$ and $\sim \mathfrak{m} = {}^{\frown}\underline{M}_2$ hold in γ, $\sim\underline{N}\phi_2(\underline{M}_2)$ also does. Hence $(\underline{x})[\underline{N}\phi_2(\underline{x}) \supset \underline{x} = {}^{\frown}\mathfrak{m}]$ holds in γ, so that on the one hand, by Def. 14.1 $(\exists_1^{\cap} \underline{x}) \underline{N}\phi_2(\underline{x})$ holds in γ. On the other hand, by (12) $\phi_2(\mathfrak{m})$, $\phi_2(\underline{M}_2)$, and $\mathfrak{m} \neq \underline{M}_2$ hold in γ, so that by Def. 11.3 $\sim(\exists_1 \underline{x}) \phi_2(\underline{x})$ also does.

Hence taking $\underline{p}$ to be $\phi_2(\underline{x})$, $(28)_3$ does not hold in γ.

Nor can $(28)_4$ hold for the same choice of $\underline{p}$, because, if it did, then by $(25)_1$, $(28)_3$ would also.

QED

THEOR. 14.2. Let the term $\underline{A}$ be free for the variable $\underline{x}$ in the matrix $\underline{F}(\underline{x}, \underline{A})$, let the matrix $\underline{F}(\underline{A}, \underline{A})$ result from $\underline{F}(\underline{x}, \underline{A})$ by replacing $\underline{x}$ by $\underline{A}$ in all its free occurrences, and let $\underline{x}$ not occur free in $\underline{A}$. Then:

(29) $$\Vdash \underline{F}(\underline{A}, \underline{A}) \supset (\exists \underline{x}) \underline{F}(\underline{x}, \underline{A}).$$

As a corollary, from $\Vdash \underline{A} = {}^{\cap}\underline{A}$ we obtain

(30) $\Vdash (\exists \underline{x}) \underline{x} = {}^{\cap}\underline{A}$ in case $\underline{x}$ does not occur free in $\underline{A}$.

Besides requirements (b) and (c) considered in N11 one can prove that the descriptions in ML^{ν} fulfill the following conditions—see Def. 11.1—where $\underline{p}$, $\underline{p}'$, and $\phi(\underline{x})$ are any matrices:[48]

(31) $\Vdash (\exists_1^{\cap} \underline{x}) \underline{x} =^{\cap} (\imath \underline{x}) \phi(\underline{x})$, $\quad \Vdash (\exists_1 \underline{x}) \underline{x} = (\imath \underline{x}) \phi(\underline{x})$,

(32) $\Vdash (\underline{x}) (\underline{p} \equiv \underline{p}') \supset (\imath \underline{x}) \underline{p} = (\imath \underline{x}) \underline{p}'$ (cf. axiom 9 in [32], p. 213],

(33) $\Vdash (\exists \underline{x}) \underline{N}\phi(\underline{x}) \underline{N}(\exists_1 \underline{x}) \phi(\underline{x}) \supset \phi [(\imath \underline{x}) \phi(\underline{x})]$,
$\Vdash \underline{N}(\exists_1 \underline{x}) \underline{p} \supset (\forall \underline{x}) [\underline{N}\underline{p}^{(ex)} \equiv \underline{x} =^{\cap} (\imath \underline{x}) \underline{p}]$.

THEOR. 14.3. For some matrix $\phi(\underline{x})$, $\underline{N}(\exists_1^{\cap} \underline{x}) \phi(\underline{x})$ does not $\underline{L}$-imply $\phi[(\imath \underline{x}) \phi(\underline{x})]$.

Indeed let $(12)_{1,2}$ hold as axioms; moreover take $\phi(\underline{x})$ to be $\underline{p}_1 \underline{x} = {}^{\cap}\underline{M}_1 \vee \sim \underline{p}_1 \underline{x} = {}^{\cap}\underline{M}_2$—cf. $(12)_3$. Hence $\Vdash \underline{N}(\exists_1^{\cap} \underline{x}) \phi(\underline{x})$, $\underline{p}_1 \Vdash (\imath \underline{x}) \phi(x) = \underline{M}_1$, and $\sim \underline{p}_1 \Vdash (\imath \underline{x}) \phi(\underline{x}) = \underline{M}_2$. So by $(12)_{1,2}$

$\Vdash \sim (\iota \underline{x}) \phi (\underline{x}) = {}^\frown \underline{M}_{\underline{i}}$ ($\underline{i} = 1, 2$). Then $\Vdash \underline{N} \sim \phi [(\iota \underline{x}) \phi (\underline{x})]$.

QED

As to the lambda operator, note that by ASs12.14–17 and Def. 11.3,

$$\text{(34)} \quad \Vdash (\exists_1 \underline{F}) (\forall \underline{x}_1, \ldots, \underline{x}_{\underline{n}}) [\underline{F}(\underline{x}_1, \ldots, \underline{x}_{\underline{n}}) \equiv \underline{p}],$$
$$\Vdash (\exists_1 \underline{f}) (\forall \underline{x}_1, \ldots, \underline{x}_{\underline{n}}) \underline{f}(\underline{x}_1, \ldots, \underline{x}_{\underline{n}}) = \Delta$$

where Δ is a term of type θ, $\underline{x}_1, \ldots, \underline{x}_{\underline{n}}$, $\underline{F}$, and $\underline{f}$ are distinct variables of the respective types $\underline{t}_1, \ldots, \underline{t}_{\underline{n}}$, $(\underline{t}_1, \ldots, \underline{t}_{\underline{n}})$, and $(\underline{t}_1, \ldots, \underline{t}_{\underline{n}} : \theta)$, and in addition $\underline{F}$ does not occur free in $\underline{p}$ nor does $\underline{f}$ in Δ.

By Ass12.14 and 15 the matrices $(\forall \underline{x}_1, \ldots, \underline{x}_{\underline{n}}) [\underline{F}(\underline{x}_1, \ldots, \underline{x}_{\underline{n}}) \equiv \underline{p}]$ and $(\forall \underline{x}_1, \ldots, \underline{x}_{\underline{n}}) \underline{f}(\underline{x}_1, \ldots, \underline{x}_{\underline{n}}) = \Delta$ are extensional with respect to $\underline{F}$ and $\underline{f}$ respectively.[49] So by (34) and the requirement (a) in N11 [Theor. 11.3] the lambda operator [Defs. 4.5 and 4.6] has the following fundamental properties:

$$\text{(35)} \quad \Vdash [(\lambda \underline{x}_1, \ldots, \underline{x}_{\underline{n}}) \underline{p}] (\underline{x}_1, \ldots, \underline{x}_{\underline{n}}) \equiv \underline{p},$$
$$\Vdash [(\lambda \underline{x}_1, \ldots, \underline{x}_{\underline{n}}) \Delta] (\underline{x}_1, \ldots, \underline{x}_{\underline{n}}) = \Delta.$$

Let us remark that in a theory of extensional logic where a nonexisting object is used, $\underline{n}$-ary functions can be defined to be the $\underline{n}$+1-ary relations that fulfill the condition $(\forall \underline{x}_1, \ldots, \underline{x}_{\underline{n}}) (\exists \underline{y}) \underline{R}(\underline{y}, \underline{x}_1, \ldots, \underline{x}_{\underline{n}})$ in $\underline{R}$. Likewise in a theory of modal logic that is like ours and uses a nonexisting object, $\underline{n}$-ary functions can be defined to be the $(\underline{n}+1)$-ary relations that (1) are modally constant and (2) fulfill the condition $(\forall \underline{x}_1, \ldots, \underline{x}_{\underline{n}}) (\exists_1^\frown \underline{y}) \underline{R}(\underline{y}, \underline{x}_1, \ldots, \underline{x}_{\underline{n}})$ in $\underline{R}$.

3

Translation in a Strong Sense of the ν-sorted Modal Language ML^{ν} into the (ν+1)-sorted Extensional Language $EL^{\nu+1}$

N15. Explicit translation rules of ML^{ν} into $EL^{\nu+1}$.

The semantical rules (δ_1) to (δ_9) for ML^{ν} induce a translation of ML^{ν} into an extensional language that may usefully be identified with $EL^{\nu+1}$. Such a translation is interesting, I think, first because it proves the so-called thesis of extensionality[50] with respect to the general modal language ML^{ν} (and in a strong sense; see fn. 20); and second because such a translation may weaken the antipathy of some physicists to the use of modal languages—cf. [13], III.9.IV, p. 896. This is of interest because ML^{ν} is intended to be applied to the foundations of mechanics set up according to certain ideas of Painlevé.

Therefore here we will make the translation explicit on the basis of rules (δ_1) to (δ_9), and in N16 we will briefly lay down obvious common semantical rules for $EL^{\nu+1}$ and enunciate a (natural) theorem relating the intensional designatum of every wff Δ in ML^{ν} to the designatum of its (extensional) translation Δ^{η} into $EL^{\nu+1}$.

Let κ be the variable $\underline{v}_{\underline{t}1}$ for $\underline{t} = \nu+1$ and let us call it the case variable.

In connection with our definition of equivalent QIs [Def. 8.1] we here formally define in $EL^{\nu+1}$ the matrix $\underline{W} \overset{t}{\underset{\kappa}{=}} \underline{Z}$ to be read as: $\underline{W}$ and $\underline{Z}$ are [objects of type $\underline{t}^{\eta}$—see (8)] equivalent in the (possible) case κ ($\underline{t} \in \tau^{\nu}$).

The following set [Def. 15.1] of formal abbreviating definitions in $EL^{\nu+1}$ is laid down recursively under the hypotheses that $\underline{W}$ and $\underline{Z}$ are arbitrary terms in $EL^{\nu+1}$ of type $\underline{t}^{\eta}$—see (8); that

$\underline{x}_1, \ldots, \underline{x}_n$ are distinct variables of the respective types $\underline{t}_1^\eta, \ldots, \underline{t}_n^\eta$ (so that they are also distinct from the case variable κ); and that $\underline{x}_i$ is the first variable of the type $\underline{t}_i$ that occurs free in neither $\underline{W}$ nor $\underline{Z}$ and is distinct from $\underline{x}_1, \ldots, \underline{x}_{i-1}$.

DEF. 15.1.

(a) $\underline{W} = \frac{t}{\kappa} \underline{Z} \equiv_D \underline{W}(\kappa) = \underline{Z}(\kappa)$ for $\underline{t} = \underline{r}$ ($\underline{r} = 1, \ldots, \nu$).

(b) $\underline{W} = \frac{t}{\kappa} \underline{Z} \equiv_D (\forall \underline{x}_1, \ldots, \underline{x}_n) [\underline{W}(\underline{x}_1, \ldots, \underline{x}_n, \kappa) \equiv \underline{Z}(\underline{x}_1, \ldots, \underline{x}_n, \kappa)]$ for $\underline{t} = (\underline{t}_1, \ldots, \underline{t}_n)$.

(c) $\underline{W} = \frac{t}{\kappa} \underline{Z} \equiv_D (\forall \underline{x}_1, \ldots, \underline{x}_n) [\underline{W}(\underline{x}_1, \ldots, \underline{x}_n) \frac{t_0}{\kappa} \underline{Z}(\underline{x}_1, \ldots, \underline{x}_n)]$ for $\underline{t} = (\underline{t}_1, \ldots, \underline{t}_n : \underline{t}_0)$.

Now in order to translate $(\exists_1 \underline{x}) \phi (\underline{x})$ [Def. 11.3] into $EL^{\nu+1}$ we set in $EL^{\nu+1}$

DEF. 15.2. $(\exists ! \alpha) \Psi (\alpha) \equiv_D (\exists \alpha) \{\Psi(\alpha)(\beta) [\Psi(\beta) \supset \alpha = \frac{t}{\kappa} \beta]\} (\underline{t} \epsilon \tau^\nu)$ where α is a variable of type $\underline{t}^\eta$, and β is the first variable of type $\underline{t}^\eta$ distinct from α and not occurring free in the matrix $\Psi(\alpha)$.

Here are the translation rules $(\underline{T}_1)$ to $(\underline{T}_9)$, which associate with every wff Δ in ML^ν its extensional translation Δ^η in $EL^{\nu+1}$. These rules correspond in order to the formation rules $(\underline{f}_1)$ to $(\underline{f}_9)$ of ML^ν considered in Def. 3.1. Analogous to the way in which the designation rules (δ_1) to (δ_8) were presented in N8 for $\underline{r} = 1, \ldots, 9$, the antecedent of rule $(\underline{T}_r)$ is the antecedent of rule $(\underline{f}_r)$. Therefore in the middle column of the next table only the main part of the antecedent of $(\underline{T}_r)$ is explicitly written.

Rule	If Δ is	then its extensional translation Δ^η in $EL^{\nu+1}$ is
$(\underline{T}_1)$	$\underline{v}_{tn}$ or $\underline{c}_{tn}$,	$\underline{v}_{\theta n}$ or $\underline{c}_{\theta n}$ respectively for $\theta = \underline{t}^\eta$—see $(\bar{8})$.

($\underline{T}_2$)	$\Delta_1 = \Delta_2$ where Δ_1, $\Delta_2 \in \mathscr{E}^{\nu}_{\underline{t}_1}$ [Def. 3.1],	$(\Delta^{\eta}_1) \overset{\underline{t}_1}{\underset{\kappa}{=}} (\Delta^{\eta}_2)$ [κ is $\underline{v}_{\theta 1}$ for $\theta = \nu+1$; see Def. 15.1].
($\underline{T}_3$)	the matrix $\underline{R}(\Delta_1, \ldots, \Delta_{\underline{n}})$,	$\underline{R}^{\eta}(\Delta^{\eta}_1, \ldots, \Delta^{\eta}_{\underline{n}}, \kappa)$.
($\underline{T}_4$)	the term $\phi(\Delta_1, \ldots, \Delta_{\underline{n}})$,	$\phi^{\eta}(\Delta^{\eta}_1, \ldots, \Delta^{\eta}_{\underline{n}})$.
($\underline{T}_5$)	$\sim\Delta_1$,	$\sim\Delta^{\eta}_1$.
($\underline{T}_6$)	$\Delta_1 \wedge \Delta_2$,	$\Delta^{\eta}_1 \wedge \Delta^{\eta}_2$.
($\underline{T}_7$)	$(\forall \underline{x})\Delta_1$,	$(\forall \underline{x}^{\eta})\Delta^{\eta}_1$.
($\underline{T}_8$)	$\underline{N}\Delta_1$,	$(\kappa)\Delta^{\eta}_1$.
($\underline{T}_9$)	$(\imath\underline{x})\phi(\underline{x})$ where $\phi(\underline{x})$ is a matrix in ML^{ν} (not necessarily containing the variable $\underline{x}$ free),	$(\imath\underline{x}^{\eta})(\kappa)\{(\exists!\underline{y}^{\eta})\phi(\underline{y})^{\eta}(\underline{y}^{\eta})[\phi(\underline{y})^{\eta} \supset \underline{x}^{\eta} \overset{\underline{t}}{\underset{\kappa}{=}} \underline{y}^{\eta}] \vee \sim(\exists!\underline{x}^{\eta})\phi(\underline{x})^{\eta}\underline{x}^{\eta} \overset{\underline{t}}{\underset{\kappa}{=}} \underline{a}^{*}_{\underline{t}\eta}\}$ [Def. 15.2] where $\underline{y}$ is the variable that has the same type as $\underline{x}$ and with $\phi(\underline{y})$ satisfies Hyp. 11.1.

Note that the translation rules imply the following theorem which, among other things, justifies the locution "modally closed matrix" [Def. 4.3]:

THEOR. 15.1. Let Δ be a term and $\underline{p}$ be a matrix in ML^{ν}. Then Δ^{η} (the extensional translation of Δ) is closed with respect to the case variable κ. Furthermore $\underline{p}^{\eta}$ is closed or open with respect to κ depending on whether $\underline{p}$ is modally closed or not.

N16. Designation rules for $EL^{\nu+1}$ and semantical justification of the translation rules.

Below are the designation rules ($\underline{d}_1$) to ($\underline{d}_8$) for EL^{ν}. For every value assignment $\underline{V}$ and model $\mathfrak{M}$ [Def. 6.1], they associate every expression Δ of type $\underline{t}$ in EL^{ν} ($\underline{t} \in \bar{\tau}^{\nu}$) with an extensional designatum $\bar{\Delta}$. For $\underline{t} \in \tau^{\nu}$, $\bar{\Delta} \in \underline{O}_{\underline{t}}^{\nu}$ [N6], and for $\underline{t} = 0$, $\bar{\Delta}$ is one of the truth-values $\underline{T}$ and $\underline{F}$. $\bar{\Delta}$ is unique, as appears from the nature of the rules. Therefore we write $\bar{\Delta} = \underline{des}_{\mathfrak{M}\underline{V}}(\Delta)$. The rules below correspond to the formation rules ($\underline{f}_1$) to ($\underline{f}_8$) in the same way the translation rules ($\underline{T}_1$) to ($\underline{T}_8$) do. Therefore the table below has been set up according to the same criteria as the one for rules ($\underline{T}_1$) to ($\underline{T}_8$). For the sake of brevity we assume that Δ and $\Delta_{\underline{i}}$ ($\underline{i} = 1, \ldots, \underline{n}$) are terms in EL^{ν}, $\underline{R}$ is an attribute, and ϕ is a functor; and the definitions

(36) $\bar{\Delta}_{\underline{i}} = \underline{des}_{\mathfrak{M}\underline{V}}(\Delta_{\underline{i}})$ ($\underline{i} = 1, \ldots, \underline{n}$), $\bar{\underline{R}} = \underline{des}_{\mathfrak{M}\underline{V}}(\underline{R})$, $\bar{\phi} = \underline{des}_{\mathfrak{M}\underline{V}}(\phi)$

are assumed to hold:

Rule	If Δ is	then $\bar{\Delta} = \underline{des}_{\mathfrak{M}\underline{V}}(\Delta)$ is
($\underline{d}_1$)	$\underline{v}_{\underline{tn}}$ or $\underline{c}_{\underline{tn}}$,	$\underline{V}(\underline{v}_{\underline{tn}})$ or $\mathfrak{M}(\underline{c}_{\underline{tn}})$ respectively.
($\underline{d}_2$)	$\Delta_1 = \Delta_2$,	$\underline{T}$ or $\underline{F}$ according to whether $\bar{\Delta}_1 = \bar{\Delta}_2$ or not.
($\underline{d}_3$)	the matrix $\underline{R}(\Delta_1, \ldots, \Delta_{\underline{n}})$,	$\underline{T}$ or $\underline{F}$ according to whether—cf. $(36)_{1,2}$—the relation $\bar{\underline{R}}$ holds for $\bar{\Delta}_1, \ldots, \bar{\Delta}_{\underline{n}}$ (i.e. whether the $\underline{n}$-tuple $\langle \bar{\Delta}_1, \ldots, \bar{\Delta}_{\underline{n}} \rangle$ belongs to the class $\bar{\underline{R}}$ of $\underline{n}$-tuples) or not.

Rule	If Δ is	then $\bar{\Delta} = \text{des}_{\mathfrak{M}\underline{V}}(\Delta)$ is
($\underline{d}_4$)	the term $\phi(\Delta_1, \ldots, \Delta_{\underline{n}})$,	$\bar{\phi}(\bar{\Delta}_1, \ldots, \bar{\Delta}_{\underline{n}})$, where $(36)_{1,3}$ hold.
($\underline{d}_5$)	$\sim\Delta_1$ [$\Delta_1 \in \underline{E}_0^{\nu}$, see N3],	the truth-value different from $\bar{\Delta}_1$—cf. $(36)_1$.
($\underline{d}_6$)	$\Delta_1 \wedge \Delta_2$,	$\underline{T}$ provided $\bar{\Delta}_1 = \bar{\Delta}_2 = \underline{T}$; otherwise $\underline{F}$.
($\underline{d}_7$)	$(\forall\underline{x})\,\Delta_1$ where $\underline{x}$ is a variable of type $\underline{t}$ and $\Delta_1 \in \underline{E}_0^{\nu}$,	$\underline{T}$ provided for every $\xi \in \underline{O}_{\underline{t}}^{\nu}$, setting $\underline{V}' = \underline{V}\binom{\underline{x}}{\xi}$ [see (11)], we have $\text{des}_{\mathfrak{M}\underline{V}'}(\Delta_1) = \underline{T}$; otherwise $\underline{F}$.
($\underline{d}_8$)	$(\iota\underline{x})\,\Delta_1$ where $\underline{x}$ is a variable of type $\underline{t}$ and $\Delta_1 \in \underline{E}_0^{\nu}$,	the element ξ of $\underline{O}_{\underline{t}}^{\nu}$ fulfilling one of the following conditions: (a) ξ is the only object of type $\underline{t}$ ($\xi \in \underline{O}_{\underline{t}}^{\nu}$) such that $\text{des}_{\mathfrak{M}\underline{V}'}(\Delta_1) = \underline{T}$ for $\underline{V}' = \underline{V}\binom{\underline{x}}{\xi}$ [see (11)]. (b) ξ is $\underline{a}_{\underline{t}}^{\nu}$ and there is not exactly one element ξ of $\underline{O}_{\underline{t}}^{\nu}$ such that $\text{des}_{\mathfrak{M}\underline{V}'}(\Delta_1) = \underline{T}$ for $\underline{V}' = \underline{V}\binom{\underline{x}}{\xi}$.

DEF. 16.1. Let $\mathcal{M}$ and $\mathfrak{M}$ be models and $\mathcal{V}$ and $\underline{V}$ value assignments for ML^{ν} and $\text{EL}^{\nu+1}$ respectively—cf. Defs. 6.1 and 6.2. Let

(37) $\mathfrak{M}(\underline{c}_{\underline{tn}}^{\eta}) = \mathcal{M}(\underline{c}_{\underline{tn}})$, $\underline{V}(\underline{v}_{\underline{tn}}^{\eta}) = \mathcal{V}(\underline{v}_{\underline{tn}})$ for $\underline{t} \in \tau^{\nu}$ ($\underline{n} = 1, 2, \ldots$),

and let $\mathfrak{M}(\underline{c}_{\theta\underline{n}}) = \underline{V}(\underline{v}_{\theta\underline{n}}) = \underline{a}_{\theta}^{\nu}$ for every type θ not having the form $\underline{t}^{\eta}$ for any $\underline{t} \in \tau^{\nu}$. Then we call $\mathfrak{M}$ and $\underline{V}$ the extensional correspondents of $\mathcal{M}$ and $\mathcal{V}$ respectively and we write $\mathfrak{M} = \mathcal{M}^{\eta}$ and $\underline{V} = \mathcal{V}^{\eta}$.

Now, on the basis of the semantical rules (δ_1) to (δ_9) [Ns8, 11] for ML^ν, the designation rules $(\underline{d}_1)$ to $(\underline{d}_8)$ for $EL^{\nu+1}$, and the translation rules $(\underline{T}_1)$ to $(\underline{T}_9)$ of ML^ν into $EL^{\nu+1}$, one can prove the following theorem:

THEOR. 16.1. We assume that Δ is a term and $\underline{p}$ is a matrix in ML^ν, and that $\mathcal{M}$ is a model and $\mathcal{V}$ a value assignment for ML^ν. Then the quasi-intensional designatum $\widetilde{\Delta}$ of Δ at $\mathcal{M}$ and $\mathcal{V}$ is the designatum of the extensional translation Δ^η of Δ into $EL^{\nu+1}$ at the extensional correspondents $\mathcal{M}^\eta$ and $\underline{V}^\eta$ of $\mathcal{M}$ and $\underline{V}$; briefly

(38) $\underline{\widetilde{des}}_{\mathcal{M}\mathcal{V}}(\Delta) = \underline{des}_{\mathfrak{M}\underline{V}}(\Delta^\eta)$ for $\mathfrak{M} = \mathcal{M}^\eta$ and $\underline{V} = \mathcal{V}^\eta$.

Furthermore, for every Γ-case γ, $\underline{p}$ holds in γ at $\mathcal{M}$ and $\mathcal{V}$ [Def. 9.6(a)] if and only if $\underline{des}_{\mathfrak{M}\underline{V}}(\underline{p}^\eta) = \underline{T}$ for $\mathfrak{M} = \mathcal{M}^\eta$ and $\underline{V} = \mathcal{V}^\eta\binom{\kappa}{\gamma}$ [see (11)].

By Theor. 16.1 our translation $\Delta \rightarrow \Delta^\eta$ of ML^ν into $EL^{\nu+1}$ is strong in the sense (considered by Carnap in [13], p. 894) that for every designator Δ in ML^ν, its translation Δ^η characterizes the intension of Δ (also when Δ is a matrix); in addition by rule $(\underline{T}_8)$ in N15 the universal quantifier $(\forall\kappa)$ can be regarded as the translation of $\underline{N}$ into $EL^{\nu+1}$.[51]

PART II

SOME USEFUL CONCEPTS DEFINABLE IN THE MODAL LANGUAGE ML^{ν}; APPLICATIONS TO QUESTIONS CONCERNING FOUNDATIONS OF CLASSICAL MECHANICS AND EVERYDAY LIFE

N17. Introduction.

The foundations of classical mechanics according to Painlevé involve Mach's definition of mass—see [29], p. 65—and Painlevé's own definitions of force and inertial frame. Following Painlevé's point of view it is natural to use a modal language, and such a language is used in the scientific or didactical works on that subject by some professors of mechanics, e.g. by Signorini in [33]. In particular, in Painlevé's theory, which is followed by Signorini, a certain counterfactual conditional is essential.[52]

These works are certainly to be considered to be rigorous relative to both their subject and the time in which they were written. However, as Painlevé himself indicated,[53] they have lower standards than the rigorous works of pure mathematics. The possibility of basing them on an explicitly determined theory of formal logic does not seem to be envisaged by their authors.

There is a certain widespread dislike among physicists for modal logic that is due, I think, to the fact that it is more complex than extensional logic, less developed than the latter, and not as well known—cf. [13], III.9.IV, p. 896. As far as I know, until recently designation rules such as Carnap's—see [12] and [13]—were set up for no modal language as general as ML^{ν}, and in particular for no language admitting the necessity sign N within the scopes of the lambda and iota operators. This has its counterpart, I think, in the axiomatic treatment of classes—or properties—and natural numbers.

Furthermore for defining mass in a general modal language [N19] such as ML^{ν}[54] and from a unified point of view—cf. requirements (a) to (e) considered in N1—certain new modal concepts [N18] need to be introduced and a certain double use of nouns needs to be taken into account [Ns20, 23]—cf. fn. 5.

As might be expected from the aversion of physicists for modal logic, the first axiomatic theories defining mass and based on an explicit theory of formal logic were (directly) based on extensional logic.[55] Such theories can be regarded, I think, only as first steps toward a rigorous systematization of foundations of mechanics according to Painlevé. Indeed as suggested by some critical remarks by Rosser—see N21—they are logically incompatible with many fundamental problems considered in classical particle mechanics, e.g. the one about two bodies attracting each other according to Newton's law under generic initial conditions. In N21—see in particular fn. 65—it is emphasized that this incompatibility feature of the above theories can by no means be attributed to a lack of skill in the author but rather are attributable to the fact that such theories are based directly on extensional logic. This disadvantageous incompatibility does not appear in the general theory of foundations of classical particle mechanics set up in [4].[56]

In [4] the ordinary modal language used in the scientific works [29] and [33] is formalized using extensional logic in a certain (indirect) way based on a device that substantially enables us to speak explicitly of the possibility of propositions and which was suggested to the author by the semantical rules given by Carnap in an extensional metalanguage for certain modal languages—see [12]. The extensional language used in [4] is unusual. Its use caused the author to add (explicitly) a certain primitive concept—see fn. 6.

The substantial insufficiency of a direct extensional formalization of some actual scientific theories (foundations of classical mechanics according to Painlevé), the formally unsatisfactory nature of a certain indirect extensional formalization of the same theories and the usefulness of a suitably general theory of modal logic not yet in existence induced the author to construct the general modal language ML^{ν} considered in part 1 and to analyze it from the semantical point of view, and moreover to propose a system of logical axioms. (In Memoir 2 it is shown that this system enables us to develop a syntactical theory of modal logic as rich as the syntactical theories for extensional logic.)

In part 2, after the formal definition of certain basic modal concepts, in particular the one of <u>absolute attribute</u> [N18], Mach's definition of mass is put into a rigorous modal form based on ML^{ν} [Def. 19.1],[57] and the same can be done with the whole theory of foundations of classical mechanics set up in [29], [33], and [4]. (The arrangement of this theory based on the procedures used in part 2 to define mass, as well as the arrangement realized in [4], is free from any restrictions such as those noted in connection with the preceding theories based directly on extensional logic.) Furthermore in part 2 other applications of the basic modal concepts are presented.

Now we describe the contents of part 2 in more detail. The basic modal concept of <u>absolute attribute</u> [N18] is such that every absolute attribute $\underline{F}$ determines its corresponding extensional concept, in that the latter is precisely the extensionalization $\underline{F}^{(\underline{e})}$

[Def. 18.9] of the former. Some absolute concepts, e.g. the concept Nn of (modally prefixed) natural number and the concept Real of (prefixed) real number, can be defined on purely logical grounds [Ns19, 27], whereas others, e.g. the concept MP of (modally prefixed) mass point, are given by experience.

Both of the absolute concepts Real and MP are practically essential in our definition of mass [Def. 19.1]. More precisely, in N20 it is shown that neither of them can be replaced in Def. 19.1 by its extensionalization—cf. fn. 4. On the other hand, in some other contexts only the extensionalizations of Real and MP can be used [N23]. So a double use of common nouns, which is not distinguished by any grammatical marks, is put in evidence [N23].[58] This double use is connected with a certain combination of modalities with existential quantification, which is responsible, I think, for the difficulties arising from defining mass rigorously in a modal language. In a certain sense this double use mirrors the difference between substances and qualities [N23]. Within the field of physics the double use is analyzed in a simple and unified way, by means of absolute attributes in N23.

In N24 the quasi-absolute attributes are briefly introduced in order to take into account (in ML^{ν}) an extension of the double use of common nouns to everyday speech where, for instance, living beings are spoken of as well.

As a further application of the absolute properties a new axiom is proposed for a modal calculus MC^{ν} (based on ML^{ν}). This axiom [A25.1] (L-true in ML^{ν}) substantially says in ML^{ν} that the individuals of a given type have the same cardinal number in every possible case.

Finally [Ns26, 27] the concepts of closure and the above concept of natural number are explicitly defined in ML^{ν}. Furthermore, we show that when $\underline{D}_1$ to $\underline{D}_{\nu}$ are infinite, Peano's axioms are $\underline{L}$-true in ML^{ν} both when the ordinary phrases "natural number" and "equal to" are replaced by the absolute concept $\underline{Nn}$ and strict identity respectively, and when they are replaced by the extensional concept $\underline{Nn}^{(e)}$ and simple identity respectively [N27].

From these results it appears, as I shall show more explicitly in Memoir 2, which has a syntactical nature, that the well-known extensional theory for natural numbers can be immediately transferred into ML^{ν} in two ways. (Both are essential for foundations of mechanics according to Painlevé in connection with a general modal language such as ML^{ν}.)

Finally, some theorems relating the theories for $\underline{Nn}$ and $\underline{Nn}^{(e)}$ are briefly considered [N27].

4

Absolute Concepts and the Definition of Mass

N18. Absolute attributes. Extensional and intensional collections.

The problem of casting Mach's definition of mass into rigorous form in a general language such as ML^{ν}, which also complies with a unified point of view, involves in a fundamental way a certain combination of modalities with existential quantification, and for translating this combination into ML^{ν} correctly it is essential to introduce certain concepts, especially that of absolute attribute [Def. 18.8]. Such concepts are also essential for dealing with a certain double use of common nouns [N23] and for laying down a new modal axiom that (is L-true in ML^{ν} and) concerns the numbers of individuals (of any given type) in the various possible cases.

In order to illustrate the concept of absolute attribute we introduce the concept of extensionalization of attributes and the concepts of extensional collection and intensional collection formed with n objects $x_1, \ldots, x_n$; furthermore we consider some properties of these concepts.

Below we write in ML^{ν} common metalinguistic definitions for the calculus of classes (or properties), assuming that $x_1, \ldots, x_n$, F, and G are distinct variables of the respective types $t_1, \ldots, t_n$, t, and t where $t = (t_1, \ldots, t_n)$; and that x_i is the first variable of type t_i different from $x_1, \ldots, x_{i-1}$ ($i = 1, \ldots, n$):

DEF. 18.1. $\bar{F} =_D (\lambda x_1, \ldots, x_n) \sim F(x_1, \ldots, x_n)$ [Def. 4.4].

DEF. 18.2. $F \cap G =_D (\lambda x_1, \ldots, x_n) [F(x_1, \ldots, x_n) \wedge G(x_1, \ldots, x_n)]$.

DEF. 18.3. $F \cup G =_D \overline{\bar{F} \cap \bar{G}}$.

DEF. 18.4. $F \subseteq G \equiv_D (\forall x_1, \ldots, x_n) [F(x_1, \ldots, x_n) \supset G(x_1, \ldots, x_n)]$.

DEF. 18.5. $F \subset G \equiv_D F \subseteq G \wedge F \neq G$.

DEF. 18.6. $\Lambda_t =_D (\lambda v_{t1}) v_{t1} \neq v_{t1}$.

Now the concept $MSep_t$ of modally separated attribute of type t is introduced (understanding obvious assumptions on y_1 to y_n):

DEF. 18.7. $MSep_t \equiv_D (\forall x_1, y_1, \ldots, x_n, y_n) [F(x_1, \ldots, x_n)$
$F(y_1, \ldots, y_n) \diamond \bigwedge_{i=1}^{n} x_i = y_i \supset \bigwedge_{i=1}^{n} x_i =^{\wedge} y_i]$
[Defs. 6.3, 6.9].

The concept of modally separated property is extensional, i.e.

(39) $\Vdash \underline{\mathrm{Ext}}_{(\underline{t})}(\underline{\mathrm{MSep}}_{\underline{t}})$ [Defs. 18.7, 6.11, AS12.14].

We shall say that the attribute $\underline{F}$ is absolute if it is both modally constant (i.e. $\underline{F}^{\frown} = \underline{F}^{\smile}$, see Defs. 6.6, 6.8, and fn. 28) and modally separated; formally:

DEF. 18.8. $\underline{\mathrm{Abs}}_{\underline{t}} = \underline{\mathrm{MConst}}_{\underline{t}} \cap \underline{\mathrm{MSep}}_{\underline{t}}$ [Defs. 13.2, 18.2, 18.7].

Hence by Defs. 18.7 and 13.2, absolute properties can be characterized as follows:

(40) $\Vdash \underline{\mathrm{Abs}}_{\underline{t}}(\underline{F}) \equiv \underline{F}^{\smile} = \underline{F}^{\frown} (\forall \underline{x}, \underline{y}) [\underline{F}(\underline{x}) \underline{F}(\underline{y}) \underline{x} =^{\smile} \underline{y} \supset \underline{x} =^{\frown} \underline{y}]\ [\underline{t} = (\underline{t}_1)]$.

By Def. 18.8 or (40), if $\underline{F}$ may be an absolute attribute, then it must be so; in other words:

(41) $\Vdash \underline{\mathrm{MConst}}_{(\underline{t})}(\underline{\mathrm{Abs}}_{\underline{t}})$ [Defs. 13.2, 18.8].

Here the extensionalization $\underline{F}^{(\underline{e})}$ of the $\underline{n}$-ary attribute $\underline{F}$ is introduced—see fn. 28:

DEF. 18.9. $\underline{F}^{(\underline{e})}(\underline{x}_1, \ldots, \underline{x}_{\underline{n}}) \equiv_D (\exists \underline{y}_1, \ldots, \underline{y}_{\underline{n}}) [\underline{F}(\underline{y}_1, \ldots, \underline{y}_{\underline{n}}) \bigwedge_{\underline{i}=1}^{\underline{n}} \underline{y}_{\underline{i}} = \underline{x}_{\underline{i}}]$

so that the matrix $\underline{F}^{(\underline{e})}(\underline{x}_1, \ldots, \underline{x}_{\underline{n}})$ is extensional (also) with respect to $\underline{x}_1, \ldots, \underline{x}_{\underline{n}}$—cf. N11.

We want to remark that if in ~~every~~ one of two distinct Γ-cases γ_1 and γ_2 the absolute property $\underline{F}$ holds for at least two objects, then there is an absolute property $\underline{G}$ different from $\underline{F}$, such that $\underline{F}^{(\underline{e})} = \underline{G}^{(\underline{e})}$. each

Before giving an example of absolute properties illustrating this point, we introduce the symbols $\epsilon^{\smile}$ and $\epsilon^{\frown}$ related to ϵ—cf. Def. 4.6—in accordance with Defs. 6.3–8 as follows:

DEF. 18.10. $\underline{x}_1, \ldots, \underline{x}_n \in^{\cup} \underline{F} \equiv_D \bigwedge_{\underline{i}=1}^{\underline{n}} \diamond \underline{F}(\underline{x}_{\underline{i}})$,

DEF. 18.11. $\underline{x}_1, \ldots, \underline{x}_n \in^{\cap} \underline{F} \equiv_D \bigwedge_{\underline{i}=1}^{\underline{n}} \underline{NF}(\underline{x}_{\underline{i}})$,

where $\underline{x}_1, \ldots, \underline{x}_n$ are terms of type $\underline{t}$ and $\underline{F}$ is a property of type $(\underline{t})$.

Here the (extensional) collection $\{\underline{x}_1, \ldots, \underline{x}_n\}$ of the objects $\underline{x}_1, \ldots, \underline{x}_n$ and their intensional collection $\{\underline{x}_1, \ldots, \underline{x}_n\}^{(\underline{i})}$ are introduced:

DEF. 18.12. $\{\underline{x}_1, \ldots, \underline{x}_n\} =_D (\lambda \underline{x}) \bigvee_{\underline{r}=1}^{\underline{n}} \underline{x} = \underline{x}_{\underline{r}}$.

DEF. 18.13. $\{\underline{x}_1, \ldots, \underline{x}_n\}^{(\underline{i})} =_D (\lambda \underline{x}) \bigvee_{\underline{r}=1}^{\underline{n}} \underline{x} =^{\cap} \underline{x}_{\underline{r}}$.

It is easy to see that

(42) $\Vdash \{\underline{x}_1, \ldots, \underline{x}_n\} \in \underline{Ext}$ [Def. 6.11], $\Vdash \{\underline{x}_1, \ldots, \underline{x}_n\}^{(\underline{i})} \in \underline{MConst}$ [Def. 13.2],

(43) $\Vdash \{\underline{x}_1, \ldots, \underline{x}_n\}^{(\underline{i})} \subseteq \{\underline{x}_1, \ldots, \underline{x}_n\}^{\cap}$ [Def. 6.6],

$\bigwedge_{\underline{j}=2}^{\underline{n}} \bigwedge_{\underline{i}=1}^{\underline{j}-1} \underline{Nx}_{\underline{j}} \neq \underline{x}_{\underline{i}} \Vdash \{\underline{x}_1, \ldots, \underline{x}_n\}^{(\underline{i})} \subset \{\underline{x}_1, \ldots, \underline{x}_n\}^{\cap}$ [Def. 18.5],

(44) $\bigwedge_{\underline{j}=2}^{\underline{n}} \bigwedge_{\underline{i}=1}^{\underline{j}-1} \underline{Nx}_{\underline{i}} \neq \underline{x}_{\underline{j}} \Vdash \{\underline{x}_1, \ldots, \underline{x}_n\}^{(\underline{i})} \in \underline{Abs}$ [Defs. 18.8, 18.13].

Furthermore, since the power of the set Γ, say $\underline{card}(\Gamma)$, is ≥ 2 [N6, paragraphs 5, 6],

(43′) $\underline{Nx}_1 \neq \underline{x}_2 \Vdash \{\underline{x}_1, \ldots, \underline{x}_n\}^{(\underline{i})} \subset \{\underline{x}_1, \ldots, \underline{x}_n\}^{\cap}$ ($\underline{n} \geq 2$).

In order to write the illustration mentioned above, let (12) hold—in particular $\underline{NM}_1 \neq \underline{M}_2$ and $\diamond \underline{p}_1 \diamond \sim \underline{p}_1$—and in addition let

(45) $\underline{M}'_1 = {}^\cap(\imath \underline{x})(\underline{p}_1\underline{x} = \underline{M}_1 \vee \sim \underline{p}_1\underline{x} = \underline{M}_2)$, $\underline{M}'_2 = {}^\cap(\imath \underline{x})(\underline{p}_1\underline{x} = \underline{M}_2 \vee \sim \underline{p}_1\underline{x} = \underline{M}_1)$, $\underline{F} = {}^\cap\{\underline{M}_1, \underline{M}_2\}^{(i)}$, $\underline{G} = {}^\cap\{\underline{M}'_1, \underline{M}'_2\}^{(i)}$

hold in the Γ-case γ at the model $\mathscr{M}$ and the value assignment $\mathscr{V}$ [Def. 9.6(a)]. Then, by the $\underline{L}$-truth of AS12.14, the matrices $\underline{NF} \neq \underline{G}$, $\underline{F}^{(e)} = {}^\cap\underline{G}^{(e)} = {}^\cap\{\underline{M}_1, \underline{M}_2\}$ [Defs. 18.9, 18.12], and $\underline{F}, \underline{G} \in \underline{Abs}$ [Def. 18.8] hold in γ at $\mathscr{M}$ and $\mathscr{V}$.(59)

DEF. 18.14. $\underline{x} \notin \underline{F} \equiv_D \sim \underline{x} \in \underline{F}$.

N19. A rigorous procedure, based on ML^ν and absolute concepts, for defining the physical magnitudes of mass points in accordance with Mach's definition of mass. Natural absolute concepts of real number and mass point.

Let us consider a particular experiment $\mathscr{E}$ and let $\underline{Exp}(\underline{M}, \rho)$ signify the real number ρ is the result of a performance of $\mathscr{E}$ on the material point $\underline{M}$.

For instance $\mathscr{E}$ may serve to determine the mass ratio of $\underline{M}$ with respect to a given mass point $\underline{M}_1$ (standard for unit mass). For example we may identify $\underline{Exp}(\underline{M}, \rho)$ with the sentence $\underline{Exp}_1(\underline{M}, \rho)$ which says that with respect to an inertial frame, $\underline{M}$ and $\underline{M}_1$ strike one another with parallel (and opposite) velocities at some instant $\underline{t}$ in such a way that

$$\Delta\underline{v} \neq 0, \quad \rho = \Delta\underline{v}_1/\Delta\underline{v}, \tag{46}$$

where $\Delta\underline{v}$ and $\Delta\underline{v}_1$ are the moduli of the velocity increments at the instant $\underline{t}$ for $\underline{M}$ and $\underline{M}_1$ respectively—cf. [16].

In some theories experiment $\mathscr{E}$ is schematized—cf. (46)—in such a way that it allows us to define a new magnitude, e.g. the mass of $\underline{M}$, by an assertion such as the following:

(a) $\mu(\underline{M})$ is the real number ρ such that (46) would hold if $\underline{M}$ and $\underline{M}_1$ hit one another with parallel velocities at an instant, $\underline{t}$.[60]

It is useful to put this assertion into the following form:

(b) $\mu(\underline{M})$ is the real number ρ such that it is possible that $\underline{\text{Exp}}_1(\underline{M}, \rho)$.

In constructing a physical theory containing for example a definition of mass based on the experiment $\mathscr{E}_1$ corresponding to the sentence $\underline{\text{Exp}}_1(\underline{M}, \rho)$, it is very important to lay down axioms that do not imply that $\mathscr{E}_1$ must take place.[61] In order to assure both the uniqueness of the real number ρ such that $\diamond\, \underline{\text{Exp}}(\underline{M}, \rho)$ hold, and the possibility of never performing the experiment $\mathscr{E}$ $[\diamond \sim(\exists\rho)\, \underline{\text{Exp}}(\underline{M}, \rho)]$, I propose to use the natural absolute concept Real of real number and the concept MP of mass point, to be intuitively described shortly, for which

(47) $\Vdash \underline{\text{Real}} \in \underline{\text{Abs}}, \quad \Vdash \underline{\text{MP}} \in \underline{\text{Abs}}$ [Def. 18.8].

In this section we show that by (47) the use of the concepts Real and MP allows us to set up rigorous theories containing definitions like that of mass; in spite of this such theories are as general (and powerful) as the corresponding theories that—as in [29] and [33]—do not claim to be based on a well-determined theory of formal logic and are not as rigorous as most mathematical works—see fn. 57.

In Ns20, 21 we show that in our definition of mass [Def. 19.1] it is practically essential to use the absolute concepts Real and MP, and that, in particular, in assertion (b) above it is understood that $\underline{M} \in \underline{\text{MP}}$ and $\rho \in \underline{\text{Real}}$—cf. fn. 5. By Real we mean precisely the most natural absolute concept of real number,[62] so that:

(48) $\Vdash 1, 2, \ldots, \frac{1}{2}, \frac{1}{3}, \frac{2}{3}, \ldots, \sqrt{2}, \pi \in \underline{\text{Real}}$.

Let ρ_{20} be defined to be the real number equal to the number of rockets flying at the end instant $\underline{t}_{20}$ of the twentieth century. Then

(49) $\Vdash \Diamond \rho_{20} = 1$, $\Vdash \Diamond \rho_{20} \neq 1$, hence $\sim \rho_{20} \in \underline{\text{Real}}$,

because Real is modally separated and $1 \in \underline{\text{Real}}$. On the other hand, by Def. 18.9,

(50) $\Vdash \underline{N}(1, 2, \ldots, \sqrt{2}, \pi \in \underline{\text{Real}}^{(\underline{e})})$, $\Vdash \underline{N}\rho_{20} \in \underline{\text{Real}}^{(\underline{e})}$.

In order to rigorously express in ordinary language (48), $(49)_3$, and $(50)_2$, for example, we may decide, for instance, to say that $1, 2, \ldots, \sqrt{2}, \pi$ are (modally) prefixed real numbers and ρ_{20} is not such a number but is necessarily a real number.

Analogously, by MP we mean precisely (modally) prefixed mass point, i.e. a very small body formed with a given matter portion, i.e. with given molecules. For instance we may consider the rockets Mida 1, Mida 2, . . . [N7] to be prefixed mass points, whereas we have to exclude from the prefixed mass points the one $\mathfrak{m}$ among those rockets, which at the instant $\underline{t}_{20}$ has the maximum distance from the earth (and the lowest index among the Midas having such a distance). However $\mathfrak{m}$ is also necessarily a rocket, i.e. a mass point. Summing up,

(51) $\Vdash$ Mida 1, Mida 2, . . . $\in \underline{\text{MP}}$, $\Vdash \sim \mathfrak{m} \in \underline{\text{MP}}$, $\Vdash \underline{N}(\mathfrak{m}$, Mida 1, . . . $\in \underline{\text{MP}}^{(\underline{e})})$.

The concept MP is of an empirical nature (usually the concept of mass point is assumed to be a primitive one), whereas the concept Real is a logical one, i.e. it is definable (e.g. in ML^{ν}) on purely logical grounds [N27].

Now, in order to prove both the uniqueness of the number ρ satisfying assertion (b) and the truth of $\Diamond \sim \underline{\text{Exp}}(\underline{\text{M}}, \rho)$, we also assume the following:

(c) For every modally prefixed material point $\underline{\text{M}}$ there is a (modally prefixed) real number ρ such that it is possible that $\underline{\text{Exp}}(\underline{\text{M}}, \rho)$; furthermore for every real number ρ', $\underline{\text{Exp}}(\underline{\text{M}}, \rho')$ (strictly) implies $\rho = \rho'$. Precisely,

(52) $\Vdash \underline{\text{M}} \in \underline{\text{MP}} \supset (\exists \rho) \{\rho \in \underline{\text{Real}} \; \Diamond \; \underline{\text{Exp}}(\underline{\text{M}}, \rho) \; (\rho') \; [\underline{\text{Exp}}(\underline{\text{M}}, \rho') \supset^{\cap} \rho' = \rho]\}$.

THEOR. 19.1. Let (47) and (52) hold as axioms.[63] Then [Defs. 9.5, 11.3]:

(53) $\Vdash \underline{\text{M}} \in \underline{\text{MP}} \supset (\exists_1 \rho) \phi(\rho)$ where $\phi(\rho) \equiv_D \rho \in \underline{\text{Real}} \; \Diamond \; \underline{\text{Exp}}(\underline{\text{M}}, \rho)$.

Indeed, assume $(53)_2$; furthermore let (a) $\underline{\text{M}} \in \underline{\text{MP}}$, (b) $\phi(\rho_1)$, and (c) $\phi(\rho_2)$ hold in the arbitrary Γ-case γ at the admissible model $\mathcal{M}$—which satisfies (47) and (52)—and at the arbitrary value assignment $\mathcal{V}$.

Then by (52), from (a) we deduce that (d) $\rho \in \underline{\text{Real}} \; \Diamond \; \underline{\text{Exp}}(\underline{\text{M}}, \rho)$ $(\rho') \; [\underline{\text{Exp}}(\underline{\text{M}}, \rho') \supset^{\cap} \rho' = \rho]$ holds in γ at $\mathcal{M}$ and $\mathcal{V}' = \mathcal{V}\binom{\rho}{\bar{\rho}}$ [see (11)] where $\bar{\rho}$ is a suitable real number.

By $(53)_2$ the matrices (b) and (c) (in ML^{ν}) $\mathcal{A}$-imply [Def. 9.4] the matrices (e) $\Diamond \underline{\text{Exp}}(\underline{\text{M}}, \rho_{\underline{i}})$ ($\underline{i} = 1, 2$). Furthermore (d) $\underline{\text{L}}$-implies the matrices $\underline{\text{Exp}}(\underline{\text{M}}, \rho_{\underline{i}}) \supset^{\cap} \rho_{\underline{i}} = \rho$ ($\underline{i} = 1, 2$) which by the truth of $(21)_1$ $\underline{\text{L}}$-imply $\Diamond \underline{\text{Exp}}(\underline{\text{M}}, \rho_{\underline{i}}) \supset \rho_{\underline{i}} =^{\cup} \rho$ [Def. 6.7] ($\underline{i} = 1, 2$). Then by (e) $\rho_{\underline{i}} =^{\cup} \rho$ ($\underline{i} = 1, 2$) also hold in γ at $\mathcal{M}$ and $\mathcal{V}'$ so that (f) $\rho_1 =^{\cup} \rho_2$ does too.

By $(53)_2$ the matrices (b) and (c) $\mathcal{A}$-imply $\rho_1, \rho_2 \in \underline{\text{Real}}$, whence by $(47)_1$ and (40), we have $\rho_1 =^{\cup} \rho_2 \supset \rho_1 =^{\cap} \rho_2$. This matrix and (f) $\mathcal{A}$-imply $\rho_1 =^{\cap} \rho_2$, hence also $\rho_1 = \rho_2$.

We conclude that (a), (b), and (c) $\mathscr{A}$-imply $\rho_1 = \rho_2$, so that by Theor. 9.2, (g) $\underline{M} \in \underline{MP} \supset [\phi(\rho_1)\phi(\rho_2) \supset \rho_1 = \rho_2]$ holds in γ at $\mathscr{M}$ and $\mathscr{V}'$. Then, since the variable ρ does not occur free in (g), by the arbitrariness of γ and $\mathscr{V}$ the matrix (g) is $\mathscr{A}$-true [Def. 9.5].

In addition, by $(53)_2$ the assumption (52) implies $\Vdash \underline{M} \in \underline{MP} \supset (\exists\rho)\phi(\rho)$. Hence by Def. 11.3, $(53)_1$ holds.

QED

Let us add that we can replace "$\rho_1 = \rho_2$" with "$\rho_1 =^{\frown} \rho_2$" in the last two paragraphs of the foregoing proof. Thus we obtain a proof of the following assertion:

(53') $\quad$ $\mathrm{M} \in \mathrm{MP} \supset (\exists_1^{\frown}\rho)\,\phi(\rho)$ if (47), (52), and $(53)_3$ hold.

Now assuming (47) and (52) for $\underline{\mathrm{Exp}} = \underline{\mathrm{Exp}}_1$ as axioms, we can turn assertion (b) above into a rigorous definition of mass in ML^{ν}:

DEF. 19.1. $\mu(\underline{M}) =_D (\imath\rho)\Phi_1(\rho)$ where $\Phi_1(\rho) \equiv_D \rho \in \underline{\mathrm{Real}} \diamond \mathrm{Exp}_1(\underline{M}, \rho)$.

By $(47)_2$ $\underline{M}' \in \underline{MP}^{(e)}$ $\mathscr{A}$-implies $(\exists_1^{\frown}\underline{M})(\underline{M} \in \underline{MP} \wedge \underline{M}' = \underline{M})$. Hence we can define the mass $\mu^{(e)}(\underline{M})$ of $\underline{M}'$ for $\underline{M}' \in \underline{MP}^{(e)}$ as follows:

DEF. 19.2. $\mu^{(e)}(\underline{M}') =_D (\imath\rho)(\exists\underline{M})[\underline{M} \in \underline{MP} \wedge \underline{M} = \underline{M}' \wedge \rho = \mu(\underline{M})]$.

This definition is satisfactory because it implies

(53*) $\quad$ $\underline{M} \in \underline{MP} \wedge \underline{M}' = \underline{M} \Vdash \mu^{(e)}(\underline{M}') = \mu(\underline{M})$,

so that in particular $\mu^{(e)}(\mathfrak{m})$ [N7] has the correct meaning, while $\mu(\mathfrak{m}) = \underline{a}^*$.

N20. Why certain absolute concepts were used in the preceding definition.

In Ns21, 22 it will be explained why a procedure based on a modal logic was chosen in N19. In this section we shall show rigorously by means of Theors. 20.1 and 20.2, respectively, that the use of nonextensional concepts of real number and mass point in the uniqueness theorem Theor. 19.1 is essential.

In N19 we identified the concept $\underline{\text{Real}}$ [$\underline{\text{MP}}$] appearing in the basic assumption (52) and the basic theorem (53) with the privileged absolute concept of (modally prefixed) real number [mass point], so that (48), $(49)_3$, and (50) [(51)] hold. We denote by $\underline{\text{Real}}_0$ [$\underline{\text{MP}}_0$] this privileged absolute concept, and (unlike in N19) we accept that $\underline{\text{Real}} = {}^\cap\underline{\text{Real}}_0$ and $\underline{\text{MP}} = {}^\cap\underline{\text{MP}}_0$ may fail to hold.

In order to explain thoroughly why in N19 we took $\underline{\text{Real}} = {}^\cap\underline{\text{Real}}_0$ and $\underline{\text{MP}} = {}^\cap\underline{\text{MP}}_0$ to hold, we consider the following natural problem: Is there a choice of $\underline{\text{Real}}$ and $\underline{\text{MP}}$ for which conditions (1) to (3) below hold, possibly under some additional reasonable assumptions?

(1) (52) holds for the new choice of $\underline{\text{Real}}$ and $\underline{\text{MP}}$ as a consequence of its validity in the sense of N19.

(2) The assertion obtained from Theor. 19.1 by replacing assumption (47) with this choice of $\underline{\text{Real}}$ and $\underline{\text{MP}}$ is a new (uniqueness) theorem.

(3) The definition of mass Def. 19.1 based on this new theorem when $\underline{\text{Exp}} = \underline{\text{Exp}}_1$ is equivalent to the one based on Theor. 19.1.

With a view to describing some "mathematically" satisfactory solutions to the above problem, we define the following concepts $\underline{\text{Real}}_1$ and $\underline{\text{Real}}_2$ of real numbers in terms of $\underline{\text{Real}}_0$.

DEF. 20.1. $\underline{\text{Real}}_1 =_D (\lambda\rho)(\exists\sigma)[\sigma \in \underline{\text{Real}}_0\ \rho = \sigma\ \underline{\text{N}}(\rho = \sigma \vee \rho = \underline{a}^*)]$ $(\Vdash \underline{\text{Real}}_0 \in \underline{\text{Abs}})$,

DEF. 20.2. $\underline{\text{Real}}_2 =_D (\lambda\rho)(\exists\sigma)[\sigma \in \underline{\text{Real}}_0\ \rho = \sigma\ \underline{\text{N}}(\rho = \sigma \vee \sim\rho \in \underline{\text{Real}}^{(\underline{e})})]$.

Let us consider the conditions

(47′) $\Vdash \underline{\text{Real}}_0 \subseteq \underline{\text{Real}}$, $\Vdash \underline{\text{Real}} \subseteq \underline{\text{Real}}_2$, $\Vdash \underline{\text{MP}} = \underline{\text{MP}}_0$

on Real and MP. Since $\Vdash \sim \underline{a}^* \in \underline{\text{Real}}^{(\underline{e})}$, $\underline{\text{Real}}_1$ fulfills conditions (47′) in the following way:

(47″) $\Vdash \underline{\text{Real}}_0 \subset \underline{\text{Real}}_1$, $\Vdash \underline{\text{Real}}_1 \subset \underline{\text{Real}}_2$.

It is not difficult to see [Appendix D] that the above problem is also solved by any choice of Real and MP fulfilling (47′). More precisely such a choice fulfills conditions (1) to (3) under the additional reasonable assumptions.

(47‴) $\Vdash \underline{\text{Exp}}(\underline{M}, \rho) \supset \underline{M} \in \underline{\text{MP}}^{(\underline{e})}$, $\Vdash \underline{\text{Exp}}(\rho, \underline{M}) \supset \rho \in \underline{\text{Real}}$, $\Vdash \underline{\text{Exp}} \in \underline{\text{Ext}}$.

Incidentally, condition $(47''')_3$ is reasonable in that it serves to present in a clear way the problem of how to choose the concepts Real and Mp. Indeed if it did not hold, by $(47''')_{1,2}$ $(\lambda\rho)\,(\exists\underline{M})\,\underline{\text{Exp}}(\underline{M}, \rho)$ $[(\lambda\underline{M})\,(\exists\rho)\,\text{Exp}\,(\underline{M}, \rho)]$ would give us an additional concept of real number [mass point] that might not be definable in terms of Real [MP].

We consider the concepts Real and MP for certain experimental purposes [N19]. Therefore they must be physically acceptable in a sense characterized by suitable and reasonable conditions on them—cf. appendix C. Included among these additional conditions are (47‴) and the one that Real be definable on purely logical grounds.

Under these additional conditions the converse of the underlined statement above (which generalizes Theor. 19.1) can be asserted in that one can prove [appendix D] that if MP and Real fulfill these conditions, (52), and (53), then they also fulfill (47′) and incidentally also certain invariance conditions—see requirement (a′) in appendix C.

For instance $\underline{\mathrm{MP}}_0$ and $\underline{\mathrm{Real}}_r$ ($\underline{r} = 1, 2$)—cf. (47″)—are physically acceptable determinations of MP and Real respectively in that they fulfill all of the above conditions. However every solution Real of (47′) different from $\underline{\mathrm{Real}}_0$ is similar to $\underline{\mathrm{Real}}_0$ and more or less awkward; furthermore $\underline{\mathrm{Real}}_0$ is the common part of all the determinations of Real that fulfill (47′), and among these determinations, first, Real is the only modally constant one, and second, it can be introduced in the simplest way.

Incidentally, the privileged absolute concept Nn of natural numbers will be defined in N27 on purely logical grounds. It is rather a matter of routine to define $\underline{\mathrm{Real}}_0$ by combining the modal tools used in N27 (to define Nn) with the customary procedures used in (extensional) mathematics to define real numbers in terms of natural numbers. (Basic existence properties of Real can be proved on the basis of AS12.19). The introduction of $\underline{\mathrm{Real}}_1$ [$\underline{\mathrm{Real}}_2$] requires the additional use of Def. 20.1 [Def. 20.2].

On the basis of the preceding considerations we conclude that ($\underline{\mathrm{MP}}_0$, $\underline{\mathrm{Real}}_0$) is the best choice of (MP, Real) solving the above natural problem. Furthermore only this choice seems to me physically acceptable and nonawkward (and relatively simple). Therefore I consider the use of $\underline{\mathrm{MP}}_0$ and $\underline{\mathrm{Real}}_0$ made in N19 as (practically speaking) essential.

In this section we prove explicitly only that in N19 we had to exclude $\Vdash \underline{\mathrm{MP}} = {}^\frown\underline{\mathrm{MP}}_0^{(e)} \vee \underline{\mathrm{Real}} = {}^\frown\underline{\mathrm{Real}}_0^{(e)}$, because this can be done without additional requirements on MP and Real and it gives us a good idea of the entire situation the problem involves.

The content of Theor. 20.1 and its proof can be intuitively expressed through the following reasoning: For some real number ρ, $\underline{\mathrm{Exp}}(\underline{\mathrm{M}}, \rho)$ must happen in some Γ-case γ. As a hypothesis for reductio ad absurdum let γ' be a Γ-case distinct from γ. Then, if "real number" is used in an extensional way, we can consider two real numbers ρ_1 and ρ_2 which coincide with ρ in γ and are different from one another in γ'. So in γ' both $\diamond$ Exp(M, ρ_i) (i = 1, 2) and $\rho_1 \neq \rho_2$ hold. By the obvious uniqueness condition $(53)_1$ we should have $\rho_1 = \rho_2$ in γ'. We conclude that at most one Γ-case can exist.

THEOR. 20.1. Let us assume—cf. Theor. 19.1—$(53)_2$, and:

(54) ‖- Ext(Real), hence ⊩ $(\exists\rho_1, \rho_2)$N$(\rho_1, \rho_2 \in$ Real $\rho_1 \neq \rho_2)$;
‖- Ext(Exp) [Def. 6.11]

(hence $(47)_1$ does not hold while $(47)_2$ may or may not hold), and ‖- NMP $\neq \Lambda$. Also let the uniqueness condition $(53)_1$ hold as either an axiom or a theorem. Then at most one Γ-case can exist, so that ML^ν cannot be an essentially modal language. In addition ‖- M $\in$ MP $\supset$ N$(\exists\rho)$ Exp (M, ρ) so that $\diamond(\rho)\sim$Exp(M, ρ) cannot be $\mathscr{A}$-true as is required—see fn. 61.

Proof. Besides the hypotheses of the theorem we assume, as a hypothesis for reductio ad absurdum, the existence of at least two Γ-cases.

Then there is an admissible model $\mathscr{M}$ [Def. 9.3], a value assignment $\mathscr{V}$, and two Γ-cases γ and γ', such that (a) M $\in$ MP, (b) Exp(M, ρ)—see (53)—and (c) N$(\rho, \rho' \in$ Real) hold in γ at $\mathscr{M}$ and $\mathscr{V}$, and moreover—see $(54)_{1,2}$—(d) $\rho' = \rho$ and (d') $\rho' \neq \rho$ hold at $\mathscr{M}$ and $\mathscr{V}$ in the Γ-cases γ and γ' respectively.

By (b), (d), and $(54)_3$, Exp(M, ρ') holds in γ at $\mathscr{M}$ and $\mathscr{V}$, so that $\diamond$ Exp(M, ρ') holds in γ' at $\mathscr{M}$ and $\mathscr{V}$. By (b), $\diamond$ Exp(M, ρ) does too. Then by (c) and $(53)_2$ $\phi(\rho)$ and $\phi(\rho')$ also hold. Since, besides $\phi(\rho)$ and $\phi(\rho')$, (d') too holds in γ' at $\mathscr{M}$ and $\mathscr{V}$, on the one hand the matrix (f) $(\exists_1 \rho)\phi(\rho)$ does not hold in γ' at $\mathcal{M}$ and $\mathcal{V}$; on the other hand, by (a) and $(53)_1$—see the hypotheses of our theorem—the matrix (f) holds in γ' at $\mathcal{M}$ and $\mathcal{V}$, which is absurd. Hence under the hypotheses of the theorem, Γ cannot have more than one element.

In order to complete our proof we remark that the above assertion about Γ implies $\Vdash \Diamond \underline{p} \supset \underline{N}\underline{p}$ for every matrix $\underline{p}$. Then (53) obviously implies $\Vdash \underline{M} \in \underline{MP} \supset \underline{N}(\exists\rho)\underline{Exp}(\underline{M}, \rho)$.

QED

THEOR. 20.2. Let us assume that $\Vdash \underline{Ext}(\underline{MP})$ [Def. 6.11], $(47)_1$, $(54)_3$, and either[64]

$$(52') \Vdash (\underline{M})\ (\underline{M} \in {}^\frown\underline{MP} \supset {}^\frown(\exists\rho)\ \{\rho \in \underline{Real}\ \Diamond\ \underline{Exp}(\underline{M}, \rho)\ (\rho')\ [\underline{Exp}(\underline{M}, \rho') \supset {}^\frown\rho' = \rho]\})$$ or

$$(53') \Vdash (\underline{M})\ (\underline{M} \in {}^\frown\underline{MP} \supset {}^\frown(\exists_1\rho)\ [\rho \in \underline{Real}\ \Diamond\ \underline{Exp}(\underline{M}, \rho)])$$ [Def. 18.10]

at the model $\mathcal{M}$, which is to be kept fixed. Furthermore, as is usually the case, let there be at least one mass point $\underline{M}_1$ and a proposition $\underline{p}$ such that the performance of the experiment $\mathcal{E}$ on $\underline{M}_1$ is compatible with both $\underline{p}$ and $\sim\underline{p}$, i.e. let

$$(55) \Vdash (\exists\underline{M}_1, \rho_1, \rho_2)\ \{\underline{M}_1 \in {}^\frown\underline{MP}\,(\rho_1, \rho_2 \in \underline{Real})\ \Diamond\ [\underline{pExp}(\underline{M}_1, \rho_1)]\ \Diamond[\sim\underline{pExp}(\underline{M}_1, \rho_2)]\}$$

hold. Then for all mass points $\underline{M}_1$ and $\underline{M}_2$, by performing the experiment $\mathcal{E}$ on $\underline{M}_1$ and $\underline{M}_2$ one always obtains the same result ρ_1. More precisely:

$$(56) \Vdash (\exists\rho_1)\ \{\rho_1 \in \underline{Real} \wedge (\rho)\ (\underline{M})\ [\underline{M} \in {}^\frown\underline{MP}\ \Diamond\ \underline{Exp}(\underline{M}, \rho) \supset {}^\frown\rho = {}^\frown\rho_1]\}$$

Intuitively this theorem says that if mass point is an extensional con cept, then under plausible assumptions all mass points have the same mass.

Proof. By $(47)_1$, $(54)_3$, (55), and either (52′) or (53′), we obtain an $\mathscr{A}$-true matrix from (55) by replacing, in (55), the last occurrence of ρ_2 with an occurrence of ρ_1 (and by deleting the other two occurrences of ρ_2). Hence:

(55′) $\underline{M}_1 \in {}^\cap \underline{MP}\ \rho_1 \in \underline{Real} \diamond [\underline{pExp}(\underline{M}_1, \rho_1)] \diamond [\sim\underline{pExp}(\underline{M}_1, \rho_1)]$

holds in γ at $\mathscr{M}$ and $\mathscr{V}$ where γ is an arbitrary Γ-case, $\mathscr{V}$ is a value assignment arbitrary except for its values $\mathscr{V}(\rho_1)$ and $\mathscr{V}(\underline{M}_1)$ in ρ_1 and $\underline{M}_1$, and $\mathscr{M}$ is an admissible model.

Intuitively the proof continues as follows: Let $\underline{M}$ be any mass point ($\underline{M} \in \underline{MP}$) so that by either (52′) or (53′) experiment $\mathscr{E}$ can be performed on $\underline{M}$ with the result ρ.

We first consider the alternative $\underline{q}$ that this performance is compatible with $\underline{p}$. In this alternative we define $\underline{M}'$ to be the mass point that coincides with $\underline{M}$ in the Γ-cases in which $\underline{p}$ holds, and with $\underline{M}_1$ in the remaining Γ-cases. So $\underline{M}'$ is necessarily a mass point, since the concept of mass point is extensional by an assumption.

The concept of the performance of $\mathscr{E}$ on a mass point is also extensional. Then, under the alternative $\underline{q}$, on the one hand $\mathscr{E}$ is performed on $\underline{M}$, hence on $\underline{M}'$, with the result ρ in a Γ-case in which $\underline{p}$ holds. On the other hand, (55′) implies that in a Γ-case in which $\sim\underline{p}$ holds, $\mathscr{E}$ is performed on $\underline{M}_1$, hence on $\underline{M}'$, with the result ρ_1. By either (52′) or (53′) the results ρ and ρ_1 of two possible performances of the experiment $\mathscr{E}$ on the mass point $\underline{M}'$ must coincide. Hence the result of some possible performances of $\mathscr{E}$ on the two mass points $\underline{M}$ and $\underline{M}_1$ coincide, if the alternative $\underline{q}$ holds.

Now we assume $\sim\underline{q}$, i.e. that the performance of $\mathscr{E}$ on $\underline{M}$ is not compatible with $\underline{p}$. Then, since this performance can happen, it must be compatible with $\sim \underline{p}$. Then the reasoning made in the alternative $\underline{q}$ holds in the alternative $\sim \underline{q}$ up to the interchange of $\underline{p}$ with $\sim \underline{p}$. So the thesis of the theorem holds.

Now we want to complete our technical proof of the theorem. To this end we suppose that (a) $\underline{M} \in {}^{\cap}\underline{MP}$ and (b) $\diamond \underline{Exp}(\underline{M}, \rho)$ hold at the model $\mathcal{M}$ and value assignment $\mathcal{V}$ (introduced in the paragraph including (55′)). Now we identify $\underline{q}$ with the matrix $\diamond [\underline{p}\underline{Exp}(\underline{M}, \rho)]$ and we set

(57) $\underline{M}' =_D (\imath \underline{x}) [(\underline{p} \equiv \underline{q}) \underline{x} = \underline{M} \vee (\sim \underline{p} \equiv \underline{q}) \underline{x} = \underline{M}_1]$ where $\underline{q} \equiv_D \diamond [\underline{p}\underline{Exp}(\underline{M}, \rho)]$.

First suppose the case that $\underline{q}$ is Γ-true. Then (57) $\underline{L}$-implies [Def. 9.4(c)] (c) $\underline{p} \supset^{\cap} \underline{M}' = \underline{M}$ and (c′) $\sim \underline{p} \supset^{\cap} \underline{M}' = \underline{M}_1$. Furthermore (55′) $\underline{L}$-implies (d) $\underline{M}_1 \in {}^{\cap} \underline{MP}$. In addition (c), (c′), (a), (d), and the matrix $\underline{Ext}(\underline{MP})$ (supposedly $\mathcal{A}$-true) $\underline{L}$-imply (e) $\underline{M}' \in {}^{\cap} \underline{MP}$.

By $(54)_3$ the matrices (c) and $\underline{q}$ [see $(57)_2$] $\underline{L}$-imply (f) $\diamond \underline{Exp}(\underline{M}', \rho)$. Analogously by $(54)_3$ the matrix (c′) and the consequence $\diamond [\sim \underline{p}\underline{Exp}(\underline{M}_1, \rho_1)]$ of (55′) $\underline{L}$-imply (f′) $\diamond \underline{Exp}(\underline{M}', \rho_1)$.

Now we consider the case that $\sim \underline{q}$ is Γ-true. Then by $(57)_2$, the matrices $\sim \underline{q}$ and (b) $\underline{L}$-imply (g) $\diamond [\sim \underline{p}\underline{Exp}(\underline{M}, \rho)]$. In addition $\sim \underline{q}$ and $(57)_1$ $\underline{L}$-imply (h) $\sim \underline{p} \supset^{\cap} \underline{M}' = \underline{M}$ and (h′) $\underline{p} \supset^{\cap} \underline{M}' = \underline{M}_1$. Then (h), (h′), (a), (d), and the matrix $\underline{Ext}(\underline{MP})$ $\underline{L}$-imply (e) $\underline{M}' \in {}^{\cap} \underline{MP}$ again. Furthermore by $(54)_3$ the matrices (g) and (h) $\underline{L}$-imply (f) $\diamond \underline{Exp}(\underline{M}', \rho)$ again, while by $(54)_3$ the matrix (h′) and the consequence $\diamond [\underline{p}\underline{Exp}(\underline{M}_1, \rho_1)]$ of (55′) $\underline{L}$-imply (f′) $\diamond \underline{Exp}(\underline{M}', \rho_1)$ again. We conclude that (e), (f), and (f′)—i.e. (l) $\underline{M}' \in \underline{MP}$ $\diamond \underline{Exp}(\underline{M}', \rho)$ $\diamond \underline{Exp}(\underline{M}', \rho_1)$—are Γ-true at $\mathcal{M}$ and $\mathcal{V}$ in every case.

The matrix (l) and either (52′) or (53′) imply $\rho = \rho_1$ in γ at $\mathcal{M}$ and $\mathcal{V}$ [Def. 9.6(b)].

We conclude that if (a) is true in γ at $\mathcal{M}$ and $\mathcal{V}$, then (b) implies $\rho = \rho_1$ in γ at $\mathcal{M}$ and $\mathcal{V}$; moreover, since the values $\mathcal{V}(\underline{M})$ and $\mathcal{V}(\rho)$ of $\mathcal{V}$ in $\underline{M}$ and ρ have been left arbitrary, $(\rho)\,(\underline{M})\,[\underline{M} \in {}^{\cap}\underline{MP} \wedge \Diamond\, \underline{Exp}(\underline{M}, \rho) \supset \rho = \rho_1]$ holds in γ at $\mathcal{M}$ and $\mathcal{V}$. By (55′) $\rho_1 \in \underline{Real}$ also holds in γ at $\mathcal{M}$ and $\mathcal{V}$, so that by the arbitrariness of γ, $\mathcal{M}$, and $\mathcal{V}$, (56) holds.

QED

N21. Certain disadvantages of existing extensional theories dealing with foundations of mechanics according to Mach, Kirchoff, and Painlevé.

Suppose that we use an extensional logic directly and that we wish to define $\mu(\underline{M})$ to be the ρ such that if we perform the experiment $\mathscr{E}$ on $\underline{M}$, then we obtain the result ρ; hence we may wish to define the magnitude $\mu(\underline{M})$ of $\underline{M}$—see p. 70—to be the ρ such that $\underline{Exp}(\underline{M}, \rho)$ holds, where $\underline{Exp}(\underline{M}, \rho)$ has the form

$$(58) \qquad (\exists \underline{t})\, [\underline{A}(\underline{M}, \underline{t}) \supset \underline{B}(\underline{M}, \underline{t}, \rho)],$$

and in case we are defining mass, $\underline{A}(\underline{M}, \underline{t})$ means that at the instant $\underline{t}$ the mass points $\underline{M}$ and $\underline{M}_1$ hit one another with parallel velocities with respect to the inertial frames, and $\underline{B}(\underline{M}, \underline{t}, \rho)$ means that (46) holds, where $\Delta\underline{v}$ and $\Delta\underline{v}_1$ are the moduli of the velocity increment at the instant $\underline{t}$ for $\underline{M}$ and $\underline{M}_1$ respectively. Under these assumptions $\underline{Exp}(\underline{M}, \rho)$ appears to be a translation into extensional logic of assertion (a) made in N19, p. 70.

Now we assume that $(\exists \underline{t})\, \underline{A}(\underline{M}, \underline{t})$ does not hold. Then $(\rho)\underline{Exp}(\underline{M}, \rho)$ —cf. (58)—holds, so that $(\exists_1 \rho)\, \underline{Exp}(\underline{M}, \rho)$ cannot hold. Hence if $\underline{Exp}(\underline{M}, \rho)$ has the form (58) and $(\exists_1 \rho)\, \underline{Exp}(\underline{M}, \rho)$ holds, then $(\exists \underline{t})\, \underline{A}(\underline{M}, \underline{t})$ also holds.

In order to define mass, $\Vdash (\exists_1 \rho) \underline{Exp}(\underline{M}, \rho)$ must be a theorem. Hence in this case (58) is equivalent—as Hermes substantially asserts in [17], p. 32—to

$$(59) \qquad (\exists \underline{t}) [\underline{A}(\underline{M}, \underline{t}) \underline{B}(\underline{M}, \underline{t}, \rho)],$$

so that one can identify $\underline{Exp}(\underline{M}, \rho)$ with the matrix (59).[65] We conclude that in the theories which are directly based on extensional logic and contain axioms enabling us to define mass, $(\exists \underline{t}) \underline{A}(\underline{M}, \underline{t})$ is a consequence of $\underline{Exp}(\underline{M}, \rho)$ in either of the above forms (58) and (59).[66] This consequence is incompatible with the hypotheses laid down to deal with most problems of mechanics. For instance when we are dealing with the problems of two bodies $\underline{M}_2$ and $\underline{M}_3$ which attract each other according to Newton's law, we fix two geometric points $\underline{P}_2$, $\underline{P}_3$ and two (vectorial) velocities $\dot{\underline{P}}_2$ and $\dot{\underline{P}}_3$, and we assume that at an instant $\underline{t}_0$, $\underline{P}_{\underline{i}}$ is the position of $\underline{M}_{\underline{i}}$ and $\dot{\underline{P}}_{\underline{i}}$ is its velocity ($\underline{i} = 2, 3$), and that $\underline{M}_2$, $\underline{M}_3$ constitute an isolated system. Then the motion is calculated.

For $\dot{\underline{P}}_3 - \dot{\underline{P}}_2$ not parallel with $\underline{P}_3 - \underline{P}_2$, one can infer that $\underline{M}_2$ and $\underline{M}_3$ do not hit any material point $\underline{M}_1$ and that at every instant the velocity intensities of $\underline{M}_2$ and $\underline{M}_3$ have zero increments. Both of these deductions are in conflict with $(\exists \underline{t}) \underline{A}(\underline{M}, \underline{t})$, hence with the theorem $(\exists_1 \rho) \underline{Exp}(\underline{M}, \rho)$ where $\underline{Exp}(\underline{M}, \rho)$ has either of the forms (58) and (59).[67]

The preceding considerations show, I think, that the incompatibility of the assumption $\underline{Exp}(\underline{M}, \rho)$, in either of the above forms, with the hypotheses laid down in most problems of mechanics is a defect due not to the particular choice of experiment $\mathscr{E}$ or to a particular form chosen for $\underline{Exp}(\underline{M}, \rho)$, but to a direct use of extensional logic (in that this use compels us to assert that experiment $\mathscr{E}$ takes place). Given this assertion, it is not surprising that in most known works on foundations of mechanics, which like Painlevé's are not explicitly based on a theory of formal logic, a logic richer than extensional logic is implicitly used—cf. fnn. 52 and 53.

Because of these considerations, some years ago (1958) I felt the need for a new work on foundations of mechanics, based on Painlevé's ideas, that would take into account on the one hand the importance of possibility in such foundations (in particular the way Painlevé and Signorini express possible propositions), and on the other hand the fact that modal logic is not as developed as extensional logic, so that some scientists feel a certain aversion to it.

The work [4] on classical particle mechanics was done in this spirit. I believe it is as rigorous as the ordinary rigorous works of mathematics; in particular a complete list of primitive concepts and axioms is given.[68]

In [4] possibility (and necessity) concepts are used, not directly (i.e. by means of a modal language), as is done in [29] and [33], but by means of an extensional language through a device related to the extensional semantical rules given by Carnap in [12] for certain modal languages. More precisely, in [4] the set CMP of the mechanically possible cases, briefly the CMP-cases, have been introduced. (They are to be related to the states considered by Wittgenstein in [39] and by Carnap in [12], hence also to the Γ-cases used here for ML^{ν}.) On the one hand absolute concepts such as mass point are considered to be classes. On the other hand ordinary contingent concepts are changed. As an example the concept $\underline{pos}^{*}_{\underline{K}}(\underline{M}, \theta)$, i.e. the concept of the position of the mass point M at the instant θ (in the inertial spatial frame K), is changed into the concept $\text{pos}_K(M, \theta, \gamma)$, which is the position of M (in K) at the instant θ in the CMP-case γ. Hence some common sentence p^* such as $P = \text{pos}_K^*(M, \theta)$ is translated into a sentence p_γ—which in the above example is $P = \text{pos}_K(M, \theta, \gamma)$—open with respect to the parameter γ ($\gamma \in$ CMP). Then $\diamond p^*$ and Np^* can obviously be translated into $(\exists\gamma)p_\gamma$ and $(\gamma)p_\gamma$ respectively.

We see that the language used in [4] is adequate[69] and extensional but not usual;[70] in particular it compels us to take the concept of CMP-case as primitive. The simple language with modal operators used by Hermes in [17] also appears to be adequate but to have some unusual features (more precisely certain limitations) —cf. fn. 5. The general language ML^ν introduced in the present work is essentially modal in that for example strict identity does not coincide with (simple) identity in it. This language seems to me both adequate for our axiomatization purposes [N22] and of a rather usual kind—in particular it includes descriptions and functors which can also express physical (contingent) concepts. The approach, based on ML^ν, that is proposed here to attain our axiomatization goals leads us to some results that are interesting in themselves. Some—e.g. AS12.19—are of a syntactical nature, others—e.g. the introduction and study of absolute and quasi-absolute attributes, cf. Ns18, 23, 24—are of a semantical nature, and one is of a rather philosophical nature—cf. N23.

N22. Usefulness of basing foundations of mechanics according to Painlevé's ideas on the modal language ML^ν.

The subject dealt with in [29] and [33] can be expressed in the modal language ML^ν, which is, I think, a natural formalization, based on a type system, of a common modal language such as the one used in [29] and [33].

On the basis of the semantical rules for ML^{ν} [Ns8, 11], ML^{ν} can be translated [N15] into the extensional language $EL^{\nu+1}$, which proves the thesis of extensionality—see [12], p. 141. On the one hand, by this translation a theory expressed in ML^{ν} can be considered to be indirectly based on the extensional language $EL^{\nu+1}$, so that the aversion of some scientists to the use of the modal language ML^{ν} may be lessened. On the other hand ML^{ν} is more efficient than $EL^{\nu+1}$. It supplies simpler forms of expressions and, consequently, simpler deductive manipulations, as Carnap suggests in [12], p. 142.

Note that the ordinary (contingent) functor "position of . . ." is extensional—as is the functor <u>pos</u>* considered in [4]. Furthermore, we know that in the modal definition [Def. 19.1] of mass, an extensional concept of mass point cannot be used [Theor. 20.2], practically an absolute one (<u>MP</u>) is required, and this fact does not appear to depend on particular features of Def. 19.1.

Hence if we base a definition of mass of the sort suggested by Mach on a general modal language $\mathscr{L}$, $\mathscr{L}$ must be able to deal with two kinds of attributes (simply related to one another): the extensional ones and the absolute ones. If we want $\mathscr{L}$ to fit this requirement in a unified way—i.e. without introducing different kinds of variables and constants for any logical type—then $\mathscr{L}$ must be able to deal with general modal properties (in particular with absolute ones) as ML^{ν} does. A unified character is advantageous because first, in conformity with the general tendency of reducing the primitive concepts, it avoids a multiplication of entities, a multiplication that is substantial especially in connection with logical types of high levels.[71] and second (and more important), it simplifies the logical axioms and the corresponding syntactical theory, as is apparent from the proposed axioms Ass12.1–23 and from Memoir 2.

5

Absolute Attributes in Connection with a Double Use of Common Nouns, with Logic, and with Natural Numbers

N23. A double use of common nouns in physics. On substances and qualities.

In N19 we spoke of (modally) prefixed real numbers (as 1, 2, . . . , π) and (modally) prefixed mass points (as Mida 1, Mida 2, . . .)—cf. (47), (48), $(51)_1$. We also considered the examples ρ_{20} and $\mathfrak{m}$, which are necessarily a real number and mass point, respectively, and are not modally prefixed—cf. (49), $(50)_2$, and $(51)_{2,3}$.

As an example, common in mechanics, of real numbers that are not modally prefixed, let $\underline{dist}(\underline{M}, \underline{M}_1, \underline{t})$ denote the distance (measured in meters) of the mass points $\underline{M}$ and $\underline{M}_1$ at the instant $\underline{t}$, so that we may assert:

(a) The distance $\underline{dist}(\underline{M}, \underline{M}_1, \underline{t})$ is necessarily a real number.

Using more technical terms—precisely, the language ML^{ν}—by $(47)_1$ and Def. 18.9 the following holds:

$$(60) \qquad \Vdash \underline{N}\, \underline{dist}(\underline{M}, \underline{M}_1, \underline{t}) \in \underline{Real}^{(\underline{e})}, \quad \sim \underline{dist}(\underline{M}, \underline{M}_1, \underline{t}) \in \underline{Real}.$$

Let us note that in ordinary speech the adjectival phrase "modally prefixed" is not used; nevertheless certain common nouns such as "real number," "mass point," "matter portion," and so on are used in two ways. For instance on the one hand, in assertion (a) above, "real number" is used extensionally—i.e. as expressing an extensional property, cf. $(60)_1$—and not in an absolute way, cf. $(60)_2$. On the other hand, "real number" is used within assertion (c) made in N19 in an absolute way and not extensionally [Theors. 19.1, 20.1]. The analogue holds for the common noun "masspoint," used extensionally in "$\mathfrak{m}$ is necessarily a mass point"—cf. $(51)_3$—and used in an absolute way in assertion (c) made in N19 [Theors. 19.1, 20.2].[72] In some contexts, e.g. "1 is a real number," the common noun "real number" may be meant in either an extensional or an absolute way at will—cf. (48) and $(50)_1$.

In N19 we took this double use of common nouns, involved in our problem of defining mass, into account on the basis of our modal language ML^{ν}. We did so in a unified and simple way, in that we showed that it suffices to assign (the translation Δ into ML^{ν} of) such nouns absolute meanings, their extensional meanings being simply the extensionalizations of the former.

Note that in certain combinations of existential quantification with modalities—see N19, assertion (c)—absolute properties serve to pick out individuals. As we showed in N20, extensional attributes cannot be used for the same purpose. That is, in certain situations it is only by means of natural absolute concepts (i.e. absolute attributes in ML^{ν}) that we can pick out a bearer of properties, in other words a subject. And this is done in a strong way in the sense that the subject being considered is "the same in all possible cases" in the most natural sense.

Note that for example the absolute concept of body—briefly body—used to define mass according to Painlevé is the only natural absolute concept of body, whereas infinitely many other absolute concepts having the same extensionalization as body can be artificially constructed. The natural concept of body can be intuitively characterized by the condition that if b is a body, then it is "the same body, i.e. the same bearer of (possible) properties, in all possible cases" in the most natural sense.[73] So a natural absolute concept is intimately and strongly related to bearers of (possible) properties, i.e. to substances. Furthermore when a common noun of an ordinary language is used in an absolute way it expresses a natural (nonartificial) absolute property.

Let us add that after having defined "the body b has the mass μ" using "body" in an absolute way, it is natural to accept that "the body$^{(e)}$ b′ has the same mass as b in case b′ happens to be the body b." (Here b′ may coincide with the rocket farthest from the earth at the end instant t_{20} of the twentieth century.) This procedure is not a mere convention because in every possible case the physical consequences of the fact that b′ has the mass μ are many, and they are tested in many possible cases where the particular experiment $\mathscr{E}_1$ considered in our definition of mass [N19] does not occur.

The role played by the natural absolute concept of body in our definition of mass is strong and practically essential in picking out bearers of properties; the concept body$^{(e)}$ also serves to express such bearers, but in a weak way, and in particular it seems to denote a set of essential properties for all bodies rather than picking out individual bodies. According to scholastics, particularly Aristotle, bearers of properties, or subjects, are (material or nonmaterial) substances. So on the one hand, (natural) absolute properties are important, even essential in certain situations, to denote things as substances.

We now show that, on the other hand, the use of absolute properties for attributing qualities is unsatisfactory, whereas extensional properties are completely suitable. For example, the most natural absolute concept of heavy coincides with the concept of heavy material body. Using "heavy" in this sense, we can satisfactorily say "Mida 6 is heavy," but not "the rocket $\mathfrak{m}$ is heavy" [N7], which is false even when $\mathfrak{m}$ happens to be Mida 6, although the same sentence appears to be true in ordinary speech. But both sentences appear true and satisfactory (i.e. endowed with virtually the same meaning as in ordinary speech) if "heavy" is used extensionally in them.

Here we have used the property of being heavy, which is a permanent property for nonliving beings such as rockets. The advantage or even the necessity of using extensional and not absolute properties is even more evident in connection with accidental properties such as that of being far from the earth at the instant $\underline{t}_{20}$. Since the scholastics, for example, call the mentioned properties qualities, the use of extensional properties (in ML^{ν}) can be said to be important, and essential in most situations, in order to attribute qualities to substances.

The preceding considerations can also be extended to living beings, provided the concept $\underline{QAbs}$ of quasi-absolute properties [N24] is used instead of $\underline{Abs}$. ($\underline{QAbs}$ is similar to and includes $\underline{Abs}$.) On the basis of these considerations it appears that <u>(quasi-)absolute attributes and extensional attributes somehow mirror the distinction between substances and qualities.</u>

Since there are innumerable extensional properties, certainly most of them do not correspond to true qualities. The analogue holds for (quasi-)absolute properties and in a stronger way, because $\underline{F}, \underline{G} \in \underline{Abs}$ and $\underline{F}^{(\underline{e})} = \underline{G}^{(\underline{e})}$ do not at all imply $\underline{F} = \underline{G}$; if $\underline{F}$ has the extension of a secondary substance such as the concept of mass point, then only one property $\underline{G}$ with $\underline{G} \in \underline{Abs}$ and $\underline{G}^{(\underline{e})} = {}^{\frown}\underline{F}^{(\underline{e})}$ constitutes the privileged (quasi-)absolute property corresponding to this substance.

The italicized statement above holds in that (in ML^{ν}) it is advisable (and important) to characterize every quality by means of an extensional property and every (secondary) substance by means of a (quasi-)absolute property, precisely by means of the (corresponding) privileged one. We may remark that privileged (quasi-)-absolute concepts of high logical levels, such as *natural number* or *real number*, are usually not considered as secondary substances. Hence the notion of a privileged (quasi-)absolute concept may be regarded as a natural extension of the notion of a secondary substance.

Incidentally, *animal* and *rational animal* can be regarded as secondary substances, but not *white animal*; in accordance with this, in ML^{ν} we have *animal* $\in$ *QAbs* [N24], *rational animal* $\in$ *QAbs*, and *white animal* $\notin$ *QAbs*.

The concepts of substance and quality have been considered by many philosophers since Aristotle and the scholastics, so that they have evolved. Indeed, several new similar concepts were determined (with varying degrees of precision) and identified with the concept of substance. Through this process the concepts of substance and quality have become enriched.

By no means does our thinking leading to the assertion above aim at analyzing in technical terms—or at giving an "explicatum" of, in Carnap's sense—the concepts of substance and quality. We intend to present neither a new instance of those concepts nor a mathematically rigorous definition of them according to the ideas of any particular philosopher. We only think that our considerations may contribute to the enrichment of the concepts of substance and quality, for example by emphasizing some possible defining features to be attributed to them and by pointing out some of their properties. For instance (1) we emphasized that in certain situations in which we pick out individuals as bearers of properties, i.e. substances, it is important that they should be "the same individuals in all possible cases" in the most natural sense of this phrase; and (2) we showed that it is possible to pick out such individuals by means of absolute attributes of the most natural kind.

N24. Extension of the double use of nouns. Quasi-absolute concepts.

In ordinary speech or, to be more technical, in the sciences of living beings, many common nouns are sometimes used extensionally and sometimes used to express certain attributes very similar to absolute ones. We shall call such attributes quasi-absolute attributes (QAbs).

As an example assume that horses are bred in a stable $\underline{S}$. At present in $\underline{S}$ there are certain male horses, say $\underline{m}_1, \ldots, \underline{m}_9$, and certain female horses, $\underline{f}_1, \ldots, \underline{f}_9$; moreover $\underline{m}_i$ and $\underline{f}_j$ have no offspring. Hence the horse breeders, who speak about the future possibilities of $\underline{S}$, may say,

(a) The oldest son $\underline{s}_{ij}$ of $\underline{m}_i$ and $\underline{f}_j$ may exist and also may not exist ($\underline{i}, \underline{j} = 1, \ldots, 9$).

By Def. 11.2 of the constant $\underline{a}^* = \underline{a}^*_t$—called the improper object—sentence (a) is translated into $\overline{\mathrm{ML}}^{\nu}$ by

(61) $$\Diamond \underline{s}_{\underline{ij}} \neq \underline{a}^* \ \Diamond \underline{s}_{\underline{ij}} = \underline{a}^* \quad (\underline{i}, \underline{j} = 1, \ldots, 9).$$

Now assume that next Sunday there will be two horse races, a first and a second, $\underline{R}_1$ and $\underline{R}_2$, respectively. Further assume that one of the two races will be assigned a larger first prize than the other, and that it will be decided on Saturday, by drawing lots, which will have the larger first prize. Let $\underline{R}$ be the race with the larger first prize.

The owner of $\underline{S}$ wishes both to win $\underline{R}$ and to sell as many horses before Saturday as is compatible with the condition of winning $\underline{R}$. Therefore he asks his horse breeders, who are well acquainted with the other stables, whether the following holds:

(b) In $\underline{S}$ there is a horse able to win race $\underline{R}$ (the one with the larger first prize) no matter how the lots are drawn.

Let us introduce (in ML^{ν}) the extensional binary relations [Def. 6.11] $\underline{W}$ and $\underline{W}^{\underline{W}}$ such that $\underline{W}(\underline{h}, \underline{R}')$ is equivalent (by definition) with $\underline{h}$ wins race $\underline{R}'$, and $\underline{W}^{\underline{W}}$ is the extensionalization of the following relation $\underline{W}^{W}$ in $\underline{h}$ and $\underline{R}'$: the owner of $\underline{S}$ wants $\underline{h}$ to win $\underline{R}'$. For the sake of simplicity we also assume that nobody makes any mistakes, so that $\underline{W}^{\underline{W}}$ can be used in ML^{ν} wherever the use of $\underline{W}^{\underline{W}}$ in ordinary speech is compatible with this assumption.[74]

Now, for the sake of simplicity, let us translate $\underline{h}$ is able to win $\underline{R}'$ into ML^{ν} by $\underline{W}^{\underline{A}}(\underline{h}, \underline{R}')$ where

(62) $$\underline{W}^{\underline{A}}(\underline{h}, \underline{R}') \equiv_D \underline{W}^{\underline{W}}(\underline{h}, \underline{R}') \supset^{\cap} \underline{W}(\underline{h}, \underline{R}').$$

We remark that the occurrence of "horses" in (b) is not extensional. Indeed let us assume that the owner of $\underline{S}$ knows that only $\underline{m}_1$ is able to win $\underline{R}_1$ and only $\underline{m}_2$ is able to win $\underline{R}_2$. Then on the one hand he sells, before Saturday, all horses in $\underline{S}$ except two, $\underline{m}_1$ and $\underline{m}_2$; furthermore he judges (b) to be false (the meaning of sentence (b) is usually such that if the owner of $\underline{S}$ judged (b) to be true, he would sell all horses in $\underline{S}$ except one).

On the other hand let us define $\underline{m}$ to be $\underline{m}_1$ or $\underline{m}_2$ according to whether $\underline{R}$ is $\underline{R}_1$ or $\underline{R}_2$. Then, under the above hypothesis on $\underline{m}_1$ and $\underline{m}_2$, $\underline{m}$ is able to win $\underline{R}$ no matter how lots are drawn, i.e. $\underline{W}^{\underline{A}}(\underline{m}, \underline{R})$ holds—see (62). We conclude that if the occurrence of "horse" in (b) were extensional, then under the last assumption on $\underline{m}_1$ and $\underline{m}_2$ (b) would be true, in contrast with before. Hence our remark holds.

We further observe that we may assert:

(c) Necessarily, $\underline{m}$ is a horse (provided $\underline{m}$ exists—cf. (61)).

In (b) the common noun horse is used in nearly an absolute way but not precisely so. Indeed the concept $\underline{Hors}$ of horse involved in (b) is, strictly speaking, not modally constant [Def. 13.2] because $\underline{s}_{11}$—see (a)—is, strictly speaking, a horse only provided $\underline{s}_{11}$ exists, i.e. $\underline{s}_{11} \neq \underline{a}^*$. Hence the concept $\underline{Hors}$ is quasi-modally constant ($\underline{QMConst}$) in accordance with the following definition which, for generality, refers to any attribute $\underline{F}$ of type $\underline{t} = (\underline{t}_1, \ldots, \underline{t}_n)$:

$$\text{DEF. 24.1.} \quad \underline{F} \in \underline{QMConst}_{\underline{t}} \equiv_D (\forall \underline{x}_1, \ldots, \underline{x}_n) \{ \Diamond \underline{F}(\underline{x}_1, \ldots, \underline{x}_n) \supset \underline{N}[\underline{F}(\underline{x}_1, \ldots, \underline{x}_n) \vee \bigvee_{\underline{i}=1}^{\underline{n}} \underline{x}_{\underline{i}} = \underline{a}^*]\}.$$

In addition the concept $\underline{Hors}$ is not modally separated [Def. 18.7] because in the possible cases where $\underline{s}_{11}$ and $\underline{s}_{12}$ exist—see (a)—they are distinct horses, but in the possible cases where they do not exist, they (are assumed in ML^ν to) coincide, precisely $\underline{s}_{11} = \underline{s}_{12} = \underline{a}^*$. However the concept Hors is quasi-modally separated (QMSep) in accordance with the following definition:

DEF. 24.2. $\underline{F} \in \underline{QMSep}_{\underline{t}} \equiv_D (\forall \underline{x}_1, \underline{y}_1, \ldots, \underline{x}_n, \underline{y}_n)[\underline{F}(\underline{x}_1, \ldots, \underline{x}_n) \underline{F}(\underline{y}_1, \ldots, \underline{y}_n) \diamond \bigwedge_{\underline{i}=1}^{\underline{n}} (\underline{x}_{\underline{i}} = \underline{y}_{\underline{i}} \neq \underline{a}^*) \supset \bigwedge_{\underline{i}=1}^{\underline{n}} \underline{x}_{\underline{i}} =^\frown \underline{y}_{\underline{i}}]$.

We conclude that the concept Hors is not absolute [Def. 18.8] but quasi absolute (QAbs) according to the following definition:

DEF. 24.3. $\underline{QAbs}_{\underline{t}} =_D \underline{QMConst}_{\underline{t}} \cap \underline{QSep}_{\underline{t}}$ [Defs. 24.1, 24.2].

By Defs. 24.1–3, 13.2, and 18.7, 8:

(63) $\Vdash \underline{MConst} \subset \underline{QMConst}$, $\Vdash \underline{MSep} \subset \underline{QMSep}$, $\Vdash \underline{Abs} \subset \underline{QAbs}$.

Now we assume that $\underline{h} \in \underline{Hors}_{\underline{S}}$ means that $\underline{h}$ is a horse of stable $\underline{S}$ and that $\underline{Hors}_{\underline{S}} \in \underline{QAbs}$ is Γ-true [Def. 9.6(c)]. Then on the basis of (62), Def. 4.6, and the preceding considerations, sentences (b) and (c) above are correctly translated into ML^ν by

(64) $(\exists \underline{h})[\underline{Hors}_{\underline{S}}(\underline{h}) \underline{W}^{\underline{A}}(\underline{h}, \underline{R})]$, $\underline{m} \neq \underline{a}^* \supset^\frown \underline{m} \in \underline{Hors}_{\underline{S}}^{(\underline{e})}$,

respectively. Observe that in neither part of (64) can $\underline{Hors}_{\underline{S}}$ and $\underline{Hors}_{\underline{S}}^{(\underline{e})}$ be interchanged.

N25. A new admissible modal axiom.

Let us remark that in our quasi-intensional semantical system for ML^ν [part 1], the objects of a given type $\underline{t}$ in a possible case γ are as numerous as those in every other possible case. This can be expressed in ML^ν by asserting the existence of an absolute property $\underline{F}$ of type $(\underline{t})$ [Def. 18.8] such that it is necessary for its extensionalization $\underline{F}^{(\underline{e})}$ to hold for every $\underline{x}$ of type $\underline{t}$. It is useful and quite possible to require, in addition, that $\underline{F}$ should hold for the "nonexisting object." Indeed the following admissible axiom is $\underline{L}$-true [Def. 9.5] in ML^{ν}:

AS25.1. $(\exists \underline{F}) [\underline{\mathrm{Abs}}_{\underline{t}} (\underline{F}) \underline{F} (\underline{a}^*) (\underline{x}) \underline{NF}^{(\underline{e})} (\underline{x})]$ $(\underline{t} = 1, \ldots, \nu)$— cf. Def. $1\bar{8}.9$.

A physical meaning can be attributed to this axiom. We assume that $\nu = 1$ holds and that a physical theory is based on MC^{ν}. Then it is natural to consider the event points as individuals—cf. [4]. Then for the power $\underline{p}$ of the set of individuals we have $\underline{p} \geq 2^{\underline{S}_{\underline{o}}}$ (continuum power) in every possible case. Maybe the material points and the (portions of) continuous bodies are also considered as individuals. However they are not more numerous than the borelian sets, so that their number is at most $2^{\underline{S}_{\underline{o}}}$. Nor can $\underline{p}$ be increased by considering animals or plants among individuals, because they are at most denumerable.

It can be added that we cannot conceive of any set of individuals (i.e. of nonsets) whose power is greater than $2^{\underline{S}_{\underline{o}}}$. So we conclude that $\underline{p} = 2^{\underline{S}_{\underline{o}}}$ holds in every possible case. This implies AS25.1.

THEOR. 25.1. Axiom AS25.1 is independent of the preceding axioms Ass12.1–23 proposed for a modal calculus MC^{ν} based on ML^{ν} and having modus ponens as its only inference rule.[75]

Before proving this theorem let us remark that the language ML^{ν} is conceived to be applied to classical physics. So, for example, $\underline{D}_1$ may be identified with the set of the event points of classical space-time, $\underline{D}_2$ with the set of mass points, and so on. Such sets have the same cardinalities in all possible cases.[76] In biology

$\underline{D}_1$ could be identified with the set of horses (living, dead, or to be born). In this case $\underline{D}_1$ obviously has different cardinalities in some different possible cases. In order to prove the theorem we give an alternative definition of quasi intensions, which (1) is compatible with the $\underline{L}$-truth of our axioms Ass12.1–23, and (2) complies with the above possible use of $\underline{D}_1$ in biology. Incidentally, in this way an easy generalization of our semantical theory for ML^ν is exhibited, and this is a reason for making the proof of Theor. 25.1 explicit.

<u>Proof</u>. In order to construct the class $\underline{QI}'_t$ of the new quasi intensions of type $\underline{t}$ ($\underline{t} \in \bar{\tau}^\nu$) mentioned above we use, instead of the individual domain $\underline{D}_r$, a function $\hat{\underline{D}}_r$ from Γ-cases to sets. From the forthcoming definition of $\underline{QI}'_t$ it will appear that when $\hat{\underline{D}}_r$ are constant functions—i.e. the set $\underline{D}_r = \hat{\underline{D}}_r(\gamma)$ is independent of γ ($\gamma \in \Gamma$)—the new quasi intensions coincide with the old quasi intensions ($\underline{QI}'_t = \underline{QI}_t$ for $\underline{t} \in \bar{\tau}^\nu$). However, to be more specific, one may assume (in accordance with any practical purpose) that the intersection $\hat{\underline{D}}_r(\gamma) \cap \hat{\underline{D}}_r(\gamma')$ is empty for every pair of Γ-cases γ and γ' with $\gamma \neq \gamma'$.

Now let us define the class $\underline{QI}'_t$ recursively by the following conditions:

(a′) For $\underline{r} = 1, \ldots, \nu$ $\underline{QI}'_r$ is the class of the functions ϕ defined on Γ and such that, for every $\gamma \in \Gamma$, $\phi(\gamma) \in \underline{D}_r(\gamma)$.

(b′) For $\underline{t}_1, \ldots, \underline{t}_n \in \tau^\nu$ and $\underline{t} = (\underline{t}_1, \ldots, \underline{t}_n)$, $\underline{QI}'_t$ is the class of the sets of ($\underline{n}$+1)-tuples $(\bar{\underline{x}}_1, \ldots, \bar{\underline{x}}_n, \gamma)$ with $\bar{\underline{x}}_1 \in \underline{QI}'_{t_1}, \ldots, \bar{\underline{x}}_n \in \underline{QI}'_{t_n}$ and $\gamma \in \Gamma$.

(c′) For $\underline{t}_0, \ldots, \underline{t}_n \in \tau^\nu$ and $\underline{t} = (\underline{t}_1, \ldots, \underline{t}_n : \underline{t}_0)$, $\underline{QI}'_t$ is the class of the functions from the Cartesian product $\underline{QI}'_{t_1} \times \ldots \times \underline{QI}'_{t_n}$ to $\underline{QI}'_{t_0}$.

(d′) For $\underline{t} = 0$, $\underline{QI}'_{\underline{t}} = \underline{QI}_0$—see (8) and (9).

Now one can see that ASs12.1–23 are $\underline{L}$-true [Def. 9.5] in connection with the new quasi intensions. On the other hand we may choose the function $\underline{D}_1(\gamma)$ of γ in such a way that for some γ_1 and γ_2 the sets $\underline{D}_1(\gamma_1)$ and $\underline{D}_1(\gamma_2)$ have different cardinalities. Then AS25.1 cannot be $\underline{L}$-true in the new sense for $\underline{t} = 1$.

QED

N26. On the concept of closure ML^ν.

In order to deal with natural numbers in ML^ν [N27] a suitable concept of closure is required. Therefore we first define in ML^ν the logical product $\bigcap\Psi$ and the logical sum $\bigcup\Psi$—cf. [32], p. 239—of the family Ψ of properties of type $\underline{t}$:

DEF. 26.1. $\bigcap\Psi =_D (\lambda \underline{x})(\forall \underline{F})[\Psi(\underline{F}) \supset F(\underline{x})]$.

DEF. 26.2. $\bigcup\Psi =_D (\lambda \underline{x})(\exists \underline{F})[\Psi(\underline{F})\,\underline{F}(\underline{x})]$.

Now we define $\underline{Her}(\beta, \underline{R})$ and $\underline{clos}(\alpha, \underline{R})$ to be read respectively as "the property β of type $(\underline{t})$ is hereditary with respect to the $(\underline{n}+1)$-ary relation $\underline{R}$ of type $(\underline{t}, \ldots, \underline{t})$" and "the closure of the property α (of type $(\underline{t})$) with respect to $\underline{R}$":

DEF. 26.3. $\underline{Her}(\beta, \underline{R}) \equiv_D (\forall \underline{x}_1, \ldots, \underline{x}_{\underline{n}}, \underline{z})[\underline{x}_1, \ldots, \underline{x}_{\underline{n}} \epsilon \beta \wedge R(\underline{x}_1, \ldots, \underline{x}_{\underline{n}}, \underline{z}) \supset \underline{z}\epsilon\beta]$ [Def. 4.6].

DEF. 26.4. $\underline{clos}(\alpha, \underline{R}) =_D \bigcap (\lambda\beta)[\alpha \subseteq \beta\ \underline{Her}(\beta, \underline{R})]$ [Defs. 26.1, 26.3, 18.4].

The $\underline{clos}(\alpha, \underline{R})$ has been defined in accordance with the way it is done in Rosser's extensional theory.[77] Therefore it is easy to see that our $\underline{clos}(\alpha, \underline{R})$ has the same properties in ML^ν as the concept of closure in extensional logic. Here we present some of them by enunciating the following semantical theorems (65) to (68) concerning ML^{ν}. They constitute, in order, the analogues for ML^{ν} of the syntactical theorems IX.5.12–16 proved in [32].[78]

(65) $\Vdash \underline{x} \in \underline{clos}(\alpha, \underline{R}) \equiv (\beta)\,[\alpha \subseteq \beta\, \underline{Her}(\beta, \underline{R}) \supset \beta(\underline{x})]$, $\Vdash \alpha \subseteq \underline{clos}(\alpha, \underline{R})$.

(66) $\Vdash (\forall \underline{x}_1, \ldots, \underline{x}_n, \underline{z})\,[\underline{x}_1, \ldots, \underline{x}_n \in \underline{clos}(\alpha, \underline{R})\, \underline{R}(\underline{x}_1, \ldots, \underline{x}_n, \underline{z}) \supset \underline{z} \in \underline{clos}(\alpha, \underline{R})]$.

(67) $\Vdash \alpha \subseteq \beta (\forall \underline{x}_1, \ldots, \underline{x}_n, \underline{z})\,[\underline{x}_1, \ldots, \underline{x}_n \in \beta \cap \underline{clos}(\alpha, \underline{R})\, \underline{R}(\underline{x}_1, \ldots, \underline{x}_n, \underline{z}) \supset \beta(\underline{z})] \supset \underline{clos}(\alpha, \underline{R}) \subseteq \beta$.

(68) $\Vdash \underline{z} \in \underline{clos}(\alpha, \underline{R}) \equiv \underline{z} \in \alpha \vee (\exists \underline{x}_1, \ldots, \underline{x}_n)\,[\underline{x}_1, \ldots, \underline{x}_n \in \underline{clos}(\alpha, \underline{R})\, \underline{R}(\underline{x}_1, \ldots, \underline{x}_n, \underline{z})]$.

The following theorem has no extensional analogue:

THEOR. 26.1. If the property α of type $\underline{t}$ and the ($\underline{n}$+1)-ary relation $\underline{R}$ of type ($\underline{t}, \ldots, \underline{t}$) are extensional, then the same holds for the $\underline{clos}(\alpha, \underline{R})$, i.e.:

(69) $\Vdash \alpha \in \underline{Ext}_{\underline{t}} \wedge \underline{R} \in \underline{Ext}_{(\underline{t}, \ldots, \underline{t})} \supset \underline{clos}(\alpha, \underline{R}) \in \underline{Ext}_{\underline{t}}$.

Following are sufficient hints for the proof of this theorem, assuming ML^{ν} as a part of our language.

Assume (a) $\alpha \in \underline{Ext}$ and (b) $\underline{R} \in \underline{Ext}$. Let β be the extensional part of $\underline{clos}(\alpha, \underline{R})$, that is:

(70) $(\underline{x})\,\{\beta(\underline{x}) \equiv_D \underline{x} \in \underline{clos}(\alpha, \underline{R})\, (\forall \underline{y})\,[\underline{y} = \underline{x} \supset \underline{y} \in \underline{clos}(\alpha, \underline{R})]\}$.

Then by $(65)_2$ and (a) we obtain (c) $\alpha \subseteq \beta$. Now suppose (d) $\underline{x}_1, \ldots, \underline{x}_n \in \beta \cap \underline{clos}(\alpha, \underline{R})$ and (e) $\underline{R}(\underline{x}_1, \ldots, \underline{x}_n, \underline{z})$. Then by (66) we have (f) $\underline{z} \in \underline{clos}(\alpha, \underline{R})$. Moreover from (b) and (e) by Def. 6.11 we have (g) $(\underline{y})\,[\underline{y} = \underline{z} \supset \underline{R}(\underline{x}_1, \ldots, \underline{x}_n, \underline{y})]$.

By (66), from (d) and (g) we obtain $(\underline{y})\,[\underline{y} = \underline{z} \supset \underline{y} \in \underline{clos}(\alpha, \underline{R})]$. Thence by (f) and (70) we obtain $\beta(\underline{z})$. We conclude that (d) and (e) imply $\beta(\underline{z})$; furthermore this implication holds for every $\underline{x}_1, \ldots, \underline{x}_n, \underline{z}$ so that (h) $(\forall \underline{x}_1, \ldots, \underline{x}_n, \underline{z})\, [\underline{x}_1, \ldots, \underline{x}_n \epsilon \beta \cap \underline{\mathrm{clos}}(\alpha, \underline{R})\, \underline{R}(\underline{x}_1, \ldots, \underline{x}_n, \underline{z}) \supset \beta(\underline{z})]$ holds. From (c) and (h), by (67) we obtain $\mathrm{clos}(\alpha, \underline{R}) \subseteq \beta$. Hence by (70), $\underline{\mathrm{clos}}(\alpha, \underline{R}) = \beta$. As a consequence by (70) and Def. 6.11, $\underline{\mathrm{clos}}(\alpha, \underline{R}) \epsilon \underline{\mathrm{Ext}}$. We conclude that Theor. 26.1 holds.

Incidentally, one could prove:

(71) $\Vdash [\underline{\mathrm{clos}}(\alpha, \underline{R})]^{\underline{(e)}} \underline{\subset} \underline{\mathrm{clos}}(\alpha^{\underline{e}}, \underline{R}^{\underline{e}})$ [Defs. 18.9, 26.4].

Furthermore in (71) "$\underline{\subset}$" can be replaced by "=" for some choices of α and $\underline{R}$—see Def. 27.5 and (82)—and by "$\subset$" [Def. 18.5] for some other choices of α and $\underline{R}$.

N27. A natural absolute concept of natural number defined on purely logical grounds in ML^{ν}. Peano's axioms in ML^{ν}.

Here we first define zero [of type $((\underline{t}))$] as the family containing only the empty property $\Lambda_{\underline{t}}$ of type $(\underline{t})$ [Def. 18.6]. Then we define 1 of type $((\underline{t}))$ as the family of the properties of type $(\underline{t})$ holding exactly for one element.

DEF. 27.1. $0 =_D \{\Lambda_{\underline{t}}\}$.

DEF. 27.2. $1 =_D (\lambda \underline{F})\, (\exists_1 \underline{x})\, \underline{F}(\underline{x})$.

It is not difficult to see that

(72) $\Vdash 0, 1 \epsilon \underline{\mathrm{Ext}}$ [Defs. 6.11, 4.6, $(42)_1$, AS12.14].

Here we define the cardinal sum $\alpha + \beta$ of the families α and β of properties of type $(\underline{t})$ and the predecessor relation $\underline{\mathrm{Pred}}$:

DEF. 27.3. $\alpha + \beta =_D (\lambda \underline{F})\, (\exists \underline{G}, \underline{H})\, [\alpha(\underline{G})\beta(\underline{H})\, \underline{G}^{\underline{(e)}} \cap \underline{H}^{\underline{(e)}} = \Lambda_{(\underline{t})}\, \underline{G} \cup \underline{H} = \underline{F}]$,

DEF. 27.4. $\underline{\text{Pred}}(\alpha,\beta) \equiv_D \alpha+1=\beta$,

where (1) $\underline{F}$, $\underline{G}$, $\underline{H}$, α, and β are distinct variables of the respective types ($\underline{t}$), ($\underline{t}$), ($\underline{t}$), (($\underline{t}$)), and (($\underline{t}$)), and where (2) $\underline{F}$, $\underline{G}$, and $\underline{H}$ have, in order, the indexes 1, 2, and 3.

Defs. 27.1–4 imply:

$$\text{(73)} \qquad \text{Pred}(\alpha,\beta) \supset 0 \neq \beta.$$

Here we define in ML^ν, on purely logical grounds, the natural absolute concept $\underline{\text{Nn}}$ of natural number by identifying $\underline{\text{Nn}}$ with the closure of $\{0\}^\cap$ with respect to the strict predecessor relation $\underline{\text{Pred}}^\cap$:

DEF. 27.5. $\underline{\text{Nn}} =_D \underline{\text{clos}}(\{0\}^\cap, \underline{\text{Pred}}^\cap)$ [$\Vdash \underline{\text{Pred}}^\cap(\alpha,\beta) \equiv \alpha+1=^\cap \beta$, see Defs. 6.3, 27.4].

Here, in order, are the analogues for ML^ν of the theorems X.1.2, X.1.4–7, and X.1.12 in [32]. The occurrences of simple identity between natural numbers in [32] are replaced here by occurrences of strict identity with an exception for the strong theorem $(74)_1$:

$$\text{(74)} \qquad \Vdash 0 \neq \underline{n}+1, \quad \Vdash 0 \in \underline{\text{Nn}}, \quad \Vdash \underline{n} \in \underline{\text{Nn}} \supset \underline{n}+1 \in \underline{\text{Nn}}.$$

$$\text{(75)} \qquad \Vdash (0 \in \beta)\,(\underline{y})\,[\underline{y} \in \beta \cap \underline{\text{Nn}} \supset \underline{y}+1 \in \beta] \supset \underline{\text{Nn}} \subseteq \beta.$$

$$\text{(76)} \quad \Vdash \underline{n} \in \underline{\text{Nn}} \equiv \underline{n} =^\cap 0 \vee (\exists \underline{m})\,(\underline{m} \in \underline{\text{Nn}}\ \underline{n} =^\cap \underline{m}+1), \quad \Vdash 1 \in \underline{\text{Nn}}.$$

Defs. 27.4 and 5 imply $(74)_1$. By Def. 27.5 and $(65)_2$, $\Vdash \{0\}^\cap \subseteq \underline{\text{Nn}}$, hence $(74)_2$.

The proofs of $(74)_3$, (75), and (76) are the semantical analogues for ML^ν of the proofs for the syntactical theorems X.1.5–7 and X.1.12 in [32]. Note that $(74)_1$ implies

$$\text{(77)} \ \Vdash \underline{n} \in \underline{\text{Nn}} \supset \text{N}0 \neq \underline{n}+1, \quad \Vdash \underline{n} \in \underline{\text{Nn}}^{(e)} \supset \sim 0 =^\cap \underline{n}+1, \quad \Vdash \underline{n} \in \underline{\text{Nn}} \supset \sim 0 =^\cap \underline{n}+1.$$

Now let each of our domains of individuals $\underline{D}_1, \ldots, \underline{D}_\nu$ [N6] have infinitely many elements. Then

(78) $\Vdash \underline{m}, \underline{n} \in \underline{Nn} \wedge \underline{m}+1 =^{\cap} \underline{n}+1 \supset \underline{m} =^{\cap} \underline{n}$, $\Vdash \underline{m}, \underline{n} \in \underline{Nn}^{(e)} \wedge \underline{m}+1 = \underline{n}+1 \supset \underline{m} = \underline{n}$ [cf. axiom 13 in [32], p. 279].

The semantical theorems $(74)_2$, $(74)_3$, $(77)_3$, $(78)_1$, and (75) say in order that Peano's axioms for the natural numbers (in the usual order) hold in ML^ν up to the replacement of the common equality sign by $=^{\cap}$. It is easy to realize that the theorems obtained from the above five by replacing $\underline{Nn}$ by $\underline{Nn}^{(e)}$ and $=^{\cap}$ by $=$ also hold in ML^ν. Among them we shall use $(78)_2$.

Let us remark that the $\underline{L}$-truth in ML^ν of Peano's axioms in the above double form (and the validity in ML^ν of modus ponens) allows us to transfer at once the ordinary theories of natural numbers into ML^ν using at will either $\underline{Nn}$ and $=^{\cap}$ or $\underline{Nn}^{(e)}$ and $=$.

THEOR. 27.1. Under the hypothesis that the individuals of any type are infinitely many we may assert [Defs. 13.2, 18.7, 18.8]:[79]

(79) $\Vdash \underline{Nn} \in \underline{MConst}$, $\Vdash \underline{Nn} \in \underline{MSep}$, $\Vdash \underline{Nn} \in \underline{Abs}$.

Here we prove $(79)_1$ and give sufficient hints for the proof of $(79)_{2,3}$ assuming ML^ν as a part of our language. Note that by $(74)_2$, $\Vdash 0 \in \underline{Nn}^{\cap}$ [Def. 6.6]; moreover $(74)_3$ and AS12.5 imply $\Vdash (\underline{n}) [\underline{n} \in \underline{Nn}^{\cap} \supset \underline{n}+1 \in \underline{Nn}^{\cap}]$. Hence by (75) we deduce $\Vdash \underline{Nn} \subseteq \underline{Nn}^{\cap}$. Furthermore, obviously $\Vdash \underline{Nn}^{\cap} \subseteq \underline{Nn}$. Then $\Vdash \underline{Nn} = \underline{Nn}^{\cap}$, whence $\Vdash \underline{Nn}^{\cup} = \underline{Nn}^{\cap}$, so that by Defs. 6.6, 6.8, and 13.2, $(79)_1$ holds.

As for $(79)_2$ let us introduce the modally separated part Ψ of $\underline{Nn}$ as follows:

(80) $(\underline{x}) \{\Psi(\underline{x}) \equiv_D \underline{x} \in \underline{Nn} (\underline{y}) [\underline{y} \in \underline{Nn} \wedge \underline{y} =^{\cup} \underline{x} \supset \underline{y} =^{\cap} \underline{x}]\}$.

(80), $(74)_2$, $(76)_1$, and $(77)_1$ imply (a) $\Psi(0)$.

Assume (b) $\underline{x} \in \Psi \cap \underline{Nn}$ and—as a hypothesis for reductio ad absurdum—(c) $\sim\Psi(\underline{x}+1)$. Then—since (b) and $(74)_3$ $\underline{L}$-imply $\underline{x}+1 \in \underline{Nn}$—(c) and (80) $\underline{L}$-imply the existence of such a $\underline{y}$ that (d) $\underline{y} \in \underline{Nn}$, (e) $\diamond \underline{y} = \underline{x}+1$, and (e′) $\diamond \underline{y} \neq \underline{x}+1$ hold. Then by $(74)_1$ we have $\diamond \underline{y} \neq 0$, i.e. $\sim \underline{y} =^{\cap} 0$, so that by (d) and $(76)_1$ there is such an $\underline{m}$ that (f) $\underline{m} \in \underline{Nn} \wedge \underline{y} =^{\cap} \underline{m}+1$ holds. From (f) and (b) we have (g) $\underline{x}, \underline{m} \in {}^{\cap}\underline{Nn}^{(e)}$.

On the one hand by $(78)_1$ [$\Vdash p \supset q$ yields $\Vdash p \supset^{\cap} q$] the matrices (f) and (e) $\underline{L}$-imply (h) $\underline{x}+1 =^{\cup} \underline{m}+1$; furthermore by $(78)_2$ (g) and (h) $\underline{L}$-imply (l) $\underline{x} =^{\cup} \underline{m}$. On the other hand (f) and (e′) $\underline{L}$-imply $\diamond \underline{x}+1 \neq \underline{m}+1$, whence (l′) $\diamond \underline{x} \neq \underline{m}$.

By (80), (f), (l), and (l′) we have $\sim\Psi(\underline{x})$, which contrasts with (b). We conclude that (c) is absurd, hence (m) $(\underline{x})[\underline{x} \in \Psi \cap \underline{Nn} \supset \Psi(\underline{x}+1)]$ holds.

By (a)—i.e. $\Psi(0)$—(m), and (75), we obtain $\underline{Nn} \subseteq \Psi$, whence by (80), $\underline{Nn} = \Psi$. Then by (80) and Def. 18.7, $(79)_2$ holds.

Theorems $(79)_{1,2}$ and Def. 18.8 obviously imply $(79)_3$. So sufficient hints have been given, I think, for a thorough proof of Theor. 27.1.

Let us consider the following theorems:

(81) $\Vdash (\lambda\, \alpha, \beta)\, (\alpha + \beta) \in \underline{Ext}$, $\Vdash \underline{Pred} \in \underline{Ext}$, $\Vdash \underline{clos}(\{0\}, \underline{Pred}) \in \underline{Ext}$ [Def. 6.11].

$(81)_{1,2}$ follow from Defs. 27.3 and 4; furthermore (69), $(42)_1$, and $(81)_2$ imply $(81)_3$.

Incidentally, one could prove, in accordance with (71), that

(82) $\Vdash \underline{Nn}^{(e)} = \underline{clos}(\{0\}, \underline{Pred})$ [Defs. 18.9, 26.4, 27.4].

Notes to Memoir 1

*The main results of this paper were presented at the Congress of Logic, Methodology, and Philosophy of Science held in Jerusalem in August 1964.

The author, an Italian mathematical physicist, met with considerable difficulties presenting this work—begun in 1960—in its final form. In connection with overcoming these difficulties, he is indebted to several professors of logic, mentioned later in this footnote, and in a particularly deep way to R. Carnap and N. Belnap.

Carnap encouraged him greatly during the first elaboration of the present work, and without such strong encouragement the author would probably not have diverted so much time and energy from his usual professional field.

Belnap spent much time improving the exposition of the present work. Most of these improvements are linguistic; the others are mostly of an expository nature and, in some cases, of a completely substantial nature. For instance he pointed out Austin's story about the donkeys in [1], which gave rise to a substantial improvement of N24.

The author became interested in logic through axiomatization problems of classical mechanics. In connection with them he pointed out certain external paradoxes—published in [4], N7, p. 82—consisting for example of an assertion p within the ordinary language of physics, which on the one hand appears false according to the customary interpretation of this language, while on the other hand we can prove the straightforward formalization of p into extensional logic—where the "if . . . , then . . ." of ordinary speech is replaced by material implication. The conclusion is that in some cases a logic richer than extensional logic—in particular modal logic—has to be used to formalize the language of physics.

In 1958 the author communicated some of these considerations somewhat related to the axiomatization of classical mechanics to B. Rosser, who advised the author to see P. Suppes at Stanford University. There, where the author worked for three months thanks to a Fulbright grant, the work [4] on foundations of classical mechanics according to P. Painlevé was begun, using a device based on some ideas of Carnap included in [12]—cf. Ns17, 19.

As far as the present work is concerned, in 1966 E. Casari, professor at the University of Pavia, Italy, recommended some improvements on the presentation of chapter 1, which deals with extensional matter. In 1967–68, following the advice of Suppes and taking advantage of Grant AFOSR 728-66, given by the Air Force Office of Scientific Research, the author decided to contact the philosophy department of the University of Pittsburgh, which is headed by A. R. Anderson and which is deeply interested in modal logic. There Belnap kindly undertook the task of casting the whole first version of this work into publishable form; his work was essential to the completion of the entire monograph.

The content of the final version of this work is essentially the same as that of the first version except that e.g. theorem $(33)_2$ is slightly strengthened and something connected with the appendices is added.

1. Such an analysis characterizes a translation (in a strong sense) of ML^{ν} into the $(\nu + 1)$-sorted extensional language $EL^{\nu+1}$.

2. MC^{ν} is based on ML^{ν} and is similar to common extensional logical calculi to a rather large extent.

3. We require this definition and more generally the foundations of classical mechanics according to Painlevé to be based on an explicitly known theory of formal logic.

4. Such difficulties arise through a certain combination of modalities with existential quantification—see Ns19–23.

5. To my knowledge [17] (presented at a Paris colloquium in 1959 and published in 1964) is the first paper that contains Mach's definition of mass and that is explicitly based on a formal system containing modal operators—on Lewis' S5 and the axiom $\diamond(\exists \underline{x})\underline{p} \equiv (\exists \underline{x})\diamond\underline{p}$, as presented in [25]. In [17] identity is used as strict identity, and functions or descriptions are not used. Functions are always replaced by relations and cannot be introduced in the usual way, if one wants to use them to express contingent concepts such as the concept of the position of the material point $\underline{M}$ at the instant $\underline{t}$ (with respect to a given space)—cf. [18], pp. 206–09.)

6. It compels us to introduce (as a primitive concept) the concept of a mechanically possible case, which is an analogue of the concept of state (of the universe) used by Wittgenstein [39], and Carnap [12], p. 9.

7. The standards of [29] are lower than those of the rigorous works of mathematics, as Painlevé himself seems to admit—see fn. 53. The same holds for [33].

8. These operators serve to build descriptions and classes respectively.

9. The semantical rules given in [12], p. 182, and mentioned above are not more general than those for $\underline{L}_2$. Moreover it is true that in $\underline{L}_3$ the modal sign $\underline{N}$ is admitted within the scope of the lambda operator (but not within that of the iota operator). However, as far as I know, semantical rules for $\underline{L}_3$ have never been published; and they must be essentially different from those for ML^{ν} because—as Carnap says in [13], III.9.IV, p. 894—they characterize no translation in a strong sense into an extensional language—cf. fnn. 20 and 21—for $\underline{L}_3$.

ML^{ν} is the first modal language presented at a congress [in 1964, cf. fn. *], that is substantially more general that $\underline{L}_2$ and combines modalities with variables in such a way that the customary inferences of the logic of quantification—in particular specification and existential generalization—remain valid. Such a combination "is, of course, of greatest importance," as Carnap says in [12], p. 196, and is realized in the present paper.

10. In ML^{ν} the lambda operator is defined by means of the iota operator [N4] in a way associated with Rosser's formalism for classes in extensional logic—see [32], p. 219.

11. One of the main aims Carnap's method achieves is to afford a solution of what may be called the antinomy of the name relation—see [12], § 31. It is easy to see that ML^{ν} solves that antinomy in substantially the same way as the modal languages constructed in [12].

12. In particular matrices (17) to (20) are true in ML^{ν} [Theor. 9.7]. These matrices correspond to the formulas (a) to (n) in [12], p. 186, considered by Carnap substantially to show that his theory fulfills (c).

13. See [32]. There the use of a type system is replaced by Rosser's stratification theory. It is very easy to transform Rosser's formal theorems for classes into their analogues for an extensional logic based on a type system. (The formal proofs become simpler.)

14. In particular, in ML^{ν} identity cannot be defined using the sign $\underline{N}$ or the signs that in extensional logic are sufficient for the same purpose—see fnn. 40, 41.

15. In Memoir 3 it will be shown that on the basis of AS12.19 a general intensional description operator can be defined in MC^{ν} and a theorem of relative completeness for MC^{ν} can be proved.

16. Explicit direct proofs of these semantical theorems are left to the reader because these theorems follow from their corresponding syntactical theorems in that the latter can be proved in MC^{ν} by means of axioms [ASs12.1–23] and inference rules that are valid in ML^{ν}.

17. Since in ML^{ν} nonextensional properties [Def. 6.11] can also be expressed, the extensions for ML^{ν} are more complex than those considered e.g. for Carnap's language $\underline{L}_2$—see [13], III.9, pp. 892, 895. In connection with nonextensional properties the extensions for ML^{ν} also involve something modal. They are called extensions because two designators having the same extension for ML^{ν} can be interchanged in every context that is usually considered to be extensional.

18. By our designation rule (δ_9) [N11] the descriptions in ML^{ν} also fulfill a stronger requirement [Theor. 11.3], discussed in N11.

19. More precisely we couple every axiom concerning the sign $\underline{N}$ with the corresponding axiom concerning $(\forall \underline{x})$ and also holding substantially in extensional logic.

20. Our translation of ML^{ν} into $EL^{\nu+1}$ is strong in that the translation Δ^{η} of every designator Δ in ML^{ν} characterizes the intension of Δ. (Moreover the universal quantifier $(\forall \kappa)$ can be considered as the translation of $\underline{N}$.)

21. To determine the QIs for the modal languages $\underline{L}_2$ and $\underline{L}_3$—see [13], III.9, p. 892—and for the corresponding semantical rules (those for $\underline{L}_3$ have never been made explicit, as far as I know) is substantially equivalent to translating $\underline{L}_2$ and $\underline{L}_3$ into an extensional language. Referring to such translations Carnap says in [13], III.9.IV, p. 894: "In the strong sense of translation, the transformation of every sentence into a synonymous one or the transformation of every designator into an $\underline{L}$-equivalent one, a translation of a modal language into an extensional one is obviously impossible since in an extensional language there can be no designator $\underline{L}$-equivalent to 'N'."

This assertion shows, I think, the significance of our strong translation of ML^{ν} into $EL^{\nu+1}$. And it makes us believe that the extensional translation understood by Carnap for $\underline{L}_3$ should be considerably different from that of ML^{ν} into $EL^{\nu+1}$.

22. The tranlation of the modal language ML^{ν} into the extensional language $EL^{\nu+1}$ is particularly useful, for as Suppes says

in [24], p. 52, "the controversial character of extensive literature on modal logic and subjective conditionals" is deplorable. Incidentally cf. Chap. 13.

23. Such a view complies e.g. with Hutten [19], p. 52.

24. Physical possibility is also understood in this sense in such purely scientific works as [5], [6], and [7], which lay down the foundations for a general theory of general relativity. This theory includes thermodynamics, electromagnetism, and constitutive equations of materials. (In [7] hereditary phenomena are taken into account.)

25. The semantical systems for modal languages set up by Carnap in [12] and [13] lead naturally, I think, to the first concept of possibility.

26. The notations used by Carnap in [12] are followed as far as possible throughout the whole of this work.

27. Let $\underline{S}$ be a nonempty closed sphere of the ordinary space. As is well known, the least open sphere containing $\underline{S}$ does not exist. However, according to our convention on descriptions, connected with condition (A), the empty set Λ turns out to coincide with this sphere:

$$\text{(a)}\ \Lambda = (\iota\sigma)\,[\underline{p}_\sigma\,(\sigma')\,(\underline{p}_{\sigma'} \supset \sigma' \underline{\subseteq} \sigma),\ \text{where}\ \underline{p}_\sigma \equiv_D \sigma \in \text{open sphere} \wedge \underline{S} \underline{\subseteq} \sigma.$$

I believe that this result is less tolerable than its analogue for condition ($\underline{A}^*$), i.e. $\underline{a}^*_t = (\iota\sigma)\,[\underline{p}_\sigma \ldots]$ where $\underline{a}^*_t$ is an unspecified object (of a suitable type) to be called the nonexisting object. This is because the empty set is a very familiar object which enters many theorems. Among these is the assertion $S \not\subseteq \Lambda$, which seems to contrast with (a).

Therefore the use of ι in the semantics based on condition ($\underline{A}$) seems to me confusing. Nevertheless such semantics are widely used.

28. As will be clear after the semantical rules are given [NN8, 11], metalinguistic definitions such as Defs. 6.6 and 6.8 may be replaced by definitions where the definiens is a lambda expression, i.e. by definitions for the form $\underline{F} \,{}^{\frown}\!\!= _D (\lambda \underline{x}_1, \ldots, \underline{x}_n)\, \underline{NF}\, (\underline{x}_1, \ldots, \underline{x}_n)$ [Def. 4.4].

29. In [12] Carnap uses $\overline{\supset}$ for our $\supset^{\frown}$ and $\overline{\equiv}$ for our $\equiv^{\frown}$. Carnap's analogue for $=^{\frown}$ would be $\equiv$, which already has a use in ML^ν different from that of $=^{\frown}$.

30. It is understood that—as Carnap intends in [13], III.9.III, IV—under the assumption that $\underline{x}$ and $\underline{y}$ are individuals, the matrix $x = \underline{y}$ holds in the possible case γ if and only if $\tilde{\underline{x}}(\gamma) = \tilde{\underline{y}}(\gamma)$ where $\tilde{\underline{x}}$ and $\tilde{\underline{y}}$ represent the intensions of $\underline{x}$ and $\underline{y}$. Something similar is also understood in the cases where the above assumption does not hold—see N8, rule (δ_2).

31. Hence for $\underline{r} = 1, \ldots, \nu$ we assume that the quasi intension of the property $\underline{F}_2$ of type $(\underline{r})$ is the (extensional) relation $\tilde{\underline{F}}_2$ of type $(\underline{r}, \nu+1)$, which holds for $\tilde{\underline{x}}$ and γ if and only if (1) $\gamma \in \Gamma$, (2) $\tilde{\underline{x}} \in QI_{\underline{r}}$, and (3) the matrix $\underline{F}_2(\overline{x})$ holds in γ for the intension of $\underline{x}$ characterized by $\tilde{\underline{x}}$—i.e. either $\gamma \in \tilde{\underline{p}}_1$ and x is $\tilde{\underline{M}}_1$ or $\gamma \in \Gamma - \tilde{\underline{p}}_1$ and $\tilde{\underline{x}} \in \{\tilde{\underline{M}}_2, \tilde{\underline{M}}_3\}$.

32. We consider the matrix $\Delta = \Delta'$ to be equivalent to $(\forall \underline{x}_1, \ldots, \underline{x}_n)[\Delta(\underline{x}_1, \ldots, \underline{x}_n) \equiv \Delta'(\underline{x}_1, \ldots, \underline{x}_n)]$ for $\underline{t} = (\underline{t}_1, \ldots, \underline{t}_n)$, and to $(\forall \underline{x}_1, \ldots, \underline{x}_n)[\overline{\Delta}(\underline{x}_1, \ldots, \underline{x}_n) = \overline{\Delta'}(\underline{x}_1, \ldots, \underline{x}_n)]$ for $\underline{t} = (\underline{t}_1, \ldots, \underline{t}_n : \underline{t}_0)$.

33. As to the propositional calculus, the above concepts of provability and $\underline{L}$-truth are equivalent, as is well known.

34. The bracketed restriction can be deleted as soon as rule (δ_9) [N11] is laid down.

35. The matrices $(17)_1$ to $(20)_4$ are the analogues of the matrices (a) to (n) considered in [12], Theor. 41.5. The $\underline{L}$-truth of $(21)_2$ will be useful in N11.

36. By Defs. 11.1 and 6.11, rules (δ_1) to (δ_8), and Def. 9.5:

$$\Vdash \underline{Ext}(\underline{F}) \equiv (\forall \underline{x}_1, \ldots, \underline{x}_n)[\underline{F}(\underline{x}_1, \ldots, \underline{x}_n) \equiv \underline{F}(\underline{x}_1, \ldots, \underline{x}_n)^{(\underline{ex}_1)} \ldots ^{(\underline{ex}_n)}].$$

37. Indeed by Def. 11.1 the resulting definiens—i.e. $(\exists x)(y)[\phi(y) \equiv \underline{x} = \underline{y}]$—$\underline{L}$-implies $(\underline{y})[\phi(\underline{y}) \equiv \phi(\underline{y})^{(\underline{ey})}]$ and hence, by Hyp. 11.1, also the matrix $(\underline{x})[\phi(\underline{x}) \equiv \phi(\underline{x})^{(\underline{ex})}]$, which asserts that $\phi(\underline{x})$ is extensional with respect to $\underline{x}$.

38. In practical life men are often able to do something (e.g. to launch a given rocket on a given day) in several ways, and they can use a theory such as classical mechanics to learn the consequences of each. The corresponding (sufficiently approximative) answers of mechanics involve various physically possible cases (i.e. states of the universe). On the one hand most of them have nothing to do with the real case $\gamma_{\mathfrak{R}}$, in that they speak of phenomena that (will) never happen in $\gamma_{\mathfrak{R}}$. But on the other hand such phenomena must also be considered in order to be able to make the best decision.

39. Here are some suggestions for proving that AS12.13 is L-true: First consider the following theorem:

THEOREM. Let Δ be a term free for x in the term $\psi(x)$, and let $\psi(\Delta)$ result from $\psi(x)$ by replacing x with Δ. Then $\Vdash x =^{\cap} \Delta \supset \psi(x) = \psi(\Delta)$.

Then prove the L-truth of AS12.13 together with the above theorem using induction, taking as inductive hypothesis the assumption that both the theorem and the L-truth of AS12.13 hold provided $\phi(x)$ and $\psi(x)$ have fewer than m symbols. Next consider both the cases where the matrix $\phi(x)$ has m symbols and any one of the forms $\sim\phi_1(x)$, $\phi_1(x) \wedge \phi_2(x)$, $\Delta_1(x) = \Delta_2(x)$, and $R(\Delta_1, \ldots, \Delta_n)$, and the cases where $\psi(x)$ has m symbols and any one of the forms $f(\Delta_1, \ldots, \Delta_n)$ and $(\iota y)p$.

40. Before Carnap made this remark, that result was considered to be surprising by some logicians—see [13], III.9.III, p. 893.

41. Simple identity can be defined in ML^{ν} as follows:

$$x = y \equiv_D (F)\,[\mathrm{Ext}(F) \supset F(x) \equiv F(y)].$$

42. ASs12.14 and 12.15 could be taken as definitions of identity of objects of type t, for every nonindividual type t. Then ASs12.10–12 would hold as theorems for this t, but they must still be postulated for every individual type. This procedure would also require a change in our formation rule (f_2) [N3].

43. AS12.18 is similar to axiom 11 in [32], p. 185. The latter corresponds only to requirement (c) [N11]. In Rosser's theory the validity of requirement (b) follows from axioms 8–10 considered in [30], pp. 184, 185. In connection with the fact that AS12.18(II) also directly implies requirement (b), in MC^{ν} the analogues of Rosser's axioms 8–10 need not be asserted as axioms.

44. Let the schemes S12.16 and S12.17 result from AS12.16 and AS12.17 respectively by substituting $\equiv^{\cap}$ for (the explicit occurrence of) $\equiv$, and $=^{\cap}$ for $=$. Then S12.16 and S12.17 can be inferred from ASs12.1–18, as will be shown in Memoir 2—cf. (46), p. 164.

45. As an example suppose that in the possible cases γ and γ' the red things are the same and that E and E' are the extensions of "red" in γ and γ' respectively. Then it is natural to take E and E' to be the same extension.

46. Indeed let $\underline{\tilde{R}}$ be an $\underline{L}$-determined QI of type $(\underline{t}_1, \ldots, \underline{t}_n)$. Then by Def. 13.1 the class $(\lambda \underline{\tilde{x}}_1, \ldots, \underline{x}_n)\underline{\tilde{R}}(\underline{\tilde{x}}_1, \ldots, \underline{\tilde{x}}_n, \gamma)$ is independent of γ, characterizes $\underline{\tilde{R}}$, and has the type $\underline{t}^{\underline{e}} = (\underline{t}_1^{\eta}, \ldots, \underline{t}_n^{\eta})$.

47. $(25)_1$ follows easily from Defs. 14.1 and 11.3.

48. The syntactical analogue of $(31)_2$ for Rosser's theory of extensional logic can easily be inferred from the instance of Theor. VIII.2.7, [32], p. 189.

49. By "the matrix $\underline{p}$ is extensional with respect to $\underline{x}$ in γ at $\mathcal{M}$ and $\mathcal{V}$" we mean that $\underline{p}^{(\underline{ex})} \equiv \underline{p}$ [Def. 11.1] holds in γ at $\mathcal{M}$ and $\mathcal{V}$—cf. N11.

50. The thesis of extensionality asserts that every nonextensional language can be translated into a suitable extensional one.

51. The possibility of such a strong translation was not obvious—cf. fn. 21.

52. In [29] Painlevé supposes, for the sake of simplicity, that the universe is formed by $\underline{n}+1$ material points $\underline{M}$, $\underline{M}_1, \ldots, \underline{M}_n$, "dont chacun reste identique à soi-même, quels que soient sa position dans l'éspace et l'instant considéré" (p. 46). On p. 65 the author says further,

> J'appelle accéleration de $\underline{M}$ causee par $\underline{M}_1$ l'accéleration qu'aurait $\underline{M}$ au même instant si ($\underline{M}$ et $\underline{M}_1$ occupant le mêmes positions avec les mêmes vitesses) tous les autres étaient écartés à l'infini.
>
> Cette terminologie admise les axiomes de la Méchanique . . . se resument ainsi.

This definition of caused acceleration, which is essential in Painlevé's axiom system, is essentially based on a counterfactual conditional. It can be taken to be not a material implication [N23] but for instance a causal implication—see [10]—so that possibility is explicitly involved.

53. Referring to the basic axioms for modern classical mechanics Painlevé says in [29], p. 64, "Je voudrais énoncer rapidement ce corps d'axiomes." This and the phrase "cette terminologie admise" quoted in fn. 52 lead us to believe that Painlevé would admit that the foundations of classical particle mechanics considered in [29], p. 65, may not be completely rigorous.

In [4], p. 106, it is pointed out that to prove a certain theorem considered in [29], p. 66, an additional axiom is needed—e.g. "Ammissione 10.2" considered in [4], p. 106.

54. The language ML^{ν} will also be used to deal with subjects that have little to do with mechanics in Ns24, 25.

55. In 1938 Hermes defined mass and force referring to continuum mechanics in special relativity—see [15]; and in 1957 he defined mass in classical particle mechanics—see [16]—using a completely rigorous procedure that was simpler than the analogue for continuum mechanics—cf. fn. 65.

56. In [4], among other things, inertial frame, mass, force, and also the spatial and temporal metrics, are defined. Hence Painlevé's ideas are realized in an evolved form, I think.

57. We prove explicitly [N19] the existence and uniqueness of mass—cf. fn. 5.

58. The same holds for the ordinary conditional "if . . . , then . . ." which—as Burks remarked in [10]—is used in some contexts for material implication and in others for causal implication.

59. Besides (42) to (44) one can see that

$$\Vdash \{\underline{x}_1, \ldots, \underline{x}_n\} = (\{\underline{x}_1, \ldots, \underline{x}_n\}^{\frown})^{(\underline{e})} = (\{\underline{x}_1, \ldots, \underline{x}_n\}^{(\underline{i})})^{(\underline{e})},$$

whereas $\Vdash \underline{F}^{(\underline{e})} = \underline{F}^{\frown(\underline{e})}$ [Def. 6.6 and fn. 28] is not true. For proving our last assertion, we may assume that the matrices $(12)_2$, (45), and $(x)\{\underline{F}(\underline{x}) \equiv^{\frown} [\underline{p}_1\underline{x} \in \{\underline{M}_1, \underline{M}_2\}^{(\underline{i})} \vee \sim\underline{p}_1\underline{x} \in \{\underline{M}'_1, \underline{M}'_2\}^{(\underline{i})}]\}$ hold in γ (at $\mathscr{M}$ and $\mathscr{V}$). Then by Def. 6.6, 18.6, and 18.13, $\underline{F}^{\frown} = \Lambda$ holds in γ. As a consequence, in γ $\underline{F}^{\frown(\underline{e})} = \Lambda$ also holds, which implies $\underline{F}^{\frown(\underline{e})} \neq \underline{F}^{(\underline{e})}$.

60. Our assertion (a) is based on a counterfactual conditional as well as on Painlevé's definition of "acceleration causée" spoken of in fn. 52.

61. Otherwise the theory does not possess any general character and is incompatible with the (physically possible) hypotheses laid down in most physical problems [N21].

62. Real can be easily defined in ML^{ν} on the basis of an obvious natural concept Nn of natural number. We shall define Nn in N27 on purely logical grounds.

63. More precisely, assume the matrices whose L-truths constitute (47) and (52) to be axioms.

64. Note that by Theor. 9.4 and AS12.9 conditions (52′) and (53′) are consequences of the uniqueness axiom (52) and the uniqueness condition (53) respectively. So, if in Theor. 20.2 we replace (52′) and (53′) with (52) and (53), then the resulting assertion also holds.

65. This procedure for defining mass—based directly on extensional logic and on the form (59) of $\underline{\text{Exp}}(\underline{M}, \rho)$—has been effectively used by Hermes in [16] (1957). More precisely, $(\exists\rho)\,\underline{\text{Exp}}\,(M, \rho)$ is substantially axiom 6 in [16].

The single-hit experiment on which this procedure for defining mass is based is simpler than the one consisting of an infinite sequence of measurements of the accelerations of $\underline{M}$ and $\underline{M_1}$ at suitable instants $\underline{t_1}, \underline{t_2}, \ldots$ where at the instant $\underline{t_n}$ the distance of $\underline{M}$ from $\underline{M_1}$ is less than one meter while the distance of $\underline{M}$ from the remaining mass points is larger than $\underline{n}$ meters ($\underline{n} = 1, 2, \ldots$).

Hermes' work [15] (1938) deals with foundations of continuum mechanics in special relativity. An analogue of the above infinite sequence of measurements is used there for defining mass.

66. Hence for such theories, in particular for [16] (1957), an analogue of the following remark made by Rosser in [31] and referring to [15] holds: "The axiom set . . . is sufficient for special relativity but not necessary. That is, a system which satisfies Hermes' axioms would be a system of special relativity, but not every system of special relativity would satisfy Hermes' axioms."

67. In [31] Rosser adds to the remark quoted in fn. 66 that

> the difficulty seems to lie mainly in axiom A8.1, which says that the corpuscles of matter behave in certain very particular fashions. It is quite possible that, in writing A8.1, Hermes really had in mind some sort of conditional statement to the effect that if the corpuscles behave in certain very peculiar fashions, then other things would happen. However, as stated, A8.1 is distinctly not conditional.

The equivalence of the forms (58) and (59) of $\underline{\text{Exp}}(\underline{M}, \rho)$ and the conditional character of the latter show that the difficulty lies, of course, in AS8.1, which is an analogue of $(\exists\underline{t})\,\underline{A}\,(\underline{M}, \underline{t})$, but a direct use of an extensional logic compels us to use axioms implying assertions like $(\exists\underline{t})\,\underline{A}\,(\underline{M}, \underline{t})$ in that in order to define mass, they must imply $(\exists_1\rho)\,\underline{\text{Exp}}\,(\underline{M}, \rho)$.

In [17] Hermes sketched a solution of the problem under consideration using a simple language with modal operators—cf. fn.

5. Presently we shall speak rather diffusely about other solutions of the problem presented in [4] and in the present work.

68. The axioms considered in [4] allow considerable economy of primitive concepts in accordance with Painlevé's ideas—see fn. 56. On the one hand these axioms are compatible with the hypotheses made in any problem of classical particle mechanics. On the other hand they are sufficient for proving any theorem considered in usual textbooks of mechanics—see [33]—and based only on the fundamental principles of mechanics.

In particular those axioms imply the following theorem (explicitly asserted in [29]): *If a space is inertial, then every space translating with respect to it with a constant vectorial velocity is inertial, and conversely*.

In [4], it is noted on p. 106 that the converse referred to here cannot be deduced from the axioms in [29], p. 65 (in accordance with Painlevé's intention of enunciating those axioms "rapidement," see p. 64) because of the lack of a possibility axiom—see Ammissione 10.2 in [4], p. 106. Furthermore, possibility axioms—constituting analogues of existence axioms in geometry—are systematically lacking both in Painlevé's book [29] and in Signorini's textbook [33]. In connection with this, the definition of *caused acceleration*—which plays an essential role in Painlevé's axioms and is based on a counterfactual conditional—is not preceded by any axiom assuring the existence of these accelerations.

One may conclude that the main improvements proposed in [4]—and which motivated the author to write the work—on the one hand have a metascientific (or linguistic) character in that they concern possibility, hence the formal logic on which foundations of mechanics can usefully be based; but on the other hand involve improvements—with respect to [29] and [33]—in the rigor of deductions, i.e. improvements that may be interesting at the purely scientific level.

69. In particular, Rosser's criticism, quoted in fn. 66, does not hold for [4].

70. Intuitive rules are given for translating the ordinary modal language used in [29] and [33] into the unusual extensional language used in [4]—see [4], p. 133.

71. Such multiplications of entities are considered to constitute a disadvantage—cf. [12], § 32, p. 137.

72. No grammatical marks distinguish the extensional use of a common noun from its absolute use. The same happens in con-

nection with material implication and causal implication—see [10] —as we said in N1.

73. In saying that "if $\underline{b}$ is a body, then it is the same . . . in all possible cases" in the most natural sense, we are speaking intuitively and we are not referring to our semantical system $(\underline{D}_1, \ldots, \underline{D}_\nu, \Gamma, \alpha^\nu)$, which could give the erroneous impression that $\underline{b}$ must have an $\underline{L}$-determined QI [Def. 13.1] and e.g. that consequently in case $\underline{b}$ is meant as a class of material points (i.e. as a property) $\underline{b}$ must be modally constant.

In connection with this case it is useful to remark that if $\underline{n} \in \underline{Nn}$ where $\underline{Nn}$ is the privileged absolute concept of natural number (defined in N27), then $\underline{n}$ is e.g. 5 in every possible Γ-case and $\underline{n}$ is not modally constant. Indeed $5 \in \underline{Nn}\ \Diamond \underline{RF}_{20} \in 5\ \Diamond \underline{RF}_{20} \notin 5$ holds where $\underline{RF}_{20}$ is the set of the rockets flying at the end instant of the twentieth century.

In case $\underline{b}$ is considered as an individual, it is rather natural for a logician to use (in connection with ML^ν) a semantical system $\underline{D}_1, \ldots, \underline{D}_\nu, \Gamma, \alpha^\nu)$ such that QI of $\underline{b}$ is $\underline{L}$-determinate. However by no means do we consider this mandatory. It is quite possible for another logician to use for the same purpose a semantical system $(\breve{\underline{D}}_1, \ldots, \breve{\underline{D}}_\nu, \Gamma, \alpha^\nu)$ such that $\breve{D}^{(e)}_{\underline{i}} = \underline{D}^{(e)}_{\underline{i}}$ and $\breve{\underline{D}}_{\underline{i}} \cap \underline{D}_{\underline{i}} = \Lambda$ hold in a suitable common modal extension of whatever his extensional metalanguage is. They will assign to an identical intension different QIs, which is quite possible. To show that the second system also may be rather natural, let us remark that for some (privileged) absolute concepts, their privileged character may be evident, while for others it may be a matter of discussion. Furthermore there exist some privileged absolute concepts $\underline{F}$ and $\underline{G}$ such that $\underline{G}^{(e)} \subset \underline{F}^{(e)}$ but $\underline{G} \cap \underline{F} = \Lambda$.

For instance let $\underline{F}$ be the privileged absolute concept of material body. Furthermore let $\underline{H}_1, \ldots, \underline{H}_n$ be $\underline{n}$ horses living today. For certain particular purposes the possibility that any of them may die at the end instant τ of tomorrow is irrelevant. Hence $\mathscr{H} =_D \{\underline{H}_1, \ldots, \underline{H}_n\}^{(i)}$—cf. Def. 18.13—yields $\mathscr{H} \in \underline{Abs}$. Our example is accomplished by setting $\underline{G} =_D \{\underline{G}_1, \ldots, \underline{G}_n\}^{(i)}$ where $\underline{G}_i$ is the body of $\underline{H}_i$ at the instant τ; i.e. $\underline{G}_i = (\imath \underline{x})$ ($\underline{x}$ is the [material] body of $\underline{H}_i$ at the instant τ).

Indeed the material body $\underline{G}_i'(\underline{G}_i' \in \underline{F})$ that happens in a particular case to be identical with $\underline{G}_i$ depends among other things on what horse $\underline{H}_i$ eats, in that case, prior to instant τ; so that in general the material body that $\underline{G}_i$ is identical to varies from case to case. Hence $\underline{G}_i \in \underline{F}$ is false ($\underline{i} = 1, \ldots, \underline{n}$). Then $\underline{G} \cap \underline{F} = \Lambda$.

Of course $\underline{G_i}$ is in every case identical with one element G_i' of $\underline{F}$. Hence $\underline{G}_i \in \underline{F}^{(e)}$. Then ($\underline{G} \subset \underline{F}^{(e)}$, hence) $\underline{G}^{(e)} \subset \underline{F}^{(e)}$.

Incidentally, the extension of horse $\underline{H_i}$ may be identified with the function $\underline{h}$ such that (1) its domain is the lifetime $\underline{T}$ of $\underline{H_i}$ and (2) for every $\tau \in \underline{T}$ $\underline{h}(\tau) = \underline{G_i'}$. (However this is not compulsory.) Furthermore, if we think of $\underline{h}$ as of a function, $\underline{h}(\gamma)$, of the Γ-case γ to which extension $\underline{h}$ is related, then we obtain the QI $\underline{\tilde{h}}$ of $\underline{H_i}$.

74. In analogy with Austin's story about the donkeys [1], p. 133, assume that the owner of $\underline{S}$ wants the horse $\underline{m_1}$ to win race $\underline{R}'$, hence he does not want the horse $\underline{m_{10}}$, which does not belong to him, to win $\underline{R}'$. Let the owner of $\underline{S}$ believe that the horse $\underline{h}$ which he sees at a certain instant is $\underline{m_{10}}$, while in fact $\underline{h}$ is $\underline{m_1}$. Then he does not want $\underline{h}$ to win $\underline{R}'$. Since it is not the case that he might not want $\underline{m_1}$ to win $\underline{R}'$ we conclude, first, that the relation $\underline{W^W}$ in $\underline{h}$ and $\underline{R}'$ that the owner of $\underline{S}$ wants $\underline{h}$ to win $\underline{R}'$ is nonextensional in $\underline{h}$; and second, for the extensionalization $\underline{W^w} = (\underline{W^W})^{(e)}$ of this relation, the condition $\underline{W^w}(\underline{h}, \underline{R}')$ holds in spite of our owner's not wanting $\underline{h}$ to win $\underline{R}'$.

For every attribute $\underline{F}$ we have $\overline{\underline{F}^{(e)}} \subseteq \bar{\underline{F}}^{(e)}$ [Defs. 18.1, 18.4, 18.9]. For $\underline{F} = \underline{W^W}$ we have $\overline{\underline{F}^{(e)}} \subset \bar{\underline{F}}^{(e)}$

The nonextensional character of the matrix $\underline{W^W}(\underline{h}, \underline{R}')$ with respect to $\underline{h}$ [Def. 11.1] and its rather unsatisfactory consequence $\underline{W^w}(\underline{h}, \underline{R}')$ mentioned above depend on a mistake by the owner of $\underline{S}$. So the exclusion of such mistakes appears natural.

75. This independence also holds for any modal calculus based on the language ML^ν, containing ASs12.1–23, and having inference rules valid in connection with both the old QIs and the new quasi intensions introduced in the proof of Theor. 25.1.

76. In some problems of mechanics an isolated system of two or three mass points is considered. This system schematizes a set of two or three bodies very far from every other body. So it is closer to reality to say that the remaining mass points—i.e. those that schematize the remaining bodies—also exist in the problem being considered but that their positions are at infinity, than to say that the set $\underline{D_1}$ of mass points has only two or three elements (in the possible case being considered). So in particle mechanics the assumption that $\underline{D_1}$ has the same cardinality in all possible cases is satisfactory from the point of view of classical physics.

77. See [32], p. 246, and observe, in particular, that Rosser's matrix $\underline{H}(\underline{A}, \beta, \underline{P})$ where $\underline{P}$ is a matrix and $\underline{A}$ and β are classes is the analogue of our matrix $\underline{A} \subset \beta \wedge \underline{\mathrm{Her}}[\beta, (\lambda \underline{x}_1, \ldots, \underline{x}_n)\underline{p}]$.

78. These semantical analogues for ML^{ν} are immediate consequences of the syntactical analogues for MC^{ν} of the syntactical theorems IX.5.12–16 in [32], p. 246. The proofs in MC^{ν} of these syntactical analogues are substantially the proofs of theorems IX.5.12–16.

79. We conjecture that the hypothesis is superfluous.

SYNTACTICAL SECTION

Memoir 2: A Modal ν-sorted Logical Calculus MC^ν Valid in the General Modal Language ML^ν

N28. Introduction*.

In part 1 of the first memoir, we introduced a general ν-sorted modal language ML^ν, where ν is an arbitrary positive integer, and in which (1) predicates and functors of every finite level are admitted, and (2) the iota operator $\imath$ (for descriptions) and Church's lambda operator (for classes) are admitted within the scope of the modal sign N. We endowed ML^ν with a semantical theory using quasi intensions in accordance with Carnap's method of extension and intension—cf. [12] and [13]. In this theory modalities are combined with variables in such a way that the customary inferences of the logic of quantification, in particular specification and existential generalization, remain valid—cf. fn. 9 in Memoir 1. On the basis of this semantical theory we translated ML^ν into a $(\nu+1)$-sorted extensional language $EL^{\nu+1}$.

In N12 we also proposed a set of axioms for a modal logical calculus MC^ν based on ML^ν, and we showed the L-truth of these axioms in ML^ν [Def. 9.5(b)].

---

* This memoir was begun in 1960. The author is indebted to R. Carnap and N. Belnap for the reasons given in Memoir 1—cf. source note to N1, p. 3.

The contents of all the memoirs in this monograph are listed at the beginning of the book. Each of the two major sections of the monograph has its own numbering for formulas. Sections and chapters are numbered consecutively throughout the monograph.

We studied the general language ML^{ν} with a view to applying modal logic to classical physics, particularly to certain axiomatization problems of mechanics for whose solution modal logic seems indispensable—see N21. To that end certain modal concepts were introduced, in particular the concept of absolute attributes—see Def. 18.8 and N20. On the basis of these concepts we pointed out a certain double use of common nouns. In connection with this and the axiomatization problems, we also considered (from the semantical point of view) two concepts of natural numbers, an absolute one, $\underline{\mathrm{Nn}}$, and its extensionalization, $\underline{\mathrm{Nn}}^{(e)}$ [Def. 18.9], and we showed how the well-known theory of natural numbers in extensional logic can immediately be transferred into the modal calculus MC^{ν} both for $\underline{\mathrm{Nn}}$ and strict identity, $=^{\cap}$, on the one hand [Def. 6.3], and for $\underline{\mathrm{Nn}}^{(e)}$ and (simple) identity, $=$, on the other hand, so that two parallel theories arise. We also proved some semantical theorems connecting these two parallel theories.

This memoir is a continuation of Memoir 1 in that the aforementioned subjects considered there from the semantical point of view are here dealt with syntactically. In particular we develop the modal calculus MC^{ν}, using nearly all of the axioms laid down in N12. More specifically, in order to reduce extensive parts of MC^{ν} to extensional logic we consider the extensional calculus $EC^{\nu+1}$ based on the extensional $(\nu+1)$-sorted language $EL^{\nu+1}$, which is the largest extensional language belonging to $ML^{\nu+1}$—see N3[1]—and we demonstrate [N31] a simple invariance property of the entailment relation under the extensional translation $MC^{\nu} \to (MC^{\nu})^{\eta}$ of MC^{ν} into $EL^{\nu+1}$ [N15].[2]

By this theorem we can at once assert a number of (formal) theorems in MC^{ν} [N31]. In addition we can quickly transfer to

MC^{ν} useful metatheorems holding in extensional logic—considered in [32]—as the generalization, duality, equivalence, and replacement theorems. In this way we also transfer to MC^{ν} some procedures based on those of Gentzen for shortening proofs in extensional logic; i.e. in N33 we transfer to MC^{ν} Rosser's theorem, VI.7.2 in [32], which enables us to use, in deductions, steps that either constitute generalizations (and are based on the so-called rule $\underline{G}$) or constitute formal analogues of an act of choice (and are based on the so-called rule $\underline{C}$).[3]

Our way of transferring formal theorems and metatheorems from the extensional calculus presented in [32] to the modal lower predicate calculus $LPC(ML^{\nu})$ for ML^{ν} is based on a translation of the $LPC(MC^{\nu})$ into (and not onto) the extensional language $EL^{\nu+1}$. Therefore the same can easily be done with the lower predicate calculus based on modal languages that are not so rich as ML^{ν} (and such are the modal languages ordinarily considered).

In our translation of ML^{ν} into $EL^{\nu+1}$ [N15] the sign $\underline{N}$ corresponds to the universal quantifier $(\forall \underline{x})$ in $EL^{\nu+1}$, where κ is the first variable of type $\nu+1$. Every quantifier $(\forall \underline{x})$ in ML^{ν} corresponds to a universal quantifier in $EL^{\nu+1}$ distinct from κ. Therefore the axioms and theorems in $EC^{\nu+1}$ concerning quantifiers have, in MC^{ν}, both an ordinary analogue (concerning ordinary quantifiers) and one or more modal analogues (concerning $\underline{N}$) [N31].

The metatheorems for $EC^{\nu+1}$ which we hinted at above and concern quantifiers have an ordinary analogue and a modal analogue for MC^{ν}.[4] However this does not hold for every metatheorem concerning quantifiers in extensional logic.[5]

We will consider axioms for (simple) identity [N34], for the iota operator [N38], and for attributes [N40]; and we will state in MC^{ν} a (single suitable) version [AS45.1] of the axiom of infinity in terms of natural numbers that mirrors, so to speak, a property of the external world.

The usual theorems on = in extensional calculi can be transferred into MC^{ν} in two ways: either by referring to strict identity, $=^{\cap}$, instead of = [N34], or referring to = and to extensional matrices [N35]. The usual theorems on descriptions in extensional logic can easily be transferred into MC^{ν} in connection with extensional matrices [N38].

We will also prove many more or less basic theorems in MC^{ν} involving matrices of any kind, and concerning identity [N36], descriptions [Ns38, 39],[6] and the existential and uniqueness operators $(\exists \underline{x})$, $(\exists_1 \underline{x})$, and $(\exists_1^{\cap} \underline{x})$ [Ns37, 39].[7] Most of them have no analogues in extensional logic.

It is not possible to define identity in MC^{ν} the way it is done in extensional logic, particularly in Rosser's book [32]—cf. fnn. 41 and 42 in Memoir 1. However, our axioms for identity and attributes enable us to set up in MC^{ν} one or more counterparts for every formal theorem considered in [32]. We set forth several of these explicitly. In particular we show that the general theorems on equivalence and substitution [32], chap. 9, sect. 2, have total analogues—cf. fn. 4—in MC^{ν} [N42] as well as, so to speak, certain extensional counterparts in MC^{ν} [N43].

Regarding the existence axioms for attributes and functions [ASs40.3, 4], let us remark that we state them in a weak—and extensional—form. Then we prove that they hold in a stronger form through an essential use of our axiom for descriptions [AS38.1].

The difference between these two forms is substantially the difference between the assertions that for each attribute F there is an attribute G equal to F, and that there is an attribute H strictly equal to F.

Let us add that we prove (in MC^{ν}) some properties of the concepts of modally constant attributes, modally separated attributes, and absolute attributes; some of them involve uniqueness operators and the iota operator [N41].

We briefly consider the foundations of the theories in MC^{ν} for the concepts Nn of absolute natural numbers and its extensionalization $Nn^{(e)}$. A few theorems (in MC^{ν}) on the concept of closure and on Nn and $Nn^{(e)}$—some of which have no analogues in extensional logic—show how to immediately reduce such theories to the well-known theory of natural numbers in extensional logic, in accordance with some results in semantics mentioned above [N27]. The possibility of the analogous reduction for real numbers seems obvious. We deal with the foundations of the theories for Nn and $Nn^{(e)}$ and with their mutual connection [Ns44, 45] in parallel with our corresponding brief semantical treatment—see Ns26, 27. We try to profit as much as possible by this treatment and we prove all theorems (in MC^{ν}) on closure and natural numbers [Ns44, 45] whose semantical analogues were not proved in Memoir 1.

Extensional matrices play an important role in several results of this work, as the preceding considerations make apparent. Therefore we shall explicitly enunciate and prove several theorems in MC^{ν} [Ns35, 43] on these matrices and related concepts [Defs. 6.11, 6.12, 18.9, 35.1, 35.2]; some of them are of a general character [Theor. 43.1].

6

Reduction of Part of the Modal Calculus MC^ν to Extensional Logic

N29. Deduction in the modal ν-sorted calculus MC^ν.

We aim at setting up a general ν-sorted modal (logical) calculus MC^ν based on ML^ν [N3]. To this end it would be practical to transfer into modal logic, in a straightforward way, extensive parts of the theory of some extensional (logical) calculus. Obviously we cannot set up the whole MC^ν in this way, even taking into account the known analogy between universal quantifiers and the modal operator, i.e. the sign N, which we shall also call the modal quantifier.

We shall make use of Rosser's theory of extensional logic [32], in which instead of a type system—such as the one for our modal language ML^ν or the extensional language EL^ν—Rosser's stratification theory is used. This difference appears only when set theory is dealt with, and it is very easy to turn Rosser's calculus into an extensional calculus—say EC^ν—based on our language EL^ν, hence also on a type system ($\nu = 1, 2, \ldots$).

As we stated in N12, on the one hand the axioms of MC^ν can be identified with AS12.r (r = 1, . . . 23) and AS25.1 (in ML^ν) and with an additional axiom for natural numbers—cf. $(78)_1$ in Memoir 1—which can be put into the form AS45.1; on the other hand the axioms of EC^ν are AS12.r in EL^ν for r = 1–4, 6, 8, 10–12, 14–18, 20–22, AS12.13′, and lastly an axiom for natural numbers—cf. $(78)_2$.

Now observe that some of our formal definitions—e.g. Defs. 4.4, 4.5—are metalinguistic (or abbreviating) definitions, i.e. they introduce metalinguistic signs such as Church's lambda, while others—e.g. Def. 18.8—are formal, i.e. they introduce new constants. Such definitions function in deductions as axioms or premises. Hence for them some considerations concerning quantifiers, similar to those made for the axioms in N12, are in order, after stating the following technical definition:

DEF. 29.1. By a system of (formal) definitions in ML^{ν} we mean a well-ordered set $\{\underline{p}_{\alpha}\}$ of sentences such that—cf. [11], p. 85—there is a similar set $\{\underline{c}_{\alpha}\}$ of distinct constants for which either $\underline{p}_{\alpha}$ has the form $\underline{c}_{\alpha} =^{\cap} \Delta$ [Def. 6.3] where (the definiens) Δ is a closed term without occurrences of $\underline{c}_{\beta}$ for $\beta \geq \alpha$, or $\underline{p}_{\alpha}$ has the form $(\forall \underline{x}_1, \ldots, \underline{x}_n) [\underline{c}_{\alpha}(\underline{x}_1, \ldots, \underline{x}_n) \equiv^{\cap} \underline{q}]$ [Def. 6.5], where (1) $\underline{x}_1, \ldots, \underline{x}_n$ are distinct variables, (2) they include all free variables in the definiens $\underline{q}$, and (3) for $\beta \geq \alpha$, $\underline{c}_{\beta}$ does not occur in $\underline{q}$.

In this memoir we do not use definitions of the form $(\forall \underline{x}_1, \ldots, \underline{x}_n) [(\underline{x}_1, \ldots, \underline{x}_n) =^{\cap} \Delta]$ where Δ fulfills the analogues of the above conditions (1) to (3) in Def. 29.1. The concept of a system of formal definitions in EL^{ν} can be defined like the one for ML^{ν} except that $=$ and $\equiv$ are to be used instead of $=^{\cap}$ and $\equiv^{\cap}$ respectively.[8]

No string of quantifiers was explicitly used in front of the definitions laid down in Memoir 1. Likewise, in this memoir it is useful to understand definitions of the forms $\Delta =_D \Delta'$ and $\Delta \equiv_D \Delta'$ to stand for definitions having, respectively, the first and second of the alternative forms considered for $\underline{p}_{\alpha}$ in Def. 29.1.

Let $\underline{L}$ be ML^{ν} or EL^{ν} or the part $\overline{EL}^{\nu+1}$ of $EL^{\nu+1}$, formed with the extensional translations of the designators in ML^{ν}. We may consider a theory $\mathscr{T}$ expressed in $\underline{L}$ as being determined by $\underline{L}$, the set $\mathscr{A}$ of (matrices assumed as) axioms, and the system $\mathscr{D}$ of the accepted (formal) definitions: $\mathscr{T} = \langle \underline{L}, \mathscr{A}, \mathscr{D} \rangle$.

$\mathscr{A}$ may not include all axiom schemes ASs12.1–23 and it may contain other (possibly nonlogical) axioms—e.g. (47) and (52) in N19.

DEF. 29.2(a). Let $\underline{p}_1, \ldots, \underline{p}_{\underline{m}}$ and $\underline{q}$ be matrices in $\underline{L}$. We say that $\underline{p}_1, \ldots, \underline{p}_{\underline{m}}$ entail (or yield) $\underline{q}$ in $\mathscr{T}$ ($\mathscr{T} = \langle \underline{L}, \mathscr{A}, \mathscr{D} \rangle$), and we write $\underline{p}_1, \ldots, \underline{p}_{\underline{m}} \vdash \underline{q}$, if in $\underline{L}$ there are $\underline{n} - \underline{m}$ (> 0) matrices $\underline{p}_{\underline{m}+1}, \ldots, \underline{p}_{\underline{n}}$, where $\underline{p}_{\underline{n}}$ is $\underline{q}$ and where for $\underline{i} = \underline{m} + 1, \ldots, \underline{n}$ one of the following alternatives (inference rules) holds:

(1) $\underline{p}_{\underline{i}}$ is an axiom or a definition in $\mathscr{T}$ ($\underline{p}_{\underline{i}} \in \mathscr{A} \cup \mathscr{D}$).

(2) $\underline{p}_{\underline{i}}$ is the same as an earlier $\underline{p}_{\underline{j}}$ ($1 \leq \underline{j} < \underline{i}$).[9]

(3) $\underline{p}_{\underline{i}}$ is derived from two earlier $\underline{p}$'s by modus ponens, i.e. there are $\underline{j}$ and $\underline{k}$ ($1 \leq \underline{j} < \underline{i}$, $1 \leq \underline{k} < \underline{i}$) such that $\underline{p}_{\underline{k}}$ is $\underline{p}_{\underline{j}} \supset \underline{p}_{\underline{i}}$.

DEF. 29.2(b). For $\underline{m} = 0$ we say that $\underline{q}$ is provable and we write $\vdash \underline{q}$—cf. [32], p. 57.

DEF. 29.2(c). Let $\underline{p}_{\underline{m}+1}, \ldots, \underline{p}_{\underline{n}}$ be as in (a). Then we say that they constitute a deduction of $\underline{q}$ from (the premises) $\underline{p}_1, \ldots, \underline{p}_{\underline{m}}$, for $\underline{m} > 0$, and a proof of $\underline{q}$, for $\underline{m} = 0$.

The above definition of $\vdash$ is sound for ML^{ν} in the following sense:

THEOR. 29.1. Let $\underline{p}_1, \ldots, \underline{p}_{\underline{m}}$ yield $\underline{q}$ in $\mathscr{T}$ ($\mathscr{T} = \langle ML^{\nu}, \mathscr{A}, \mathscr{D} \rangle$). Then [by Def. 29.1 and Theor. 9.2] $\underline{p}_1, \ldots, \underline{p}_{\underline{m}}$ $\mathscr{A}$-imply $\underline{q}$ [Def. 9.4(b)], so that (1) if $\underline{p}_1, \ldots, \underline{p}_{\underline{m}}$ are $\mathscr{A}$-true [Def. 9.5(a)], then the same holds for $\underline{q}$ [Theor. 9.3] and (2) if $\underline{p}_1, \ldots, \underline{p}_{\underline{m}}$ hold in the Γ-case γ [N6] at the value assignment $\mathscr{V}$ and at the admissible model $\mathscr{M}$ [Defs. 9.3, 9.6(a)], then the same holds for $\underline{q}$ [Theor. 9.5].

N30. On the extensional translations of the axioms of the modal calculus MC^{ν}.

Let $\overline{EL}^{\nu+1}$ be that part of the extensional language $EL^{\nu+1}$ formed with the extensional translations of the designators in ML^{ν} according to rules $(\underline{T}_1)$ to $(\underline{T}_9)$ [N15], briefly

(1) $$\overline{EL}^{\nu+1} = (ML^{\nu})^{\eta}.$$

Furthermore by the extensional translation $(MC^{\nu})^{\eta}$ of MC^{ν} we mean—cf. fn. 2—the extensional calculus based on the language $EL^{\nu+1}$ whose axioms are the extensional translations [N15] of the axioms of MC^{ν}. (The analogue will also be assumed to hold for $LPC(ML^{\nu})^{\eta}$ and $LPC(ML^{\nu})$.)

For the ease of the reader we now list the axioms of the lower predicate calculi $LPC(ML^{\nu})$ and $LPC(EL^{\nu+1})$ based on the languages ML^{ν} and $EL^{\nu+1}$ respectively. As a preliminary we state the following conventions:

(a) If a wff is given as $\underline{A}(\underline{x})$ and Δ is a term—in particular a variable—of the same type as the variable $\underline{x}$, then by $\underline{A}(\Delta)$ we mean the result of replacing all free occurrences of $\underline{x}$ in $\underline{A}(\underline{x})$ with (occurrences of) Δ.

(b) If after giving a wff as $\underline{A}(\underline{x})$ we use $\underline{A}(\Delta)$, then we understand that Δ is free for $\underline{x}$ in $\underline{A}(\underline{x})$, i.e. no free occurrence of $\underline{x}$ in $\underline{A}(\underline{x})$ belongs to the scope of any operator $(\forall\underline{z})$ or $(\iota\,\underline{z})$ where $\underline{z}$ is any variable occurring free in Δ.

Similar conventions hold in case a wff is given e.g. as $\underline{A}(\underline{x}, \underline{y}, \underline{z})$:

(c) The matrix $\Phi(\underline{x})$ $[\Phi(\underline{y})]$ results from $\Phi(\underline{y})$ $[\Phi(\underline{x})]$ by replacing the free occurrences of $\underline{y}$ $[\underline{x}]$ with occurrences of $\underline{x}$ $[\underline{y}]$.

Rosser's extensional lower predicate calculus is based on six axiom schemes—see (1) to (6) in [32], p. 101—which are the extensional analogues of the following six axiom schemes:[10]

AS30.1. $\underline{(N)}\,\underline{p} \supset \underline{pp}$,

AS30.2. $\underline{(N)}\,\underline{pq} \supset \underline{p}$,

AS30.3. $\underline{(N)}\,(\underline{p} \supset \underline{q}) \supset [\sim(\underline{pr}) \supset \sim(\underline{rq})]$,

AS30.4. $\underline{(N)}\,(\underline{x})\,(\underline{p} \supset \underline{q}) \supset [(\underline{x})\underline{p} \supset (\underline{x})\underline{q}]$,

AS30.5. $\underline{(N)}\,\underline{p}_1 \supset (\underline{x})\underline{p}_1$,

where $\underline{p}_1$ has no free occurrences of $\underline{x}$, and where $\underline{(N)}$ is any string of $\underline{N}$'s and universal quantifiers whose scope is the entire subsequent formula—cf. N12, $(24)_2$.

AS30.6. $\underline{(N)}\,(\underline{x})\,\Phi\,(\underline{x}, \underline{y}) \supset \Phi\,(\underline{y}, \underline{y})$

where $\underline{x}$ and $\underline{y}$ are not necessarily distinct variables and they have, of course, the same logical type [N2].

The modal analogues—discussed later—of the axiom schemes ASs30.4–6 are, in order,

AS30.7. $\underline{(N)}\,\underline{N}\,(\underline{p} \supset \underline{q}) \supset (\underline{Np} \supset \underline{Nq})$,

AS30.8. $\underline{p}_1 \supset \underline{Np}_1$,

where $\underline{p}_1$ is modally closed [Def. 4.3], and

AS30.9. $\underline{(N)}\,\underline{Np} \supset \underline{p}$.

The axioms of the LPC($EL^{\nu+1}$) are ASs30.1–6 (in $EL^{\nu+1}$); those of the LPC(ML^{ν}) are ASs30.1–9.

On the basis of the translation rules $(\underline{T}_1)$ to $(\underline{T}_9)$ [N15] we may assert the following theorem:

THEOR. 30.1. For $\underline{r} = 1, 2, 3$ the matrix $\underline{p}$ in ML^{ν} is an instance of AS30.$\underline{r}$ (in ML^{ν}) if and only if $\underline{p}^{\eta}$ is an instance of

AS30.$\underline{r}$ in $EL^{\nu+1}$; furthermore $\underline{p}$ is an instance of AS30.$\underline{r}+6$ [of AS30.$\underline{r}+3$] in ML^{ν} if and only if (1) $\underline{p}^{\eta}$ is an instance of AS30.$\underline{r}+3$ in $EL^{\nu+1}$ and (2) the variable $\underline{x}$ explicitly written (quantified) in this instance coincides [does not coincide] with κ.

The extensional translation $LPC(ML^{\nu})^{\eta}$ of the $LPC(ML^{\nu})$—whose axioms are by definition the extensional translations of the axioms of the $LPC(ML^{\nu})$—is the lower predicate calculus $LPC(\overline{EL}^{\nu+1})$ subordinated by the $LPC(EL^{\nu+1})$ on $\overline{EL}^{\nu+1}$. We cannot assert, more generally, that $(CM^{\nu})^{\eta}$ is the calculus $\overline{EC}^{\nu+1}$ subordinated by $EC^{\nu+1}$ on $\overline{EL}^{\nu+1}$. Indeed, if for $\underline{r} = 10$ to 12 $\underline{p}$ is an instance of AS12.$\underline{r}$ in ML^{ν}, then $\underline{p}^{\eta}$ is not any axiom of $EC^{\nu+1}$; e.g. for $\underline{r} = 10$ $\underline{p}^{\eta}$ is $\underline{x} \overset{t}{\underset{\kappa}{=}} \underline{x}$ [Def. 15.1]; however $\underline{p}^{\eta}$ is only a theorem in $EC^{\nu+1}$—cf. the forthcoming formula (2).

More generally the following theorem holds:

THEOR. 30.2. If $\underline{p}$ is an axiom of MC^{ν}, then $\underline{p}^{\eta}$ is a theorem in $EC^{\nu+1}$ (and in some cases an axiom).

The proof of Theor. 30.2 is rather long but natural and easy. We need only remark explicitly, first, that by Def. 15.1 of $\overset{t}{\underset{\kappa}{=}}$ in $EL^{\nu+1}$ and the analogy between the rules $(\underline{T}_1)$ to $(\underline{T}_9)$ of the extensional translation of ML^{ν} [N15] on the one hand, and our designation rules (δ_1) to (δ_9) for ML^{ν} [Ns8, 11] on the other hand, Theor. 10.2 suggests that for all designators Δ_1 and Δ_2 in $\overline{EL}^{\nu+1}$ of the same type $\underline{t} \epsilon \tau^{\nu+1}$ we have

$$\text{(2)} \qquad \vdash (\kappa)\Delta_1 \overset{t}{\underset{\kappa}{=}} \Delta_2 \equiv \Delta_1 = \Delta_2 \text{ in } EC^{\nu+1}.$$

Second, for proving Theor. 30.2 in case $\underline{p}$ is an instance of AS12.22 or AS12.23 (on $\underline{a}_{\underline{t}}^{*}$), by Def. 11.2 $\underline{a}_{\underline{t}}^{*}$ is $(\iota\underline{x})\Phi(\underline{x})$ where $\Phi(\underline{x})$ is $\underline{v}_{11} \neq \underline{v}_{11}$. Then by rules $(\underline{T}_2)$ and $(\underline{T}_5)$ in N15, $\Phi(\underline{x})^{\eta}$ is $\underline{x}^{\eta} \neq \underline{x}^{\eta}$, which by rule ($\underline{T}_9$) [N15] has the form $(\imath\, \underline{x}^\eta)\, \Psi(\underline{x}^\eta)$ where $\Psi(\underline{x}^\eta)$ yields $(\kappa)\underline{x}^\eta \overset{t}{\underset{\kappa}{=}} \underline{a}^*_{\underline{t}\eta}$ (in $EC^{\nu+1}$). Then we have $(\exists_1 \underline{x}^\eta)\, \Psi(\underline{x}^\eta)$ and $(\kappa)(\underline{a}^*_{\underline{t}})^\eta \overset{t}{\underset{\kappa}{=}} \underline{a}^*_{\underline{t}\eta}$. By (2) this yields

(3) $$\vdash (\underline{a}^*_{\underline{t}})^\eta = \underline{a}^*_{\underline{t}\eta} \quad \text{in } EC^{\nu+1}.$$

N31. Invariance properties of the entailment relation for the extensional translation. Some consequences within the lower predicate calculus.

On the basis of our definition of deduction [Def. 29.2], rules ($\underline{T}_1$) to ($\underline{T}_9$) of extensional translation, and Theor. 30.1, it is easy to prove that in case $\underline{p}_1, \ldots, \underline{p}_n$ are matrices in ML^ν, $\underline{p}_1, \ldots, \underline{p}_n$ is a proof of $\underline{p}_0$ in the LPC($\overline{ML}^\nu$) if and only if $\underline{p}_1^\eta, \ldots, \underline{p}_n^\eta$ is a proof of $\underline{p}_0^\eta$ in the LPC($\overline{EL}^{\nu+1}$).

So as far as the lower predicate calculus is concerned, the invariance properties mentioned in the title can be expressed by the following theorem:

THEOR. 31.1. For any matrices $\underline{p}_0, \ldots, \underline{p}_m$ in ML^ν, $\underline{p}_1, \ldots, \underline{p}_m \vdash \underline{p}_0$ in the LPC(ML^ν) or in MC^ν if and only if $\underline{p}_1^\eta, \ldots, \underline{p}_m^\eta \vdash \underline{p}_0^\eta$ in the (LPC(ML^ν)$^\eta$, hence in the) LPC($\overline{EL}^{\nu+1}$) or in $(\overline{MC}^\nu)^\eta$ respectively.

The part of Theor. 31.1 that concerns MC^ν—and $(MC^\nu)^\eta$—and Theor. 30.2 obviously yield the following theorem:

THEOR. 31.2. If $\underline{p}_1, \ldots, \underline{p}_m \vdash \underline{p}_0$ in MC^ν, then $\underline{p}_1^\eta, \ldots, \underline{p}_m^\eta \vdash \underline{p}_0$ in $EC^{\nu+1}$.

We want to state the main formal theorems of the LPC(ML^ν), mostly using Theor. 31.1. To begin with we remark that if $\mathscr{E}(\underline{p}_1, \ldots, \underline{p}_n)$ is a matrix built up from the arbitrary matrices $\underline{p}_1, \ldots, \underline{p}_n$ using only $\sim$ and $\wedge$—i.e. the formation rules ($\underline{f}_5$) and ($\underline{f}_6$)

in N3—and if $\mathscr{E}(\underline{p}_1, \ldots, \underline{p}_n)$ is a theorem in $EC^{\nu+1}$ for every choice of the matrices $\underline{p}_1, \ldots, \underline{p}_n$ in $EL^{\nu+1}$, then $\mathscr{E}(\underline{p}_1, \ldots, \underline{p}_n)$ is a theorem in MC^{ν} for every choice of the matrices $\underline{p}_1, \ldots, \underline{p}_n$ in ML^{ν}. Such theorems constitute the extensional part of the propositional calculus for MC^{ν}.

Obviously the truth-value theorem—cf. Theor. IV.5.2 in [32], p. 70—and the deduction theorem—i.e. if $\underline{p}_1, \ldots, \underline{p}_n \vdash \underline{p}_0$, then $\underline{p}_2, \ldots, \underline{p}_n \vdash \underline{p}_1 \supset \underline{p}_0$ [Theor. 33.2]—also hold for MC^{ν}.

In the forthcoming Theor. 31.3, which is a preliminary for Theor. 31.4, we assert that certain well-known theorems of the ordinary extensional lower predicate calculus, particularly of the LPC($EL^{\nu+1}$), hold in the LPC($\overline{EL}^{\nu+1}$). Then, on the basis of Theor. 31.1, we can assert the validity of the same theorems in the LPC(ML^{ν}).

Theor. 31.3 concerns the LPC($\overline{EL}^{\nu+1}$) which differs from for example the LPC($EL^{\nu+1}$) in that in the former there is only one variable of the same type as the variable κ. (There are other differences but they are inessential.) However the validity of Theor. 31.3 can quickly be seen from the following remark about any matrix $\underline{p}$ in ML^{ν} for which $\vdash \underline{p}^{\eta}$ in the LPC($EL^{\nu+1}$). The matrix $\underline{p}^{\eta}$ certainly has a proof $\underline{p}'_1, \ldots, \underline{p}'_n$, where for $\underline{i} = 1, \ldots, \underline{n}$, $\underline{p}'_i$ is formed with submatrices of $\underline{p}^{\eta}$, so that (1) $\underline{p}'_i$ has the form $\underline{q}_i^{\eta}$ where $\underline{q}_i$ is a matrix in ML^{ν}, and (2) $\underline{p}'_i$ does not include any variable that does not occur in $\underline{p}^{\eta}$. This remark leads us to assert that every theorem of the ordinary extensional lower predicate calculus is provable in the LPC($\overline{EL}^{\nu+1}$), hence [Theor. 31.1] in the LPC(MC^{ν}).

Those among the theorems in the LPC(MC^{ν}) that will be used in the remainder of the monograph are substantially those listed in Theor. 31.3. Let $\underline{p}$ be one of them. Then the existence of the above $\underline{p}'_1, \ldots, \underline{p}'_n$ appears immediately and directly—on the basis of our previous remark involving $p'_1, \ldots, p'_n$—from any common textbook of extensional logic such as [32].

THEOR. 31.3. Let the matrix r be closed with respect to x; furthermore let x and the matrices $\Phi(x)$ and $\Phi(y)$ in $\overline{EL}^{\nu+1}$ fulfill condition (c) in N30. Then—see conventions (a) and (b) in N30—the following hold in the LPC($\overline{EL}^{\nu+1}$):[11]

(4) $\vdash (x)p \supset p$, $\vdash (\exists x)p \equiv \sim(x)\sim p$, $\vdash (x)p \equiv \sim(\exists x)\sim p$,

(5) $\vdash (x)(pq) \equiv (x)p(x)q$, $\vdash (x)r \equiv r$, $\vdash (x)(pr) \equiv (x)pr$,

(6) $\vdash (x)(p \vee r) \equiv (x)p \vee r$, $\vdash (x)(p \supset r) \supset [(\exists x)p \supset r]$, $\vdash (x)(r \supset p) \supset [r \supset (x)p]$,

(7) $\vdash (\exists x)(y)p \supset (y)(\exists x)p$, $\vdash (x)\Phi(x) \equiv (y)\Phi(y)$, $\vdash \Phi(y, y) \supset (\exists x)\Phi(x, y)$,

(8) $\vdash p \supset (\exists x)p$, $\vdash (\exists x)(pq) \supset (\exists x)p(\exists x)q$, $\vdash (\forall x, y)p \equiv (\forall y, x)p$,

(9) $\vdash (x)(p \supset q) \supset [(\exists x)p \supset (\exists x)q]$, $\vdash (\exists x)(p \supset q) \supset [(x)p \supset (\exists x)q]$, $\vdash (\exists x)\Phi(x) \equiv (\exists y)\Phi(y)$.

Now let p, q, and r be matrices in ML^{ν} fulfilling condition (c) in N30, let r be closed with respect to x, and let p_i be the i-th of the eighteen matrices $(4)_1$ to $(9)_3$. Then if the variable x or y occurs in p_i, it is different from κ. In addition p_i^{η} [N15] is an instance of just the i-th of the above theorems $(4)_1$ to $(9)_3$ in the LPC($\overline{EL}^{\nu+1}$). Hence by Theor. 31.1 p_i is a theorem in the LPC(ML^{ν}), hence in MC^{ν}. We may call p_i (an instance of) the <u>ordinary analogue</u> for MC^{ν} of the i-th theorem in $EC^{\nu+1}$ mentioned above.

Now let us consider a matrix q_i in ML^{ν} such that q_i^{η} is an instance of the above i-th theorem $\mathscr{T}_i$ in $EC^{\nu+1}$, with x identical to κ —cf. rule $(\underline{T}_8)$ in N15. Then we may say that $\underline{q}_i$ is an instance of the modal analogue of $\mathscr{T}_i$. For instance, for $i = 1, 2, 3$ $\mathscr{T}_i$ is $(4)_i$ so that $\underline{q}_1$ is $\underline{Np} \supset \underline{p}$ (hence $\underline{q}_1 \in$ AS30.9), $\underline{q}_2$ is $\Diamond \underline{p} \equiv \sim\underline{N}\sim\underline{p}$, i.e. the definition of $\Diamond$, and $\underline{q}_3$ is $\underline{Np} \equiv \sim\Diamond\sim\underline{p}$.

Note that, on the one hand, if in $\mathscr{T}_i$ the variable $\underline{x}$ is κ and $\underline{y}$ is supposed to be interchangeable with $\underline{x}$, then $\underline{y}$ must also be κ. On this account $\mathscr{T}_i$ may become useless, e.g. when $\mathscr{T}_i$ is $(7)_2$; furthermore $\underline{q}_i$ may lose some interest without becoming wholly useless, e.g. when $\mathscr{T}_i$ is $(7)_3$. Indeed in this case $\underline{q}_i$ is an instance of both $(7)_3$ and $(8)_1$; in addition $\underline{q}_i$ is $\underline{p} \supset \Diamond\, \underline{p}$.

On the other hand, when $\mathscr{T}_i$ is $(7)_1$, it has a second modal analogue that is obtained assuming κ to be $\underline{y}$ and not $\underline{x}$—see $(7')_{1,2}$ below. We obtain a third modal analogue of $(7)_1$ by taking both $\underline{x}$ and $\underline{y}$ to be κ —see $(7')_3$. The second modal analogue of $(8)_3$ in the above sense is obviously equivalent to the first—see $(8')_3$ below.

Theors. 31.1, 31.3, and the preceding considerations imply the following theorem:

THEOR. 31.4(a). Let $\underline{p}$, $\underline{q}$, $\underline{r}$, $\Phi(\underline{x}, \underline{y})$, and $\Phi(\underline{x})$ be matrices in ML^{ν}; let $\underline{r}$ be closed with respect to $\underline{x}$; and let condition (c) in N30 hold. Then $(4)_1$ to $(9)_3$ hold in the LPC(MC^{ν}).

THEOR. 31.4(b). If in addition $\underline{r}$ is modally closed, then we also have in the LPC(MC^{ν}):[12]

(5′) $\vdash \underline{N}(\underline{pq}) \equiv \underline{NpNq}$, $\vdash \underline{Nr} \equiv \underline{r}$, $\vdash \underline{N}(\underline{pr}) \equiv \underline{Npr}$,

(6′) $\vdash \underline{N}(\underline{p} \vee \underline{r}) \equiv \underline{Np} \vee \underline{r}$, $\vdash \underline{N}(\underline{p} \supset \underline{r}) \supset (\Diamond\underline{p} \supset \underline{r})$, $\vdash \underline{N}(\underline{r} \supset \underline{p})$
$\supset (\underline{r} \supset \underline{Np})$,

(7′) $\vdash \Diamond\, (\underline{y})\underline{p} \supset (\underline{y})\, \Diamond\, \underline{p}$, $\vdash (\exists\underline{x})\,\underline{Np} \supset \underline{N}(\exists\underline{x})\underline{p}$, $\vdash \Diamond\underline{Np} \supset \underline{N}\Diamond\,\underline{p}$,

(8′) $\vdash \underline{p} \supset \Diamond\underline{p}$, $\vdash \Diamond\, (\underline{pq}) \supset \Diamond\,\underline{p}\Diamond\,\underline{q}$, $\vdash \underline{N}(\underline{y})\underline{p} \equiv (\underline{y})\,\underline{Np}$,

(9′) $\vdash \underline{N}(\underline{p} \supset \underline{q}) \supset (\Diamond \underline{p} \supset \Diamond \underline{q})$, $\vdash \Diamond (\underline{p} \supset \underline{q}) \supset (\underline{Np} \supset \Diamond \underline{q})$,
$\vdash \Diamond \underline{r} \equiv \underline{Nr}$.

N32. The generalization, duality, equivalence, and replacement theorems in MC^ν.

The results of N31 and in particular of Theor. 31.1 allow us quickly to transfer into the modal calculus MC^ν some well-known metatheorems of the extensional lower predicate calculus. First we enunciate the latter for the LPC($\overline{EL}^{\nu+1}$), using almost the same words as [32].

The following theorem is the generalization theorem for the LPC($\overline{EL}^{\nu+1}$). It is the analogue for $\overline{EL}^{\nu+1}$ of Theor. VI.4.2 in [32], p. 106, and can be proved with substantially the same words even when $\underline{x}$ is the variable κ which cannot be replaced by any other variable in $\overline{EL}^{\nu+1}$.

THEOR. 32.1. If $\underline{p}_1, \ldots, \underline{p}_n$ and $\underline{q}$ are matrices in $\overline{EL}^{\nu+1}$, $\underline{x}$ is a variable which has no free occurrences in any of $\underline{p}_1, \ldots,$ $\underline{p}_n$, and $\underline{p}_1, \ldots, \underline{p}_n$ entail $\underline{q}$ in the LPC($\overline{EL}^{\nu+1}$), then $\underline{p}_1, \ldots,$ $\underline{p}_n$ entail $(\underline{x})\underline{q}$ in the LPC($\overline{EL}^{\nu+1}$).

Here are the ordinary and modal analogues for MC^ν of Theor. 32.1, i.e. its total analogue for MC^ν.

THEOR. 32.2. Let $\underline{p}_1, \ldots, \underline{p}_n$ and $\underline{q}$ be matrices in MC^ν and let $\underline{p}_1, \ldots, \underline{p}_n$ entail $\underline{q}$ in the LPC(ML^ν). Furthermore let $\underline{p}_1, \ldots, \underline{p}_n$ be either all closed with respect to $\underline{x}$, or all modally closed [Def. 4.3]. Then $\underline{p}_1, \ldots, \underline{p}_n$ entail $(\underline{x})\underline{q}$ or $\underline{Nq}$ respectively in the LPC(MC^ν).

Proof. Under our hypotheses, by Theor. 31.1 the matrices $\underline{p}_1^\eta$, $\ldots, \underline{p}_n^\eta$ in $\overline{EL}^{\nu+1}$ [see N15] entail $\underline{q}^\eta$ in the LPC($\overline{EL}^{\nu+1}$), and by rules ($\underline{T}_7$) and ($\underline{T}_8$) in N15, $\underline{p}_1^\eta, \ldots, \underline{p}_{\underline{n}}^\eta$ are closed with respect to $\underline{x}^\eta$ or κ (respectively). Then by Theor. 32.1 $\underline{p}_1^\eta, \ldots, \underline{p}_{\underline{n}}^\eta$ entail, in the LPC($\overline{\text{EL}}^{\nu+1}$), $(\underline{x}^\eta)\underline{q}^\eta$ or $(\kappa)\underline{q}^\eta$ respectively. By rules ($\underline{T}_7$) and ($\underline{T}_8$) in N15 these matrices are, in order, $[(\underline{x})\underline{q}]^\eta$ and $(\underline{N}\underline{q})^\eta$. Hence by Theor. 31.1 $\underline{p}_1, \ldots, \underline{p}_{\underline{n}}$ entail in the LPC(MC$^\nu$) $(\underline{x})\underline{q}$ or $\underline{N}\underline{q}$ respectively.

QED

In order to transfer to MC$^\nu$ the equivalence and substitution theorems—i.e. Theors. VI.5.4 and VI.5.6 in [32], pp. 109–11—let us consider the following hypotheses suggested by [32], which refer to either ML$^\nu$ or $\overline{\text{EL}}^{\nu+1}$ at will:

HYP. 32.1 [32.2]. $\underline{p}_1, \ldots, \underline{p}_{\underline{n}}, \underline{A}, \underline{B}$ are matrices (statements) and $\underline{x}_1, \ldots, \underline{x}_{\underline{a}}$ are variables. The matrix $\underline{W}$, i.e. $\underline{F}(\underline{p}_1, \ldots, \underline{p}_{\underline{n}}, \underline{A}, \underline{B})$, is built up out of some or all of the $\underline{p}$'s and $\underline{A}$ and $\underline{B}$ by means of $\sim$, $\wedge$ [by means of $\sim$, $\wedge$, and $\underline{N}$], and by means of $(\underline{x})$, where each time $(\underline{x})$ is used, $\underline{x}$ is one of $\underline{x}_1, \ldots, \underline{x}_{\underline{a}}$, and where one may use each $\underline{p}$ or each $\underline{x}$ or $\underline{A}$ or $\underline{B}$ more than once if desired. In addition $\underline{V}$ is the result of replacing some or none of the $\underline{A}$'s in $\underline{W}$ by $\underline{B}$'s.

Here is the equivalence theorem for the LPC($\overline{\text{EL}}^{\nu+1}$), to be proved substantially the way Theor. VI.5.4 is in [32]:

THEOR. 32.3. Let Hyp. 32.1 hold in connection with $\overline{\text{EL}}^{\nu+1}$, so that $\underline{p}_1, \ldots, \underline{p}_{\underline{n}}, \underline{A}, \underline{B}, \underline{x}_1, \ldots, \underline{x}_{\underline{a}}$ are wffs in $\overline{\text{EL}}^{\nu+1}$. In addition let $\underline{y}_1, \ldots, \underline{y}_{\underline{b}}$ be variables of $\overline{\text{EL}}^{\nu+1}$ such that there are no free occurrences of the $\underline{x}$'s in $(\forall \underline{y}_1, \ldots, \underline{y}_{\underline{b}})(\underline{A} \equiv \underline{B})$. Then $\vdash (\forall \underline{y}_1, \ldots, \underline{y}_{\underline{b}})(\underline{A} \equiv \underline{B}) \supset (\underline{W} \equiv \underline{V})$ in the LPC($\overline{\text{EL}}^{\nu+1}$).

The ordinary and modal analogues of Theor. 32.3 are the parts of the following theorem corresponding to Hyps. 32.1 and 32.2 respectively.

THEOR. 32.4. Let Hyp. 32.2 hold and let $\underline{p}_1, \ldots, \underline{p}_n$, $\underline{A}$, $\underline{B}$ ($\underline{V}$ and $\underline{W}$) be matrices in ML^ν. Furthermore let $\underline{y}_1, \ldots, \underline{y}_b$ be variables in ML^ν such that there are no free occurrences of the $\underline{x}$'s in $(\forall \underline{y}_1, \ldots, \underline{y}_b)(\underline{A} \equiv \underline{B})$.

Then first, $\vdash (\forall \underline{y}_1, \ldots, \underline{y}_b)(\underline{A} \equiv^\cap \underline{B}) \supset (\underline{W} \equiv^\cap \underline{V})$ in MC^ν; and second, if the stronger hypothesis [Hyp. 32.1] holds we can also assert that $\vdash (\forall \underline{y}_1, \ldots, \underline{y}_b)(\underline{A} \equiv \underline{B}) \supset (\underline{W} \equiv \underline{V})$ in MC^ν.

Proof. Since Hyp. 32.2 holds, by rules $(\underline{T}_1)$ to $(\underline{T}_9)$ in N15 the analogue of Hyp. 32.1 holds for $\underline{p}_1^\eta, \ldots, \underline{p}_n^\eta, \underline{A}^\eta, \underline{B}^\eta, \underline{W}^\eta, \underline{V}^\eta$, and for the variables $\underline{x}_1^\eta, \ldots, \underline{x}_a^\eta$ and for κ (on building $\underline{W}^\eta$ up out of $\underline{p}_1^\eta, \ldots, \underline{p}_n^\eta, \underline{A}^\eta$, and $\underline{B}^\eta$ we may use $(\forall\xi)$, where ξ is any one of those variables).

By our hypothesis on $\underline{y}_1, \ldots, \underline{y}_b$ the variables $\underline{x}_1^\eta, \ldots, \underline{x}_a^\eta$ and κ do not occur free in $(\forall \underline{y}_1^\eta, \ldots, \underline{y}_b^\eta, \kappa)(\underline{A}^\eta \equiv \underline{B}^\eta)$. Hence by Theors. 32.2, 32.3, and $(6)_3$ the matrix $(\forall \underline{y}_1^\eta, \ldots, \underline{y}_b^\eta, \kappa)(\underline{A}^\eta \equiv \underline{B}^\eta) \supset (\kappa)(\underline{W}^\eta \equiv \underline{V}^\eta)$ is provable in the $LPC(\overline{EL}^{\nu+1})$. By rules $(\underline{T}_1)$ to $(\underline{T}_9)$ in N15 this matrix is the extensional translation of $(\forall \underline{y}_1, \ldots, \underline{y}_b)(\underline{A} \equiv^\cap \underline{B}) \supset (\underline{W} \equiv^\cap \underline{V})$. Hence by Theor. 31.1 the last matrix is provable in MC^ν.

Now let Hyp. 32.1 also hold. Hence the sign $\underline{N}$ is not used in constructing $\underline{W}$, i.e. $\underline{F}(\underline{p}_1, \ldots, \underline{p}_n, \underline{A}, \underline{B})$, out of $\underline{p}_1, \ldots, \underline{p}_n$, $\underline{A}$, and $\underline{B}$, so that by rules $(\underline{T}_1)$ to $(\underline{T}_9)$ the quantifier (κ) is not used in constructing $\underline{W}^\eta$ out of $\underline{p}_1^\eta, \ldots, \underline{p}_n^\eta, \underline{A}^\eta$, and $\underline{B}^\eta$; more precisely, (at most) the quantifiers $(\underline{x}_1^\eta), \ldots, (\underline{x}_a^\eta)$ are used to construct $\underline{W}^\eta$. Furthermore by our hypothesis on $\underline{y}_1, \ldots, \underline{y}_b$ and by our translation rules, the variables $\underline{x}_1^\eta, \ldots, \underline{x}_a^\eta$ do not occur free in $(\forall \underline{y}_1^\eta, \ldots, \underline{y}_b^\eta)(\underline{A}^\eta \equiv \underline{B}^\eta)$. Hence by Theor. 32.3 the matrix $(\forall \underline{y}_1^\eta, \ldots, \underline{y}_b^\eta)(\underline{A}^\eta \equiv \underline{B}^\eta) \supset (\underline{W} \equiv \underline{V})$ is provable in the $LPC(\overline{EL}^{\nu+1})$. Hence by rules $(\underline{T}_1)$ to $(\underline{T}_9)$ and Theor. 31.1 the matrix $(\forall \underline{y}_1, \ldots, \underline{y}_b)(\underline{A} \equiv \underline{B}) \supset (\underline{W} \equiv \underline{V})$ is provable in MC^ν.

QED

The following replacement theorem for MC^ν could be proved from its analogue in the $LPC(\overline{EL}^{\nu+1})$ in the same way Theor. 32.4 was. But it is simpler to prove it directly.

THEOR. 32.5. Assume Hyp. 32.1 [Hyp. 32.2] in connection with ML^ν and let $\vdash \underline{A} \equiv \underline{B}$ hold in the $LPC(ML^\nu)$. Furthermore let $\vdash \underline{W}$ hold in the theory $\mathcal{T}$ including the $LPC(ML^\nu)$.[13] Then $\vdash \underline{V}$ holds in $\mathcal{T}$.

Proof. By Theor. 32.2 our hypotheses imply that $\vdash (\forall \underline{y}_1, \ldots, \underline{y}_b)(\underline{A} \equiv^{\frown} \underline{B})$ in the $LPC(MC^\nu)$. Hence by Theor. 32.4 our theorem follows.

QED

Let $\underline{W}^*$ be the dual of the matrix $\underline{W}$ built up out of (some or all of) the matrices $\underline{p}_1, \ldots, \underline{p}_n$ in ML^ν by the use of $\sim$, $\wedge$, $\vee$, $(\forall \underline{x}_i)$, $(\exists \underline{x}_i)$ $(\underline{i} = 1, \ldots \underline{r})$, $\underline{N}$, and $\diamond$. Naturally this means that $\underline{W}^*$ is obtained from $\underline{W}$ by simultaneously performing the following changes:

(1) If an occurrence of $\underline{p}_i$ has a $\sim$ attached, remove it; otherwise attach one.

(2) Replace $\wedge$ by $\vee$, $(\forall \underline{x}_i)$ by $(\exists \underline{x}_i)$ $(\underline{i} = 1, \ldots, \underline{r})$, $\underline{N}$ by $\diamond$, and vice versa.

Obviously (for $\nu = 1, 2, \ldots$) the above definition holds when $\underline{p}_1, \ldots, \underline{p}_n$, $\underline{W}$, and $\underline{W}^*$ are wffs in $EL^{\nu+1}$ or $\overline{EL}^{\nu+1}$. By rules $(\underline{T}_1)$ to $(\underline{T}_9)$, $\underline{W}^{*\eta}$ is $(\underline{W}^\eta)^*$, i.e. the extensional translation of the dual is the dual of the extensional translation.

Here is the duality theorem and its corollary [Theor. 32.7] for ML^ν. They can be proved by referring to their extensional counterparts in $\overline{EL}^{\nu+1}$—see [32], Theor. VI.6.3, p. 117, and Theor. VI.4, p. 118. But a direct proof of Theor. 32.7 is also very brief.

THEOR. 32.6. If $\underline{W}^*$ is the dual of $\underline{W}$, then $\vdash \sim\underline{W} \equiv \underline{W}^*$ holds in MC^ν.

THEOR. 32.7. Let $\underline{W}$ and $\underline{V}$ be built up out of some or all the matrices $\underline{p}_1, \ldots, \underline{p}_n$ in ML^ν by use of $\sim$, $\wedge$, $(\forall \underline{x}_i)$, $(\exists \underline{x}_i)$ $(\underline{i} = 1, \ldots, \underline{r})$, $\underline{N}$, and $\diamond$, where we may use each $\underline{p}_h$ and each variable $\underline{x}_i$, $\underline{N}$, and $\diamond$ as often as desired (possibly never). Let $\underline{X}$ and $\underline{Y}$ be the results of replacing $\wedge$ by $\vee$, $\underline{N}$ by $\diamond$, $(\forall \underline{x}_i)$ by $(\exists \underline{x}_i)$ and vice versa $(\underline{i} = 1, \ldots, \underline{r})$ in $\underline{W}$ and $\underline{V}$ respectively. If in addition $\vdash \underline{W} \equiv \underline{V}$ in the LPC(MC^ν) and if this would continue to hold if we replace $\underline{p}_i$ by $\sim\underline{p}_i$, then $\vdash \underline{X} \equiv \underline{Y}$ in the LPC(MC^ν) (so that $\vdash \underline{X}$ implies $\vdash \underline{Y}$).

By Theors. 32.6 and 32.7 we can immediately prove the dual forms of some of the theorems $(4)_1$ to $(9)_3$ and $(5')_1$ to $(9')_3$. We denote such duals by an asterisk.

THEOR. 32.8. Let $\underline{r}$ be closed with respect to $\underline{x}$ and let $\underline{r}_1$ be modally closed [Def. 4.3]. Then

(5)* $\vdash (\exists \underline{x})(\underline{p} \vee \underline{q}) \equiv (\exists \underline{x})\underline{p} \vee (\exists \underline{x})\underline{q}$, $\vdash (\exists \underline{x})\underline{r} \equiv \underline{r}$, $\vdash (\exists \underline{x})(\underline{p} \vee \underline{r}) \equiv (\exists \underline{x})\underline{p} \vee \underline{r}$,

(5′)* $\vdash \diamond(\underline{p} \vee \underline{q}) \equiv \diamond\underline{p} \vee \diamond\underline{q}$, $\vdash \diamond\underline{r}_1 \equiv \underline{r}_1$, $\vdash \diamond(\underline{p} \vee \underline{r}_1) \equiv \diamond\underline{p} \vee \underline{r}_1$.

As to the prenex form we remark that, for instance, $(\underline{p} \supset \underline{Nq})^\eta$ is $\underline{p}^\eta \supset (\kappa)\underline{q}^\eta$, which has the form $\Phi(\kappa) \supset (\kappa)\Psi(\kappa)$. Its prenex form is the matrix $(\kappa')[\Phi(\kappa) \supset \Psi(\kappa')]$ where κ' has the same logical type as κ but is distinct from κ. Hence this matrix in $EL^{\nu+1}$ is the extensional translation of no matrix in ML^ν. Therefore the modal (or total) analogue of the usual metatheorem on the prenex form holding for $EC^{\nu+1}$ does not hold for ($\overline{EL}^{\nu+1}$ or) MC^{ν}. Using our extensional translation (in the same way as we did to prove e.g. Theor. 32.4) we immediately see that the ordinary analogue for MC^{ν} of this metatheorem for $EC^{\nu+1}$ holds in the following form:

THEOR. 32.9. Let $\underline{W}$ be built up out of matrices $\underline{p}_1, \ldots, \underline{p}_m$ in ML^{ν} (which may contain $\underline{N}$) using $\sim$, $\wedge$, $(\underline{x}_1), \ldots, (\underline{x}_a)$ as often as desired. Then there are matrices $\underline{q}_1, \ldots, \underline{q}_n$ which are the same as the $\underline{p}$'s or are obtained from them by replacing some free variables with other variables, and there is a matrix $\underline{V}$ obtained from the $\underline{q}$'s using (only) $\sim$ and $\wedge$—i.e. only rules $(\underline{f}_5)$ and $(\underline{f}_6)$ in N3—and a string $(\underline{Q})$ of (ordinary) universal quantifiers, such that $\underline{W} \equiv (\underline{Q})\,\underline{V}$.

N33. Theorems for shortening deductions in MC^{ν}. Rules G and C.

Here we transfer to ML^{ν} two theorems for shortening proofs in the extensional logical calculus $EC^{\nu+1}$. They are a theorem [Theor. 33.1] based on some procedures due to Gentzen—cf. [32], Theor. VI.7.2, p. 130—and two versions [Theors. 33.2, 33.3] of the deduction theorem—cf. fnn. 3, 4.

We first define a new concept of deduction in the theory $\mathscr{T} = \langle ML^{\nu}, \mathscr{A}, \mathscr{D} \rangle$—see Def. 29.2—and precisely (the concept of) a deduction of $\underline{q}$ from $\underline{p}_1, \ldots, \underline{p}_m$ made in $\mathscr{T}$ using rules including the generalization rule (rule $\underline{G}$) and the formal analogue of an act of choice (rule $\underline{C}$).

DEF. 33.1. Assume that $\underline{p}_1, \ldots, \underline{p}_m, \underline{q}_1, \ldots, \underline{q}_n$ are matrices in ML^{ν} ($\underline{m} \geq 0$); in addition let $\underline{r}_1, \ldots, \underline{r}_a, \underline{s}_1, \ldots, \underline{s}_b$ ($\underline{a} \geq 0$, $\underline{b} \geq 0$) be $\underline{a} + \underline{b}$ (distinct) positive integers with $\underline{r}_1 < \underline{r}_2 <$

$\dots \underline{r}_{\underline{a}} \le \underline{n}$ and $\underline{s}_1 < \underline{s}_2 < \dots < \underline{s}_{\underline{b}} \le \underline{n}$; furthermore let each of $\underline{x}_1, \dots, \underline{x}_{\underline{a}}, \underline{y}_1, \dots, \underline{y}_{\underline{b}}$ be either a variable of ML^ν or the variable κ (which does not belong to ML^ν).

Under these assumptions we say that the sequence $\underline{q}_1, \dots, \underline{q}_{\underline{n}}$ constitutes a deduction of $\underline{q}_{\underline{n}}$ from $\underline{p}_1, \dots, \underline{p}_{\underline{m}}$ made in the theory $\mathscr{T}$ ($\mathscr{T} = \langle ML^\nu, \mathscr{A}, \mathscr{D} \rangle$) using rule $\underline{G}$ at the steps $\underline{r}_1, \dots, \underline{r}_{\underline{a}}$ (in connection) with the variables $\underline{x}_1, \dots, \underline{x}_{\underline{a}}$, and using rule $\underline{C}$ at the steps $\underline{s}_1, \dots, \underline{s}_{\underline{b}}$ (in connection) with the variables $\underline{y}_1, \dots, \underline{y}_{\underline{b}}$, if the following requirements hold:

(a) For $\underline{i} = 1, \dots, \underline{a}$ ($\underline{x}_{\underline{i}}$ may be the same as some others of $\underline{x}_1, \dots, \underline{x}_{\underline{a}}$ but), if $\underline{x}_{\underline{i}}$ is different from κ, then it does not occur free in $\underline{p}_1 \wedge \dots \wedge \underline{p}_{\underline{m}}$ or any $\underline{C}$-step earlier than $\underline{q}_{\underline{r}_{\underline{i}}}$, i.e. any $\underline{q}_{\underline{s}_\rho}$ with $1 \le \underline{s}_\rho < \underline{r}_{\underline{i}}$.

(b) Either none of $\underline{x}_1, \dots, \underline{x}_{\underline{a}}$ is κ or exactly one, $\underline{x}_1$, of them is κ and all of $\underline{p}_1, \dots, \underline{p}_{\underline{m}}$ and all $\underline{C}$-steps earlier than $\underline{q}_{\underline{r}_1}$ (i.e. all $\underline{q}_{\underline{s}_\rho}$ with $1 \le \underline{s}_\rho < \underline{r}_1$) are modally closed.

(c) For $\underline{i} = 1, \dots, \underline{b}$ the variable $\underline{y}_{\underline{i}}$ does not occur free in $\underline{p}_1 \wedge \dots \wedge \underline{p}_{\underline{m}}$ or any $\underline{C}$-step earlier than $\underline{q}_{\underline{s}_{\underline{i}}}$ (i.e. any $\underline{q}_{\underline{s}_{\underline{j}}}$ with $\underline{j} < \underline{i}$) or in any axiom of the theory $\mathscr{T}$ whose closure with respect to $\underline{y}_{\underline{i}}$ is not an axiom of $\mathscr{T}$.

(d) Either none of $\underline{y}_1, \dots, \underline{y}_{\underline{b}}$ is κ or (1) exactly one $\underline{y}_{\underline{i}}$ of them is κ, (2) $\underline{p}_1, \dots, \underline{p}_{\underline{m}}$ and the $\underline{C}$-steps earlier than $\underline{q}_{\underline{s}_{\underline{i}}}$ (i.e. the steps $\underline{q}_{\underline{s}_\rho}$ with $\rho < \underline{i}$) are modally closed [Def. 4.3], and (3) the modal closures of the axioms of $\mathscr{T}$ are axioms of $\mathscr{T}$.[14]

(e) For every $\underline{i}$ with $1 \le \underline{i} < \underline{n}$ one of the following conditions (rules) holds:

(1) The index $\underline{i}$ is different from $\underline{r}_1, \dots, \underline{r}_{\underline{a}}, \underline{s}_1, \dots, \underline{s}_{\underline{b}}$; and $\underline{q}$ is an axiom or definition in $\mathscr{T}$ ($\underline{q}_{\underline{i}} \in \mathscr{A} \cup \mathscr{D}$).

(2) i is as in (1) and either q_i is one of $p_1, \ldots, p_m$ or q_i is q_j for some $j < i$.

(3) i is as in (1) and there are j and k less than i and such that q_k is $q_j \supset q_i$.

(4) (Ordinary Rule G) $i = r_l$ (for some l with $1 \leq l \leq a$), x_l is not κ, and there is a j less than i such that q_i is $(\forall x_l) q_j$.

(4′) (Modal Rule G) $i = r_l$ (for some l with $1 \leq l \leq a$), x_l is κ, and there is a j less than i such that q_i is Nq_j.

(5) (Ordinary Rule C) $i = s_h$ (for some h with $1 \leq h \leq b$), y_h is not κ, and there is a j less than i and a variable z_h such that q_j is $(\exists z_h) \Phi(z_h)$ where $\Phi(z_h)$ is the result of replacing all free occurrences of y_h in q_i by occurrences of z_h, and where q_i is the result $\Phi(y_h)$ of replacing the free occurrences of z_h in $\Phi(z_h)$ by occurrences of y_h.

(5′) (Modal Rule C) $i = s_h$ (for some h with $1 \leq h \leq b$), y_h is κ, and there is a j less than i such that q_i is $\diamond q_j$.

Suppose that among the matrices $p_1, \ldots, p_m$ there are both p and $(\exists y_h) p$ [or $\diamond p$] where y has no free occurrences in p [p is modally closed]. Then we may arrange a deduction from the premises $p_1, \ldots, p_m$ in such a way that for some h and for $i = s_h$, q_i is p.

Then it is clear that in our deduction the h-th use of rule C is not essential in that if we take $i = s_h$ off from our $s_1, \ldots, s_b$, q_i would result from an earlier rule than rule (5) (or rule (5′)), namely rule (2). In an analogous situation we may say that the l-th use of rule G—see rules (4) and (4′) in Def. 33.1—is not essential and we may decide that the same should hold for modus ponens, i.e. rule (3).

Observe that if in the deduction spoken of in Def. 33.1 some G- or C-steps are not essential, then by taking off some of the integers

$\underline{r}_1, \ldots, \underline{r}_a, \underline{s}_1, \ldots, \underline{s}_b$ we may keep only the essential $\underline{G}$- and $\underline{C}$-steps.

As we already hinted at, for $\underline{l} = 1, \ldots, \underline{a}$ we may say that the $\underline{l}$-th $\underline{G}$-step $\underline{q}_{\underline{r}_l}$ is a modal $\underline{G}$-step—see rule (4′)—or an ordinary $\underline{G}$-step—see rule (4)—according to whether $\underline{x}_l$ is κ or not; furthermore for $\underline{h} = 1, \ldots, \underline{b}$ we may say that $\underline{q}_{\underline{s}_h}$ is a modal $\underline{C}$-step—see rule (5′)—or an ordinary $\underline{C}$-step—see rule (5)—according to whether $\underline{y}_h$ is κ or not.

We understand an obvious analogue of Def. 33.1 for $\overline{EL}^{\nu+1}$ (or EL^{ν}):

DEF. 33.2(a) Let $\underline{L}$ be either ML^{ν} or $\overline{EL}^{\nu+1}$ (or else EL^{ν}). Then we say that $\underline{p}_1, \ldots, \underline{p}_m$ entail $\underline{p}_0$ in the theory $\mathcal{T} = \langle \underline{L}, \mathcal{A}, \mathcal{D} \rangle$ using rules including rule $\underline{C}$ (and rule $\underline{G}$)—and we write $\underline{p}_1, \ldots, \underline{p}_m \vdash_c \underline{p}_0$ (in $\mathcal{T}$)—if there are $\underline{n}$ matrices $\underline{q}_1, \ldots, \underline{q}_n$ and $\underline{a}+\underline{b}$ integers $\underline{r}_1, \ldots, \underline{r}_a, \underline{s}_1, \ldots, \underline{s}_b$ fulfilling the assumption made in Def. 33.1 and such that $\underline{q}_n$ is $\underline{p}_0$ and in addition that $\underline{q}_1, \ldots, \underline{q}_n$ constitute a deduction of $\underline{p}_0$ from $\underline{p}_1, \ldots, \underline{p}_m$ in $\mathcal{T}$ made using rules $\underline{G}$ and $\underline{C}$ at the steps $\underline{r}_1, \ldots, \underline{r}_a$ and $\underline{s}_1, \ldots, \underline{s}_b$ respectively.

DEF. 33.2(b) If $\underline{b} = 0$, i.e. rule $\underline{C}$ is not used, then we also write $\underline{p}_1, \ldots, \underline{p}_m \vdash_{\underline{G}} \underline{p}_0$ (in $\mathcal{T}$).

Here is the theorem on $\vdash_c$ for the cases $\underline{L} = EL^{\nu}$, $\underline{L} = \overline{EL}^{\nu+1}$, and $\underline{L} = ML^{\nu}$. It includes the (total) analogues for MC^{ν} of the theorems on rules $\underline{G}$ and $\underline{C}$ in extensional logic—see [32], Theors. VI.7.1 and VI.7.2, pp. 125, 130.

THEOR. 33.1. Let us consider a theory $\mathcal{T} = \langle \underline{L}, \mathcal{A}, \mathcal{D} \rangle$ where $\underline{L}$ is either EL^{ν} or $\overline{EL}^{\nu+1}$ or ML^{ν} and $\mathcal{A}$ includes the axio of the lower predicate calculus for EL^{ν} or $\overline{EL}^{\nu+1}$ or ML^{ν} respec tively. Moreover let $\underline{p}_1, \ldots, \underline{p}_m \vdash_{\underline{c}} \underline{q}$ in $\mathscr{T}$. Furthermore let there be a deduction $\underline{q}_1, \ldots, \underline{q}_n$ of $\underline{q}$ from $\underline{p}_1, \ldots, \underline{p}_m$ made in $\mathscr{T}$ using rules $\underline{G}$ and $\underline{C}$, and let none of the variables $\underline{y}_1, \ldots, \underline{y}_b$ introduced by rule $\underline{C}$ (in particular none of those different from κ) occur free in $\underline{q}$; in addition either let none of $\underline{y}_1, \ldots, \underline{y}_b$ be κ or let $\underline{q}$ be modally closed [Def. 4.3]. Then also $\underline{p}_1, \ldots, \underline{p}_m \vdash \underline{q}$ holds in $\mathscr{T}$ [Def. 29.2].

Proof. The part of our theorem for the case $\underline{L} = EL^{\nu}$ or the case $\underline{L} = \overline{EL}^{\nu+1}$ is in effect the analogue for EL^{ν} or $\overline{EL}^{\nu+1}$ respectively of Theor. VI.7.2 in [32], p. 130. This part differs from the last theorem in [32] only in that in Theor. VI.7.2 all $\underline{G}$- and $\underline{C}$-steps are assumed to be essential. By our remark on nonessential $\underline{G}$- and $\underline{C}$-steps made after Def. 33.1 this part—which concerns EL^{ν} and $\overline{EL}^{\nu+1}$—can be proved in the LPC(EL^{ν}) or the LPC($\overline{EL}^{\nu+1}$) with substantially the same words used in [32] to prove Theor. VI.7.2. Now let $\mathscr{T} = \langle \overline{EL}^{\nu+1}, \mathscr{A}^{\eta}, \mathscr{D}^{\eta} \rangle$ hold, where $\mathscr{A}^{\eta}$ is the class of the extensional translations into $EL^{\nu+1}$ of the axioms forming $\mathscr{A}$ [N15] and $\mathscr{D}^{\eta}$ is a class of definitions equivalent in the LPC($\overline{EL}^{\nu+1}$) to the extensional translations into $\overline{EL}^{\nu+1}$ of the definitions forming $\mathscr{D}$.[15] Incidentally, by Theor. 30.1 $\mathscr{A}^{\eta}$ includes the axioms of the LPC($\overline{EL}^{\nu+1}$).

The variable κ is not a wff in ML^{ν}, hence thus far κ^{η} can be considered to be meaningless. Therefore we may assume here that κ^{η} should mean κ itself, which will be useful shortly.

Now let $\underline{r}_1, \ldots, \underline{r}_a, \underline{s}_1, \ldots, \underline{s}_b$ be integers such that the deduction $\underline{q}_1, \ldots \underline{q}_n$ mentioned among the hypotheses of Theor. 33.1 is a deduction of $\underline{q}$ from $\underline{p}_1, \ldots, \underline{p}_m$ made in $\mathscr{T}$ using rules $\underline{G}$ and $\underline{C}$ at the steps $\underline{r}_1, \ldots, \underline{r}_a$ and $\underline{s}_1, \ldots, \underline{s}_b$ respectively in connection with the variables $\underline{x}_1, \ldots, \underline{x}_a$ and $\underline{y}_1, \ldots, \underline{y}_b$ respectively.

Then by our translation rules ($\underline{T}_1$) to ($\underline{T}_9$) in N15, and by the analogue of Def. 33.1 for $\overline{EL}^{\nu+1}$, the matrices $\underline{q}_1^{\eta}, \ldots, \underline{q}_{\underline{n}}^{\eta}$ constitute a deduction of $\underline{q}^{\eta}$ from $\underline{p}_1^{\eta}, \ldots, \underline{p}_{\underline{m}}^{\eta}$ made in $\mathscr{T}^{\eta}$ using rules $\underline{G}$ and $\underline{C}$ at the steps $\underline{r}_1, \ldots, \underline{r}_{\underline{a}}$ and $\underline{s}_1, \ldots, \underline{s}_{\underline{b}}$ respectively in connection with the variables $\underline{x}_1^{\eta}, \ldots, \underline{x}_{\underline{a}}^{\eta}$ and $\underline{y}_1^{\eta}, \ldots, \underline{y}_{\underline{b}}^{\eta}$ respectively.

By hypothesis, $\underline{y}_1, \ldots \underline{y}_{\underline{b}}$ do not occur free in $\underline{q}$. Hence if $\underline{y}_{\underline{i}}$ is not κ, by rules ($\underline{T}_1$) and ($\underline{T}_9$), $\underline{y}_{\underline{i}}^{\eta}$ does not occur free in $\underline{q}^{\eta}$ ($\underline{i}$ = 1, . . . , $\underline{b}$). Also by hypothesis, if $\underline{y}_{\underline{i}}$ is κ, then $\underline{q}$ is modally closed, so that by Theor. 15.1 and our convention that κ^{η} is κ, the variable $\underline{y}_{\underline{i}}^{\eta}$, i.e. κ, does not occur free in $\underline{q}^{\eta}$. We conclude that none of $\underline{y}_1^{\eta}, \ldots, \underline{y}_{\underline{b}}^{\eta}$ occurs free in $\underline{q}^{\eta}$. Hence by the part of Theor. 33.1 for the case $\underline{L} = \overline{EL}^{\nu+1}$ we can assert that $\underline{p}_1^{\eta}, \ldots, \underline{p}_{\underline{m}}^{\eta} \vdash \underline{q}^{\eta}$ in $\mathscr{T}^{\eta}$.

As a consequence we can consider a deduction $\underline{q}_1', \ldots \underline{q}_{\underline{s}}'$ of $\underline{q}^{\eta}$ from $\underline{p}_1^{\eta}, \ldots, \underline{p}_{\underline{m}}^{\eta}$ in $\mathscr{T}^{\eta}$. Then we may choose in $\mathscr{A} \cup \mathscr{D}$ $\underline{r}-\underline{m}$ matrices $\underline{p}_{\underline{m}+1}, \ldots, \underline{p}_{\underline{r}}$ such that $\underline{q}_1', \ldots, \underline{q}_{\underline{s}}'$ constitute a deduction of $\underline{q}^{\eta}$ from $\underline{p}_1^{\eta}, \ldots, \underline{p}_{\underline{r}}^{\eta}$ in the LPC($\overline{EL}^{\nu+1}$). Then by Theor. 31.1 there is a deduction of $\underline{q}$ from $\underline{p}_1, \ldots, \underline{p}_{\underline{r}}$ in the LPC(MC^{ν}).

Hence by Def. 29.2 and the way in which $\underline{p}_{\underline{m}+1}, \ldots, \underline{p}_{\underline{r}}$ were introduced, there is a deduction of $\underline{q}$ from $\underline{p}_1, \ldots, \underline{p}_{\underline{m}}$ in $\mathscr{T}$.

QED

According to the above procedure based on our translation of ML^{ν} into $EL^{\nu+1}$ [N15] and on Theor. 31.1, one can easily prove the following deduction theorems—cf. [32], Theor. IV.6.1, p. 75, and Theor. VI.8.1, p. 148:

THEOR. 33.2. Let $\underline{p}_1, \ldots, \underline{p}_{\underline{n}}, \underline{q} \vdash \underline{r}$ hold in $\mathscr{T} = \langle ML^{\nu}, \mathscr{A}, \mathscr{D} \rangle$ where $\mathscr{A}$ contains AS30.$\overline{1}$–3 of the extensional propositional calculus for MC^{ν}. Then also $\underline{p}_1, \ldots, \underline{p}_{\underline{n}} \vdash \underline{q} \supset \underline{r}$ holds in $\mathscr{T}$.

THEOR. 33.3. Let $\underline{p}_1, \ldots, \underline{p}_n, \underline{q} \vdash_{\underline{c}} \underline{r}$ hold in $\mathscr{T} = \langle ML^{\nu}, \mathscr{A}, \mathscr{D} \rangle$ where $\mathscr{A}$ contains AS30.1–9 of the LPC(MC^{ν}). Then also $\underline{p}_1, \ldots, \underline{p}_n \vdash_{\underline{c}} \underline{q} \supset \underline{r}$ holds in $\mathscr{T}$.

N34. Axioms and basic theorems for identity in MC^{ν}.

Let $\underline{x}$ and $\underline{y}$ be free for $\underline{z}$ in the matrix $\Phi(\underline{x}, \underline{y}, \underline{z})$, and recall conventions $(24)_2$ in N12 and (a) and (b) in N30. Then the axioms (axiom schemes) for identity in MC^{ν}—cf. AS12.10–13—are:[16]

AS34.1. $(\underline{N})\underline{x} = \underline{x}$,

AS34.2. $(\underline{N})\underline{x} = \underline{y} \supset \underline{y} = \underline{x}$,

AS34.3. $(\underline{N})\underline{x} = \underline{y}\ \underline{y} = \underline{z} \supset \underline{x} = \underline{z}$,

AS34.4. $(\underline{N})\underline{x} =^{\frown} \underline{y} \supset [\Phi(\underline{x}, \underline{y}, \underline{x}) \equiv \Phi(\underline{x}, \underline{y}, \underline{y})]$.

We add to Def. 11.3 of $(\exists_1 \underline{x})$ and Def. 14.1 of $(\exists_1^{\frown} \underline{x})$ the following definitions of $(\exists^{(1)} \underline{x})$ and $(\exists^{(1)\frown} \underline{x})$ where $(\exists^{(1)} \underline{x})\underline{p}$ is to be read as there is at most one $\underline{x}$ such that $\underline{p}$ and $(\exists^{(1)\frown} \underline{x})\underline{p}$ is to be read as there is at most a strictly unique $\underline{x}$ such that $\underline{p}$.

DEF. 34.1. $(\exists^{(1)} \underline{x}) \Phi(\underline{x}) \equiv_D (\forall \underline{x}, \underline{y}) [\Phi(\underline{x}) \Phi(\underline{y}) \supset \underline{x} = \underline{y}]$,

DEF. 34.2. $(\exists^{(1)\frown} \underline{x}) \Phi(\underline{x}) \equiv_D (\forall \underline{x}, \underline{y}) [\Phi(\underline{x}) \Phi(\underline{y}) \supset \underline{x} =^{\frown} \underline{y}]$,

where the usual Hyp. 11.1 holds [N11].

The first three of the following theorems in MC^{ν} are straightforward consequences of the above definitions:

(10) $\vdash (\underline{x})(\underline{p} \supset \underline{q}) \supset [(\exists^{(1)} \underline{x})\underline{q} \supset (\exists^{(1)} \underline{x})\underline{p}]$,

$\vdash (\underline{x})(\underline{p} \supset \underline{q}) \supset [(\exists^{(1)\frown} \underline{x})\underline{q} \supset (\exists^{(1)\frown} \underline{x})\underline{p}]$,

(11) $\vdash (\exists^{(1)\frown} \underline{x})\underline{p} \supset (\exists^{(1)} \underline{x})\underline{p}$, $\vdash \underline{N}(\exists^{(1)} \underline{x})\underline{p} \supset \underline{N}(\exists^{(1)\frown} \underline{x})\underline{Np}$.

In order to prove $(11)_2$ we now start with $\underline{N}(\exists^{(1)} \underline{x}) \Phi(\underline{x})$ and assume that $\Phi(x)$ and $\Phi(y)$ fulfill condition (c) in N30. Then by Def.

34.1, AS30.7, and $(5')_1$ we deduce the modally closed matrix $(\forall \underline{x}, \underline{y}) [\underline{N}\Phi(\underline{x}) N\Phi(\underline{y}) \supset {}^\frown \underline{x} = {}^\frown \underline{y}]$, which by $(5')_2$ and Def. 34.2 yields $\underline{N}(\exists^{(1)\frown} \underline{x}) \underline{N}\Phi(\underline{x})$. By Theor. 33.2 we conclude that $(11)_2$ holds.

By convention $(24)_2$ on $(\underline{N})$ [N12] and by AS30.7 and $(5')_1$, if we replace $=$ by $=^\frown$ and, if we like, $(\underline{N})$ by $(\ldots)$ in AS34.1–3, we obtain theorems in MC^ν. In particular, from AS34.1 we obtain $(\ldots)$ $\underline{x} =^\frown \underline{x}$, which together with AS34.4 tells us that the axioms for equality laid down in [32], pp. 163–64 hold in MC^ν up to the replacement of $=$ by $=^\frown$.

Since the axioms of the lower predicate calculus in [32] also hold in MC^ν, all theorems on identity proved in [32] are also provable in MC^ν, up to the above replacement. (In addition, by the modal part of the generalization theorem, i.e. Theor. 32.2, those theorems can be put into the form $(\underline{N}) \ldots .$)

Hence in particular the following holds—cf. Theors. VII.1.4 and 5 in [32], pp. 165–66:

(12) $$\vdash \Phi(\underline{x}, \underline{y}, \underline{x}) \sim \Phi(\underline{x}, \underline{y}, \underline{y}) \supset \sim \underline{x} =^\frown \underline{y},$$

(13) $$\vdash \Phi(\underline{y}) \equiv (\exists \underline{x}) [\underline{x} =^\frown \underline{y}\, \Phi(\underline{x})], \quad \vdash \Phi(\underline{y}) \equiv (\underline{x}) [\underline{x} =^\frown \underline{y} \supset \Phi(\underline{x})].$$

Furthermore by Def. 34.2, from Def. 14.1 of "there is a strictly unique $\underline{x}$ such that . . . ," and from Theor. VII.2.1 in [32], p. 167, we obtain the following chain of four equivalences (the first of them is Def. 14.1 again):

(14) $$\vdash (\exists_1^\frown \underline{x}) \Phi(\underline{x}) \equiv (\exists \underline{x}) \{\Phi(\underline{x}) (\underline{y}) [\Phi(\underline{y}) \supset \underline{y} =^\frown \underline{x}]\} \equiv$$
$$(\exists \underline{x}) (\underline{y}) [\underline{x} =^\frown \underline{y} \equiv \Phi(\underline{y})] \equiv (\exists \underline{x}) \Phi(\underline{x}) \wedge (\exists^{(1)\frown} \underline{x}) \Phi(\underline{x}) \equiv$$
$$(\exists \underline{x}) \Phi(\underline{x}) \sim (\exists \underline{x}, \underline{y}) [\Phi(\underline{x}) \Phi(\underline{y}) \sim \underline{x} =^\frown \underline{y}].$$

At last, from Theor. VII.2.2 in [32], p. 169, we have:

(15) $$\vdash (\underline{y}) (\exists_1^\frown \underline{x}) \underline{x} =^\frown \underline{y} \qquad \text{[see also AS34.1 and } (7)_3].$$

7

Modal Theorems on Extensional Matrices, Identity, and Descriptions

N35. On extensional matrices in MC^{ν}. How to reduce theorems on identity in MC^{ν} for these matrices to theorems in extensional logic.

In accordance with Carnap's concept of extensional propertie[s]— cf. Defs. 6.11, 6.12—we define the extensionalities of the matri[x] $\phi(\underline{x}_1, \ldots, \underline{x}_n)$ and the term $\Psi(\underline{x}_1, \ldots, \underline{x}_n)$ in ML^{ν} with respec[t] to the distinct variables $\underline{x}_1, \ldots, \underline{x}_n$ to be the matrices $(\underline{\text{ext}}\ \underline{x}_1, \ldots, \underline{x}_n)\, \phi(\underline{x}_1, \ldots, \underline{x}_n)$ and $(\underline{\text{ext}}\ \underline{x}_1, \ldots, \underline{x}_n)\, \Psi(\underline{x}_1, \ldots, \underline{x}_n)$—closed with respect to $\underline{x}_1, \ldots, \underline{x}_n$—introduced as follows (see Def. 6.9):

DEF. 35.1. $(\underline{\text{ext}}\ \underline{x}_1, \ldots, \underline{x}_n)\, \phi(\underline{x}_1, \ldots, \underline{x}_n) \equiv_D (\forall \underline{x}_1, \underline{y}_1, \ldots, \underline{x}_n, \underline{y}_n)\, [\phi(\underline{x}_1, \ldots, \underline{x}_n)\ \bigwedge_{\underline{i}=1}^{\underline{n}} \underline{x}_{\underline{i}} = \underline{y}_{\underline{i}} \supset \phi(\underline{y}_1, \ldots, \underline{y}_n)]$,

DEF. 35.2. $(\underline{\text{ext}}\ \underline{x}_1, \ldots, \underline{x}_n)\, \Psi(\underline{x}_1, \ldots, \underline{x}_n) \equiv_D (\forall \underline{x}_1, \underline{y}_1, \ldots, \underline{x}_n, \underline{y}_n)\, [\bigwedge_{\underline{i}=1}^{\underline{n}} \underline{x}_{\underline{i}} = \underline{y}_{\underline{i}} \supset \Psi(\underline{x}_1, \ldots, \underline{x}_n) = \Psi(\underline{y}_1, \ldots, \underline{y}_n)]$,

where $\underline{x}_1, \ldots, \underline{x}_n$ are $\underline{n}$ distinct variables and where for $\underline{i} = 1, \ldots, \underline{n}$ $\underline{y}_{\underline{i}}$ is the first variable distinct from $\underline{x}_1, \ldots, \underline{x}_n, \underline{y}_1, \ldots, \underline{y}_{\underline{i}-1}$ and not occurring free in $\phi(\underline{x}_1, \ldots, \underline{x}_n)$ or $\Psi(\underline{x}_1, \ldots, \underline{x}_n)$ respectively.

Hence the matrix $(\underline{\text{ext}}\ \underline{x})\, \Phi(\underline{x})$ is true if and only if $\Phi(\underline{x})$ is extensional with respect to $\underline{x}$—cf. Def. 6.11; in addition a syntactical analogue of this constitutes the thesis $(16)_2$ of the following theorem on extensional matrices and terms:

THEOR. 35.1. Let $\underline{x}, \underline{y}, \underline{x}_1, \underline{y}_1, \ldots, \underline{x}_n, \underline{y}_n, \underline{F}$, and $\underline{f}$ be distinct variables of the respective types $\underline{t}, \underline{t}, \underline{t}_1, \underline{t}_1, \ldots, \underline{t}_n, \underline{t}_n, (\underline{t}_1, \ldots, \underline{t}_n)$, and $(\underline{t}_1, \ldots, \underline{t}_n : \underline{t})$ [N2]. Furthermore let $\Phi(\underline{x})$ and

$\phi(\underline{x}_1, \ldots, \underline{x}_{\underline{n}})$ be matrices and let $\psi(\underline{x})$ be a term in ML^{ν}, all being without any occurrence of $\underline{y}, \underline{y}_1, \ldots, \underline{y}_{\underline{n}}$. Then:

$$(16) \quad \vdash \Phi(\underline{x}) \supset \Phi(\underline{x})^{(\underline{ex})}, \quad \vdash (\underline{\text{ext}}\ \underline{x})\,\Phi(\underline{x}) \equiv (\underline{x})\,[\Phi(\underline{x}) \equiv \Phi(\underline{x})^{(\underline{ex})}],$$

$$(17) \quad \vdash (\underline{\text{ext}}\ \underline{x})\,\Phi(\underline{x})^{(\text{ex})}, \quad \vdash \Phi(\underline{x})^{(\underline{ex})(\underline{ex})} \equiv \Phi(\underline{x})^{(\underline{ex})},$$

$$(18) \quad \vdash \phi(\underline{x}_1, \ldots, \underline{x}_{\underline{n}})^{(\underline{ex}_1)\ldots(\underline{ex}_{\underline{n}})} \equiv (\exists \underline{y}_1, \ldots, \underline{y}_{\underline{n}})\,[\textstyle\bigwedge_{\underline{i}=1}^{\underline{n}} \underline{x}_{\underline{i}} = \underline{y}_{\underline{i}}$$
$$\phi(\underline{y}_1, \ldots, \underline{y}_{\underline{n}})],$$
$$\vdash \underline{F}^{(\underline{e})}(\underline{x}_1, \ldots, \underline{x}_{\underline{n}}) \equiv \underline{F}(\underline{x}_1, \ldots, \underline{x}_{\underline{n}})^{(\underline{ex}_1)\ldots(\underline{ex}_{\underline{n}})},$$

$$(19) \quad \vdash \underline{p}^{(\underline{ex})(\underline{ey})} \equiv \underline{p}^{(\underline{ey})(\underline{ex})}, \quad \vdash (\underline{\text{ext}}\,\underline{x}_1, \ldots, \underline{x}_{\underline{n}})\,\underline{p} \equiv \textstyle\bigwedge_{\underline{i}=1}^{\underline{n}} (\underline{\text{ext}}\ \underline{x}_{\underline{i}})\,\underline{p},$$

$$(20) \quad \vdash (\underline{\text{ext}}\ \underline{x})\,\Phi(\underline{x}) \equiv (\forall \underline{x}, \underline{y})\,\{\underline{x} = \underline{y} \supset [\Phi(\underline{x}) \equiv \Phi(\underline{y})]\}, \qquad [\text{Def. 35.1}],$$
$$\vdash (\underline{\text{ext}}\ \underline{x})\,\psi(\underline{x}) \equiv (\forall \underline{x}, \underline{y})\,[\underline{x} = \underline{y} \supset \psi(\underline{x}) = \psi(\underline{y})] \qquad [\text{Def. 35.2}],$$

$$(21) \quad \vdash \underline{\text{Ext}}(\underline{F}) \equiv (\underline{\text{ext}}\ \underline{x}_1, \ldots, \underline{x}_{\underline{n}})\,\underline{F}(\underline{x}_1, \ldots, \underline{x}_{\underline{n}}),$$
$$\vdash \underline{\text{Ext}}(\underline{f}) \equiv (\underline{\text{ext}}\ \underline{x}_0, \ldots, \underline{x}_{\underline{n}})\,\underline{x}_0 = \underline{f}(\underline{x}_1, \ldots, \underline{x}_{\underline{n}}),$$

$$(22) \quad \vdash \underline{\text{Ext}}(\underline{F}) \equiv (\forall \underline{x}_1, \underline{y}_1, \ldots, \underline{x}_{\underline{n}}, \underline{y}_{\underline{n}})\,\{\textstyle\bigwedge_{\underline{i}=1}^{\underline{n}} \underline{x}_{\underline{i}} = \underline{y}_{\underline{i}} \supset [\underline{F}(\underline{x}_1, \ldots,$$
$$\underline{x}_{\underline{n}}) \equiv \underline{F}(\underline{y}_1, \ldots, \underline{y}_{\underline{n}})]\},$$

$$(23) \quad \vdash \text{Ext}(\text{f}) \equiv (\forall \underline{x}_1, y_1, \ldots, x_{\underline{n}}, y_{\underline{n}})\,[\textstyle\bigwedge_{\underline{i}=1}^{\underline{n}} \underline{x}_{\underline{i}} = \underline{y}_{\underline{i}} \supset \underline{f}(\underline{x}_1, \ldots,$$
$$\underline{x}_{\underline{n}}) = \underline{f}(\underline{y}_1, \ldots, \underline{y}_{\underline{n}})].$$

Proof. (I) Assume $\Phi(\underline{x})$. Then by AS34.1, $\underline{x} = \underline{x}\,\Phi(\underline{x})$. Hence by $(7)_3$ $(\exists \underline{y})\,[\underline{y} = \underline{x}\,\Phi(\underline{y})]$. Then by Def. 11.1, $\Phi(\underline{x})^{(\underline{ex})}$. So by the deduction theorem [Theor. 33.2] $(16)_1$ is proved.

(II) Now we first assume (a) $(\underline{\text{ext}}\ \underline{x})\,\Phi(\underline{x})$ and (b) $\Phi(\underline{x})^{(\underline{ex})}$. By Def. 11.1, from (b) we deduce $(\exists \underline{y})\,[\underline{y} = \underline{x}\,\Phi(y)]$, whence by rule $\underline{C}$, $\underline{y} = \underline{x}\,\Phi(\underline{x})$, so that by (a) and Def. 35.1 we have $\Phi(\underline{x})$. Hence by Theor. 33.3, (a) entails—using rule $\underline{C}$—the matrix $\Phi(\underline{x})^{(\underline{ex})} \supset \Phi(\underline{x})$; and $\underline{x}$ does not occur free in (a). Then by $(16)_1$ and rule $\underline{G}$ we have (c) $(\underline{x})\,[\Phi(\underline{x}) \equiv \Phi(\underline{x})^{(\underline{ex})}]$. Since $\underline{y}$ does not occur free in (c), by our theorem on rules $\underline{G}$ and $\underline{C}$ [Theor. 33.1]], (a) entails (c).

Now we want to prove that (c) entails (a). To this end we assume only (c) and (d) $\Phi(\underline{x})\underline{x} = \underline{y}$. The latter entails $(\exists \underline{x})\,[\Phi(\underline{x})\underline{x} = \underline{y}]$, whence by Def. 11.1, $\Phi(\underline{y})^{(\underline{ey})}$. Thence by (c) we deduce (e) $\Phi(\underline{y})$. Since (c), (d) $\vdash$ (e), then by Theor. 33.2 (c) entails $\Phi(\underline{x})\underline{x} = \underline{y} \supset \Phi\,(\underline{y})$. Then by rule $\underline{G}$ and Def. 35.1 we have $(\underline{\text{ext}}\ \underline{x})\ \Phi(\underline{x})$, i.e. (a).

Hence (a) and (c) entail one another. Then by Theor. 33.2, $(16)_2$ also holds.

(III) To prove $(17)_1$ let $\overline{\Phi}\,(\underline{x})$ be $\Phi(\underline{x})^{(\underline{ex})}$ and assume (a) $\overline{\Phi}(\underline{x})$ and (b) $\underline{x} = \underline{z}$, where $\underline{z}$ is a variable distinct from $\underline{x}$ and $\underline{y}$. By rule $\underline{C}$, from (a) and Def. 11.1 we deduce (c) $\underline{x} = \underline{y}\,\Phi\,(\underline{y})$. By AS34.2 and 3, (b) and (c) yield $\underline{z} = \underline{y}\,\Phi(\underline{y})$, whence by $(7)_3$ we deduce $(\exists \underline{y})\,[\underline{z} = \underline{y}\,\Phi\,(\underline{y})]$, which by Def. 11.1 is $\overline{\Phi}(\underline{z})$. By Theor. 33.3 we conclude $\vdash_{\underline{c}} \overline{\Phi}\,(\underline{x})\underline{x} = \underline{z} \supset \overline{\Phi}\,(\underline{z})$. Thence we obtain $(17)_1$ by rule $\underline{G}$, Theor. $3\overline{3}.1$, and Def. 35.1.

(IV) From $(16)_2$ and $(17)_1$ we deduce $(17)_2$. In addition $(18)_1$ easily follows from Def. 11.1, $(18)_2$ follows from $(18)_1$ and Def. 18.9, $(19)_1$ follows from $(18)_1$, and $(19)_2$ follows from Def. 35.1.

$(20)_{1,2}$ are easy consequences of Defs. 35.1 and 35.2 respectively.

$(21)_{1,2}$ follow at once from Defs. 6.11, 6.12, 35.1, 35.2; (22) follows from Def. 6.11 and (23) from Def. 6.12.

QED

Now we show [Theor. 35.2] how to reduce to matter of extensional logic (EC^{ν}) all theorems on identity in MC^{ν} concerning only matrices that are extensional [Def. 11.1] with respect to every variable explicitly mentioned (or with respect to certain of such variables).

THEOR. 35.2. $(\text{ext } \underline{z})\ \Phi(\underline{x},\underline{y},\underline{z}) \vdash \underline{x} = \underline{y} \supset [\Phi(\underline{x},\underline{y},\underline{x}) \equiv \Phi(\underline{x},\underline{y},\underline{y})]$ [Def. 35.1] where $\underline{x}, \underline{y}, \underline{z}$ are distinct variables.

Proof. Assume (a) $(\text{ext } \underline{z})\ \Phi(\underline{x},\underline{y},\underline{z})$. Thence by $(20)_1$ we deduce (b) $(\forall \underline{z}, \underline{u})\ \{\underline{u} = \underline{z} \supset [\Phi(\underline{x},\underline{y},\underline{z}) \equiv \Phi(\underline{x},\underline{y},\underline{u})]\}$, where $\underline{u}$ is a variable not occurring in $\underline{x} = \underline{y}\, \Phi(\underline{x},\underline{y},\underline{z})$. (b) yields $\underline{x} = \underline{y} \supset [\Phi(\underline{x},\underline{y},\underline{x}) \equiv \Phi(\underline{x},\underline{y},\underline{y})]$.

QED

The scheme $(\ldots)\ \underline{x} = \underline{x}$ which belongs to AS34.1 and the matrix on the R.H.S. (right-hand side) of the sign $\vdash$ in Theor. 35.2 are substantially Rosser's axioms 7A and 7B for identity in [32], pp. 163, 164. Hence under suitable hypotheses such as the premise $(\text{ext } \underline{z})\ \Phi(\underline{x},\underline{y},\underline{z})$ considered in Theor. 35.2, all theorems on identity proved in the extensional theory [32]—in particular the analogues of (12) to (15) for simple identity—can be proved in MC^ν in the same way as in [32].

By AS34.1–3, $\vdash (\text{ext } \underline{x}, \underline{y})\, \underline{x} = \underline{y}$. Hence one can prove the analogues $(24)_1$, $(24)_2$, (25), and (26) for simple identity of the theorems (12), (15), (13), and (14) respectively. They are justified by the corresponding theorems in [32]—i.e. Theors. VII.1.4, 5 and VII.2.1—and by Theor 35.2, Def. 34.1, and the deduction theorem, Theor. 33.2. We assume that $\underline{x}, \underline{y}, \underline{z}$ are distinct variables, that $\underline{y}$ does not occur in $\Phi(\underline{x})$, and that condition (c) in N30 holds. Then:

$$(24)\quad \vdash (\text{ext } \underline{z})\ \Phi(\underline{x},\underline{y},\underline{z}) \wedge \Phi(\underline{x},\underline{y},\underline{x}) \sim \Phi(\underline{x},\underline{y},\underline{y}) \supset \underline{x} \neq \underline{y},$$
$$\vdash (\underline{y})\ (\exists_1 \underline{x})\, \underline{x} = \underline{y},$$

$$(25)\quad (\text{ext } \underline{x})\ \Phi(\underline{x}) \vdash \Phi(\underline{y}) \equiv (\exists \underline{x})\ [\underline{x} = \underline{y}\, \Phi(\underline{x})],$$
$$(\text{ext } \underline{x})\ \Phi(\underline{x}) \vdash \Phi(\underline{y}) \equiv (\underline{x})\ [\underline{x} = \underline{y} \supset \Phi(\underline{x})],$$

$$(26)\quad (\text{ext } \underline{x})\, \Phi(\underline{x}) \vdash (\exists_1 \underline{x})\, \Phi(\underline{x}) \equiv (\exists \underline{x})\ \{\Phi(\underline{x})\, (\underline{y})\ [\Phi(\underline{y}) \supset \underline{y} = \underline{x}]\}$$
$$\equiv (\exists \underline{x})\ (\underline{y})\ [\underline{x} = \underline{y} \equiv \Phi(\underline{y})] \equiv (\exists \underline{x})\ \Phi(\underline{x}) \wedge (\exists^{(1)} \underline{x})\ \Phi(\underline{x})$$
$$\equiv (\exists \underline{x})\ \Phi(\underline{x}) \sim (\exists \underline{x}, \underline{y})\ [\Phi(\underline{x})\ \Phi(\underline{y})\, \underline{x} \neq \underline{y}] \qquad \text{[Def. 34.1]}$$

Note that the first equivalence in (26) is Def. 11.3 and that, by the second equivalence in (26), under the assumption $(\text{ext}\ \underline{x})\ \Phi(\underline{x})$ Def. 11.3 is equivalent to the simpler definition $(\exists_1 \underline{x})\ \Phi(\underline{x}) \equiv_D (\exists \underline{x})\ (\underline{y})\ [\Phi(\underline{y}) \equiv \underline{y} = \underline{x}]$ used in [32]. In N36 it will be shown [Theor. 36.2] that in MC^ν $(\exists_1 \underline{x})$ cannot be defined in the latter way—cf. also Memoir 1, fn. 37.

N36. Theorems on identity in MC^ν for matrices of any kind.

In the last two sections on identity in MC^ν we considered particular cases, i.e. that of strict identity and that of extensional matrices. Now we consider the general case. Some of the forthcoming theorems, e.g. Theor. 36.1, are the analogues of some theorems of extensional logic, while others—e.g. Theor. 36.2—show that the analogues for MC^ν of some theorems of extensional logic cannot be theorems in MC^ν.

THEOR. 36.1. $\vdash (\exists_1 \underline{x})\ \Phi(\underline{x}) \equiv (\exists \underline{x})\ \Phi(\underline{x})\ (\exists^{(1)} \underline{x})\ \Phi(\underline{x}) \equiv (\exists \underline{x})\ \Phi(\underline{x}) \sim (\exists \underline{x}, \underline{y})\ [\Phi(\underline{x})\ \Phi(\underline{y})\ \underline{x} \neq \underline{y}]$ [Defs. 11.3, 34.1] where condition (c) in N30 holds and where y does not occur in $\Phi(\underline{x})$ and is distinct from $\underline{x}$.

Proof. Assume (a) $(\exists_1 \underline{x})\ \Phi(\underline{x})$. Then by Def. 11.3 we have $(\exists \underline{z})\ \{\Phi(\underline{z})\ (\underline{y})\ [\Phi(\underline{y}) \supset \underline{y} = \underline{z}]\}$, where $\underline{z}$ and $\underline{y}$ are distinct variables not occurring in $\Phi(\underline{x})$. By rule $\underline{C}$ this yields (b) $\Phi(\underline{z})$ and (c) $(\underline{y})[\Phi(\underline{y}) \supset \underline{y} = \underline{z}]$. By $(7)_3$ (b) yields (d) $(\exists \underline{x})\ \Phi(\underline{x})$.

Now we add the assumptions (e) $\Phi(\underline{x})$ and (f) $\Phi(\underline{y})$. Then by (c) we have $\underline{x} = \underline{z}$ and $\underline{y} = \underline{z}$, whence by AS34.2 and 34.3, $\underline{x} = \underline{y}$. Hence by Theor. 33.3, (a) yields (using rule $\underline{C}$) $\Phi(\underline{x})\ \Phi(\underline{y}) \supset \underline{x} = \underline{y}$, whence by rule $\underline{G}$, (g) $(\forall \underline{x}, \underline{y})\ [\Phi(\underline{x})\ \Phi(\underline{y}) \supset \underline{x} = \underline{y}]$. Since (g) contains no free occurrences of $\underline{z}$, by Theor. 33.1 (a) yields (g) as well as (d).

Now, conversely, we assume (d) $(\exists \underline{x})\ \Phi(\underline{x})$ and (g) $(\forall \underline{x}, \underline{y})\ [\Phi(\underline{x})\ \Phi(\underline{y}) \supset \underline{x} = \underline{y}]$ instead of (a). Moreover we add the assumption

(h) $\Phi(\underline{y})$. By rule $\underline{C}$ with $\underline{x}$, (d) yields (d′) $\Phi(\underline{x})$; whence by (h) and by our assumption (g) we deduce (i) $\underline{y} = \underline{x}$. Hence by Theor. 33.3 we have (d), (g) $\vdash_{\underline{c}} \Phi(\underline{y}) \supset \underline{y} = \underline{x}$. Then, by rule $\underline{G}$, (d) and (g) yield, using rule $\underline{C}$, (1) $(\underline{y})\, [\Phi(\underline{y}) \supset \underline{y} = \underline{x}]$.

By $(7)_3$ (1) and (d′) yield (m) $(\exists \underline{x})\, \{\Phi(\underline{x})\, (\underline{y})\, [\Phi(\underline{y}) \supset \underline{y} = \underline{x}]\}$.

Since $\underline{x}$ does not occur free in (m), by Theor. 33.1 we get (d), (g) $\vdash$ (m). By Def. 11.3 the matrix (m) is $(\exists_1 \underline{x})\, \Phi(\underline{x})$, i.e. (a).

We conclude that $\vdash$ (a) $\equiv$ (d) $\wedge$ (g), and since, by Def. 34.1, (g) is $(\exists^{(1)} \underline{x})\, \Phi(\underline{x})$, the first equivalence of Theor. 36.1 holds. The second equivalence of Theor. 1 follows easily from Def. 34.1.

QED

From the forthcoming theorem it is apparent that in MC^{ν}, $(\exists_1 \underline{x})\, \Phi(\underline{x})$ cannot be defined to be $(\exists \underline{x})\, (\underline{y})\, [\Phi(\underline{y}) \equiv \underline{y} = \underline{x}]$ as is often (usefully) done in extensional logic—e.g. in [32].

THEOR. 36.2. $\vdash (\exists \underline{y})\, (\underline{x})\, [\Phi(\underline{x}) \equiv \underline{x} = \underline{y}] \equiv (\exists_1 \underline{x})\, \Phi(\underline{x})\, (\underline{\text{ext}}\ \underline{x})\, \Phi(\underline{x})$ where $\underline{y}$ does not occur free in $\Phi(\underline{x})\underline{x} = \underline{y}$.

Proof. Assume (a) $(\exists \underline{y})\, (\underline{x})\, [\Phi(\underline{x}) \equiv \underline{x} = \underline{y}]$, (b) $\Phi(\underline{z})$, and (c) $\underline{x} = \underline{z}$, where the variable $\underline{z}$ does not occur in $\Phi(\underline{x})\underline{x} = \underline{y}$ and where $\Phi(\underline{x})$ $\Phi(\underline{z})$ fulfill condition (c) in N30 on $\Phi(\underline{x})$ and $\Phi(\underline{y})$. By rule $\underline{C}$, from (a) we have (d) $(\underline{x})\, [\Phi(\underline{x}) \equiv \underline{x} = \underline{y}]$, which together with (b) yields $\underline{z} = \underline{y}$. Then (c) and ASs34.2 and 34.3 yield $\underline{x} = \underline{y}$, so that by (d) we have (e) $\Phi(\underline{x})$. Hence by Theor. 33.3 and rule $\underline{G}$ (a) yields $(\forall \underline{x}, \underline{z})\, [\Phi(\underline{z})\, \underline{z} = \underline{x} \supset \Phi(\underline{x})]$, which by Def. 35.1 is (f) $(\underline{\text{ext}}\ \underline{x})\, \Phi(\underline{x})$.

In addition (d) yields $(\underline{y})\, [\Phi(\underline{y}) \supset \underline{y} = \underline{x}]$, whence by (e) and $(8)_1$ we have $(\exists \underline{x})\, \{\Phi(\underline{x})\, (\underline{y})\, [\Phi(\underline{y}) \supset \underline{y} = \underline{x}]\}$. By Def. 11.3 this yields (g) $(\exists_1 \underline{x})\, \Phi(\underline{x})$. We conclude (using Theor. 33.1) that (a) $\vdash$ (f) $\wedge$ (g)—i.e. (a) yields (f) $\wedge$ (g).

Now conversely, we assume (f) and (g). Then by (26) we have (a).

By Theor. 33.2 we conclude that Theor. 36.2 holds.

QED

N37. Some theorems in MC^{ν} on $(\exists x)$, $(\exists_1 x)$, and $(\exists_1^{\cap} x)$ that have no extensional analogues.

THEOR. 37.1. The following syntactical analogues of the semantical theorems $(25)_{1,2}$, $(26)_2$, and $(27)_2$ in N14 hold in MC^{ν}:[17]

$$(27)\quad \vdash (\exists_1^{\cap}\underline{x})\underline{p} \supset (\exists_1\underline{x})\underline{p} \quad \text{[Defs. 11.3, 14.1]},$$

$$\vdash \Diamond (\exists_1^{\cap}\underline{x})\underline{Np} \supset \underline{N}(\exists_1^{\cap}\underline{x})\underline{Np},$$

$$(28)\quad \vdash (\exists_1\underline{x})\underline{p}(\exists\underline{x})\underline{Np} \supset (\exists_1\underline{x})\underline{Np},$$

$$\vdash \underline{N}(\exists_1\underline{x})\underline{p} \Diamond (\exists\underline{x})\underline{Np} \supset \underline{N}(\exists_1^{\cap}\underline{x})\underline{Np}.$$

Furthermore, assuming that the matrices $\Phi_{\underline{i}}(\underline{x}_{\underline{i}})$ and $\Phi_{\underline{i}}(\underline{y}_{\underline{i}})$ fulfill condition (c) in N30 (and using conventions (a) and (b) in N30), we have:

$$(29)\quad \vdash (\exists\underline{x}_1, \ldots, \underline{x}_{\underline{n}})[\underline{R}(\underline{x}_1, \ldots, \underline{x}_{\underline{n}}) \prod_{\underline{i}=1}^{\underline{n}} \Phi_{\underline{i}}(\underline{x}_{\underline{i}})] \prod_{\underline{i}=1}^{\underline{n}} [\Phi_{\underline{i}}(\underline{y}_{\underline{i}}) \,.\, (\exists_1^{\cap}\underline{x}_{\underline{i}})\Phi_{\underline{i}}(\underline{x}_{\underline{i}})] \supset \underline{R}(\underline{y}_1, \ldots, \underline{y}_{\underline{n}}).$$

Proof. (I) By AS30.9, $\vdash \underline{x} =^{\cap} \underline{y} \supset \underline{x} = \underline{y}$. Hence $(27)_1$ is an obvious consequence of $(14)_{1,2,3}$, $(11)_1$, and Theor. 36.1.

(II) Hereafter we denote $\underline{p}$ by $\Phi(\underline{x})$ and we assume that $\underline{y}$ is free for $\underline{x}$ in $\Phi(\underline{x})$. Note that by $(14)_{1,2}$, $(\exists_1^{\cap}\underline{x})\underline{Np}$ is (strictly equivalent to) $(\exists\underline{x})(\underline{y})[\underline{N}\Phi(\underline{y}) \equiv \underline{x} =^{\cap} \underline{y}]$, which is modally closed [Def. 4.3]. As a consequence AS30.8 implies $(27)_2$.

(III) To prove $(28)_1$ let us assume (a) $(\exists_1\underline{x})\Phi(\underline{x})$ and (b) $(\exists\underline{x})\underline{N}\Phi(\underline{x})$. By AS30.9 we have (c) $\vdash (\forall\underline{x}, \underline{y})[\underline{N}\Phi(\underline{x})\underline{N}\Phi(\underline{y}) \supset \Phi(\underline{x})\Phi(\underline{y})]$.[18] By Theor. 36.1 and Def. 34.1, (a) yields (d) $(\forall\underline{x}, \underline{y})[\Phi(\underline{x})\Phi(\underline{y}) \supset \underline{x} = \underline{y}]$. By Def. 34.1, from (c) and (d) we deduce (e)

$(\exists^{(1)}\underline{x})\underline{N}\Phi(\underline{x})$. By Theor. 36.1, (b) and (e) yield $(\exists_1\underline{x})\underline{N}\Phi(\underline{x})$. Hence by Theor. 33.2 we have $(28)_1$.

(IV) Since $(\exists\underline{x})\underline{Np}$ is modally closed, (a) $\vdash \Diamond(\exists\underline{x})\underline{Np} \equiv \underline{N}(\exists\underline{x})\underline{Np}$ follows from $(9')_3$. Now we assume (b) $\underline{N}(\exists_1\underline{x})\underline{p}$ and (c) $\Diamond(\exists\underline{x})\underline{Np}$. By (a), from (c) we deduce (d) $\underline{N}(\exists\underline{x})\underline{Np}$. By Theor. 36.1 (b) yields $\underline{N}(\exists^{(1)}\underline{x})\underline{p}$, whence by $(11)_2$, (e) $\underline{N}(\exists^{(1)\frown}\underline{x})\underline{Np}$. By $(14)_{1,2,3}$, from (d) and (e) we deduce $\underline{N}(\exists_1^{\frown}\underline{x})\underline{Np}$. Then by Theor. 33.2, $(28)_2$ holds.

(V) To prove the implication $\underline{p} \supset \underline{q}$ written in (29) we assume $\underline{p}$ and use rule $\underline{C}$ with $\underline{x}_1, \ldots, \underline{x}_n$. Then, using Def. 34.2 and AS34.4, we easily deduce $\underline{q}$. We conclude that (29) holds.

QED

N38. On descriptions in MC^ν.

Our axioms for descriptions are AS12.18(I), (II), i.e.—cf. Memoir 1, $(24)_2$—

AS38.1(I) $(\underline{N})(\exists_1\underline{x})\underline{p}\ \underline{p} \supset \underline{x} = (\iota\underline{x})\underline{p}$,

(II) $(\underline{N}) \sim (\exists_1\underline{x})\underline{p} \supset (\iota\underline{x})\underline{p} = \underline{a}^*$ [Defs. 11.2, 3].

Now we want the complete instantiation principle AS12.8:

AS38.2. $(\underline{N})(\underline{x})\Phi(\underline{x}) \supset \Phi(\Delta)$ where Δ is free for $\underline{x}$ in $\Phi(\underline{x})$,

its special case AS30.6 being insufficient.

Here we prove some formal theorems on descriptions concerning matrices of any kind. Furthermore we show, among other things, that the analogues for ML^ν of the theorems on descriptions proved e.g. in Rosser's extensional calculus [32] are theorems in MC^ν provided we use only extensional matrices. To this end we show that the analogues for MC^ν of Rosser's axioms for descriptions hold in connection with these matrices.

THEOR. 38.1. Under condition (c) considered in N30,

$$(30)\ \vdash (\underline{x})(\underline{p} \equiv \underline{q}) \supset (\iota \underline{x})\underline{p} = (\iota \underline{x})\underline{q}, \quad \vdash (\iota \underline{x})\Phi(\underline{x}) = (\iota \underline{y})\Phi(\underline{y}),$$
$$\vdash (\exists_1 \underline{x})\underline{p}\,(\exists \underline{x})\underline{Np} \supset (\iota \underline{x})\underline{p} = (\iota \underline{x})\underline{Np},$$

$$(31)\ \vdash (\exists_1 \underline{x})\underline{p} \supset (\underline{x})[\underline{x} = (\iota \underline{x})\underline{p} \equiv \underline{p}^{(\underline{ex})}], \quad \vdash (\underline{ext}\ \underline{x})\Phi(\underline{x})(\exists_1 \underline{x})$$
$$\Phi(\underline{x}) \supset \Phi[(\iota \underline{x})\Phi(\underline{x})],$$

$$(32)\ \vdash (\exists \underline{x})\underline{p} \equiv (\exists \underline{x})\underline{p}^{(\underline{ex})}, \quad \vdash (\exists_1 \underline{x})\underline{p} \equiv (\exists_1 \underline{x})\underline{p}^{(\underline{ex})},$$

$$(33)\ \vdash (\iota \underline{x})\underline{p} = (\iota \underline{x})\underline{p}^{(\underline{ex})}, \quad \vdash (\underline{x})[\underline{p}^{(\underline{ex})} \equiv \underline{q}^{(\underline{ex})}] \supset (\iota \underline{x})\underline{p} = (\iota \underline{x})\underline{q}.$$

Proof. (I) In order to prove $(30)_1$ we first assume (a) $(\underline{x})(\underline{p} \equiv \underline{q})$ and (b) $(\exists_1 \underline{x})\underline{p}$. Then by Def. 11.3 and Theor. 32.5 we easily obtain (b′) $(\exists_1 \underline{x})\underline{q}$. By Theor. 36.1, (b) yields $(\exists \underline{x})\underline{p}$, whence by rule $\underline{C}$ with $\underline{x}$, we get $\underline{p}$. Thence by (a) and AS30.6 we obtain $\underline{q}$.

By AS38.1(I), from $\underline{p}, \underline{q}$, (b), and (b′) we obtain both $\underline{x} = (\iota \underline{x})\underline{p}$ and $\underline{x} = (\iota \underline{x})\underline{q}$. Thence by ASs34.2, 34.3, and 38.2 we deduce the matrix (c) $(\iota \underline{x})\underline{p} = (\iota \underline{x})\underline{q}$, in which $\underline{x}$ does not occur free. Hence by Theor. 33.1, (c) is entailed by (a) and (b).

Now we assume (a) and (d) $\sim (\exists_1 \underline{x})\underline{p}$. Thence (d′) $\sim (\exists_1 \underline{x})\underline{q}$. By AS38.1(II), (d) and (d′) yield $(\iota \underline{x})\underline{p} = \underline{a}^*$ and $(\iota \underline{x})\underline{q} = \underline{a}^*$. Thence (c) holds again.

Since (d) is $\sim$ (b) [also (a), (b) $\vdash$ (c) and (a), (d) $\vdash$ c)], (a) yields (c), so that by Theor. 33.2, $(30)_1$ holds.

(II) With regard to $(30)_2$, by condition (c) in N30, $(9)_3$ holds, whence by Theor. 36.1 we easily obtain (a) $\vdash (\exists_1 \underline{x})\Phi(\underline{x}) \equiv (\exists_1 \underline{y})\Phi(\underline{y})$—cf. fn. 18.

Now we first assume (b) $(\exists_1 \underline{x})\Phi(\underline{x})$. Then by (a) we have (b′) $(\exists_1 \underline{y})\Phi(\underline{y})$. By rule $\underline{C}$, from (b) and (b′) we obtain $\Phi(\underline{x})$ and $\Phi(\underline{y})$. Thence by (b), (b′), and AS38.1(I) we deduce (c) $\underline{x} = (\iota \underline{x})\Phi(\underline{x})$ and (c′) $\underline{y} = (\iota \underline{y})\Phi(\underline{y})$.

By Theor. 36.1 and Def. 34.1, from (b) we deduce (d) $(\forall \underline{x}, \underline{y})$ $[\Phi(\underline{x})\Phi(\underline{y}) \supset \underline{x} = \underline{y}]$. So (d), $\Phi(\underline{x})$, and $\Phi(\underline{y})$ yield $\underline{x} = \underline{y}$. Thence by (c), (c′), and ASs34.2, 34.3 we have (e) $(\imath\underline{x})\Phi(\underline{x}) = (\imath\underline{y})\Phi(\underline{y})$.

Now we assume (f) $\sim(\exists_1\underline{x})\Phi(\underline{x})$ instead of (b). Thence by (a) we have (f′) $\sim(\exists_1\underline{y})\Phi(y)$. By AS38.1(II), (f), (f′) and ASs38.2, 34.2, 3, we easily deduce (e) again. So both (b) and (f) yield (e). Hence since (f) is $\sim$ (b), $(30)_2$ holds.

(III) To prove $(30)_3$ we take $\underline{p}$ to be $\Phi(\underline{x})$, as in the sequel; moreover we assume (a) $(\exists_1\underline{x})\Phi(\underline{x})$ and (b) $(\exists\underline{x})\underline{N}\Phi(\underline{x})$. Then by $(28)_1$ we obtain (c) $(\exists_1\underline{x})\underline{N}\Phi(\underline{x})$. By rule $\underline{C}$, (b) yields (d) $\underline{N}\Phi(\underline{x})$, whence (e) $\Phi(\underline{x})$.

By AS38.1(I), from (a) and (e) we deduce (f) $\underline{x} = (\imath\underline{x})\Phi(\underline{x})$, and from (c) and (d) we deduce (f′) $\underline{x} = (\imath\underline{x})\underline{N}\Phi(\underline{x})$. By ASs34.2, 34.3, and 38.2, (f) and (f′) yield the matrix (g) $(\imath\underline{x})\Phi(\underline{x}) = (\imath\underline{x})\underline{N}\Phi(\underline{x})$, where $\underline{x}$ does not occur free. Hence by Theor. 33.1 we have (a), (b) $\vdash$ (g), whence by Theor. 33.2 we obtain $(30)_3$.

(IV) With regard to $(31)_1$ let us first assume (a) $(\exists_1\underline{x})\Phi(\underline{x})$ and (b) $(\imath\underline{x})\Phi(\underline{x}) = \underline{x}$.

Introducing $\underline{y}$ by rule $\underline{C}$ through Def. 11.3, from (a) we obtain (c) $\Phi(\underline{y})$. By ASs38.1(I) and 30.6 we have $\vdash (\exists_1\underline{x})\Phi(\underline{x})\Phi(\underline{y}) \supset \underline{y} = (\imath\underline{x})\Phi(\underline{x})$. So by (a) and (c) we obtain (d) $\underline{y} = (\imath\underline{x})\Phi(\underline{x})$. By ASs34.3 and 38.2, (d) and (b) yield (e) $\underline{y} = \underline{x}$.

From (c) and (e) we deduce (f) $(\exists\underline{y})[\underline{y} = \underline{x}\,\Phi(\underline{y})]$, which by Def. 11.1 is (equivalent to) (g) $\Phi(\underline{x})^{(\underline{ex})}$, where $\underline{y}$ does not occur free. Hence by Theor. 33.1, (a) and (b) yield (g).

Now conversely, we assume (a) $(\exists_1\underline{x})\Phi(\underline{x})$ and (g) $\Phi(\underline{x})^{(\underline{ex})}$. Thence—by Def. 11.1—(f) holds, so that by rule $\underline{C}$ we obtain (h) $\underline{y} = \underline{x}\,\Phi(\underline{y})$. Then (c) $\Phi(\underline{y})$ holds again, and from (a) and (c) we can deduce (d) $\underline{y} = (\imath\underline{x})\Phi(\underline{x})$ as before. (d) and (h) yield (b) $(\imath\underline{x})$

$\Phi(\underline{x}) = \underline{x}$. Since $\underline{y}$ does not occur free in (b), by Theor. 33.1 we have (a), (g) $\vdash$ (b).

Since (a), (b) $\vdash$ (g) also holds, by Theor. 33.2 we easily obtain $(31)_1$.

(V) To prove $(31)_2$ we assume (a) $(\underline{\text{ext}}\ \underline{x})\Phi(\underline{x})$ and (b) $(\exists_1\underline{x})\Phi(\underline{x})$. By rule $\underline{C}$ with $\underline{y}$ through Def. 11.3, (b) yields (c) $\Phi(\underline{y})$. (b), (c), ASs38.1(I) and 30.6 yield (d) $\underline{y} = (\imath\underline{x})\Phi(\underline{x})$. By Def. 35.1 and AS38.2, from (a) we deduce (e) $\Phi(\underline{y})\underline{y} = (\imath\underline{x})\Phi(\underline{x}) \supset \Phi[(\imath\underline{x})\Phi(\underline{x})]$. (c), (d), and (e) yield $\Phi[(\imath\underline{x})\Phi(\underline{x})]$. Hence by Theor. 33.2, $(31)_2$ holds.

(VI) As to $(32)_1$ we first remark that the matrices $(16)_1$ and $(9)_1$ yield (a) $\vdash (\exists\underline{x})\Phi(\underline{x}) \supset (\exists\underline{x})\Phi(\underline{x})^{(\underline{ex})}$.

Now we assume (b) $(\exists\underline{x})\Phi(\underline{x})^{(\underline{ex})}$. Hence by Def. 11.1 and rule $\underline{C}$ with $\underline{x}$ we obtain (c) $(\exists\underline{y})[\underline{y} = \underline{x}\Phi(\underline{y})]$. Thence by $(9)_1$ and by $\vdash (\underline{y})[\underline{y} = \underline{x}\Phi(\underline{y}) \supset \Phi(\underline{y})]$, we have $(\exists\underline{y})\Phi(\underline{y})$, whence (d) $(\exists\underline{x})\Phi(\underline{x})$ by $(9)_3$. Then by Theor. 33.2, $\vdash$ (b) $\supset$ (d) so that, by (a), $(32)_1$ holds.

(VII) With regard to $(32)_2$ we denote $\Phi(\underline{x})^{(\underline{ex})}$ by $\Psi(\underline{x})$. We first assume (a) $(\exists_1\underline{x})\Phi(\underline{x})$ and (b) $\Psi(\underline{x})\Psi(\underline{y})$. By Def. 34.1 and Theor. 36.1, from (a) we deduce (c) $(\forall\underline{u},\underline{v})[\Phi(\underline{u})\Phi(\underline{v}) \supset \underline{u} = \underline{v}]$, where the distinct variables $\underline{u}$ and $\underline{v}$ do not occur in $\Phi(\underline{x})\underline{x} = \underline{y}$. By Def. 11.1, (b) yields, introducing $\underline{u}$ and $\underline{v}$ by rule $\underline{C}$, (d) $\Phi(\underline{u})\underline{u} = \underline{x}\Phi(\underline{v})\underline{v} = \underline{y}$. This yields $\Phi(\underline{u})\Phi(\underline{v})$, whence by (c) we deduce $\underline{u} = \underline{v}$. Thence by (d) and ASs34.2, 34.3 we obtain (e) $\underline{x} = \underline{y}$. By Theor. 33.3, from (a) we can deduce (b) $\supset$ (e), using rule $\underline{C}$. Hence by rule $\underline{G}$, (a) yields $(\forall\underline{x},\underline{y})[\Psi(\underline{x})\Psi(\underline{y}) \supset \underline{x} = \underline{y}]$, which by Def. 34.1 yields (f) $(\exists^{(1)}\underline{x})\Psi(\underline{x})$. By Theor. 36.1 and $(32)_1$, (a) yields (g) $(\exists\underline{x})\Psi(\underline{x})$. Again by Theor. 36.1, (f) and (g) yield (h) $(\exists_1\underline{x})\Psi(\underline{x})$ where $\underline{u}$ and $\underline{v}$ do not occur free. Hence by Theor. 33.1 we conclude that (a) $\vdash$ (h)—cf. fn. 18.

Now we assume, conversely, (h), i.e. $(\exists_1\underline{x})\Psi(\underline{x})$. Thence by Theor. 36.1 we deduce (i) $(\exists\underline{x})\Phi(\underline{x})^{(\underline{ex})}$ and (l) $(\exists^{(1)}\underline{x})\Phi(\underline{x})^{(\underline{ex})}$.

$(32)_1$ and (i) yield (m) $(\exists \underline{x}) \Phi(\underline{x})$; furthermore $(10)_1$, $(16)_1$, and (1) yield (n) $(\exists^{(1)} \underline{x}) \Phi(\underline{x})$. By Theor. 36.1, from (m) and (n) we deduce (a) $(\exists_1 \underline{x}) \Phi(\underline{x})$. Hence (h) $\vdash$ (a). Since (a) $\vdash$ (h) also holds, by Theor. 33.2 we have $(32)_2$.

(VIII) In order to prove $(33)_1$ we first assume (a) $(\exists_1 \underline{x}) \Phi(\underline{x})$. Thence by $(32)_2$, we deduce (a′) $(\exists_1 \underline{x}) \Psi(\underline{x})$, where $\Psi(\underline{x})$ is $\Phi(\underline{x})^{\underline{(ex)}}$.

By AS38.1(I), from (a) and (a′) we obtain (b) $(\underline{x}) [\Phi(\underline{x}) \supset \underline{x} = (\imath \underline{x}) \Phi(\underline{x})]$ and (b′) $(\underline{x})[\Psi(\underline{x}) \supset \underline{x} = (\imath \underline{x}) \Psi(\underline{x})]$ respectively.

By Theor. 36.1 and rule $\underline{C}$, from (a) we obtain (c) $\Phi(\underline{x})$, which by $(16)_1$ yields (d) $\Psi(\underline{x})$. From (b), (b′), (c), and (d) we deduce $\underline{x} = (\imath \underline{x}) \Phi(\underline{x})$ and $\underline{x} = (\imath \underline{x}) \Psi(\underline{x})$, whence (e) $(\imath \underline{x}) \Phi(\underline{x}) = (\imath \underline{x}) \Psi(\underline{x})$. By Theor. 33.1 we conclude that (a) $\vdash$ (e).

Now we assume (f) $\sim (\exists_1 \underline{x}) \Phi(\underline{x})$. Then by $(32)_2$ we have (f′) $\sim (\exists_1 \underline{x}) \Psi(\underline{x})$.

By AS38.1(II), from (f) and (f′) we deduce $(\imath \underline{x}) \Phi(\underline{x}) = \underline{a}^*$ and $(\imath \underline{x}) \Psi(\underline{x}) = \underline{a}^*$, whence (e) again. We conclude that (f) $\vdash$ (e), besides (a) $\vdash$ (e). Furthermore (f) is $\sim$ (a). Hence $(33)_1$ holds.

(IX) $(33)_2$ easily follows from $(30)_1$, $(33)_1$, and ASs34.2, 34.3.

QED

In Rosser's theory of extensional logic [32], four axiom schemes for descriptions are used: axiom schemes 8–11, pp. 184–85. The analogues for MC^ν of axiom schemes 8–10 are, in order, AS38.2, $(30)_1$, and $(30)_2$.

The analogue for ML^ν (but not for MC^ν) of axiom scheme 11 in [32] is the matrix $(\underline{N})$ $(\exists_1 \underline{x})\underline{p} \supset (\underline{x}) [(\imath \underline{x})\underline{p} = \underline{x} \equiv \underline{p}]$, which is true provided $\underline{p}$ is extensional with respect to $\underline{x}$. However it cannot be asserted in general. More precisely one can easily prove

(34) $\vdash (\exists_1 \underline{x})\underline{p} \supset \{(\underline{\text{ext}}\ \underline{x})\underline{p} \equiv (\underline{x}) [(\imath \underline{x})\underline{p} = \underline{x} \equiv \underline{p}]\}$ [Def. 35.1].

On the basis of some considerations in N11 we choose AS38.1(I) as the counterpart in MC^{ν} of axiom scheme 11 in [32]. AS38.1(II) enables us to derive the analogues of Rosser's axiom schemes 9 and 10 as theorems (a similar procedure is used by Rosser in connection with restricted variables, see [32], p. 192).

The theorems $(30)_1$, (31), (32), and (33) exhibit useful extensional properties of our iota operator.

Theorem $(31)_2$, which refers to extensional matrices, is the counterpart for MC^{ν} of Theor. VIII.2.2 in [32], p. 187 (in extensional logic). Here are theorems in MC^{ν} that can be regarded as counterpart of $(31)_2$ for matrices of any kind.

THEOR. 38.2. $\vdash (\exists_1 \underline{x})\Phi(\underline{x}) \supset \Psi[(\iota\underline{x})\Phi(\underline{x})]$ where $\Psi(\underline{x})$ is $\Phi(\underline{x})^{(ex)}$.

Proof. Let us assume (a) $(\exists_1 \underline{x})\Phi(\underline{x})$. Then by AS38.1(I) we have (b) $(\underline{x})[\Phi(\underline{x}) \supset \underline{x} = (\iota\underline{x})\Phi(\underline{x})]$. Furthermore by Theor. 36.1, (a) yields $(\exists\underline{x})\Phi(\underline{x})$, whence by rule $\underline{C}$, we obtain (c) $\Phi(\underline{x})$. Thence by (b) we deduce $\Phi(\underline{x})\underline{x} = (\iota\underline{x})\Phi(\underline{x})$, which by $(7)_3$ yields (d) $(\exists\underline{y})[\Phi(\underline{y})\underline{y} = (\iota\underline{x})\Phi(\underline{x})]$, where $\underline{y}$ does not occur in $\Phi(\underline{x})$.

By Def. 11.1, (d) is $\Psi[(\iota\underline{x})\Phi(\underline{x})]$ and $\underline{x}$ does not occur free in it. Hence by Theors. 33.1. 2 we obtain our theorem.

QED

THEOR. 38.3. Let $\Psi(\underline{x})$ be $(\exists_1\underline{x})\Phi(\underline{x})\Phi(\underline{x})^{(ex)} \vee \sim(\exists_1\underline{x})\Phi(\underline{x})\underline{x} = \underline{a}^*$.

Then:[19]

$$\vdash (\exists\underline{x})[\Phi(\underline{x})\underline{N}\Psi(\underline{x})] \equiv \Phi[(\iota\underline{x})\Phi(\underline{x})],$$

$$\vdash (\exists\underline{x})\underline{N}\Phi(\underline{x})\underline{N}(\exists^{(1)}\underline{x})\Phi(\underline{x}) \supset \Phi[(\iota\underline{x})\Phi(\underline{x})],$$

$$\vdash (\underline{x})[\Diamond\Phi(\underline{x}) \equiv \underline{N}\Phi(\underline{x})]\underline{N}(\exists_1\underline{x})\Phi(\underline{x}) \supset \Phi[(\iota\underline{x})\Phi(\underline{x})],$$

$$\vdash \underline{N}(\exists_1\underline{x})\underline{N}\Phi(\underline{x}) \supset \underline{N}\Phi[(\iota\underline{x})\underline{N}\Phi(\underline{x})], \tag{35}$$

$$\vdash (\exists_1^{\frown}\underline{x})\underline{N}\Phi(\underline{x}) \supset \underline{N}\Phi[(\iota\underline{x})\Phi(\underline{x})],$$

$$\vdash (\underline{x})[\Diamond\Phi(\underline{x}) \equiv \underline{N}\Phi(\underline{x})](\exists_1^{\frown}\underline{x})\Phi(\underline{x}) \supset \Phi[(\iota\underline{x})\Phi(\underline{x})].$$

In spite of its general character, $(35)_1$ is not very useful in practice, whereas $(35)_2$ and its "corollaries" $(35)_{3,4,6}$ are.

<u>Proof of Theor. 38.3.</u> (I) With regard to $(35)_1$, from $(32)_2$ we easily obtain (a) $\vdash \underline{N}(\exists_1\underline{x})\Psi(\underline{x})$ and, by AS38.1, also (b) $\vdash \underline{N}\Psi[(\iota\underline{x})\Phi(\underline{x})]$. By $(7)_3$ and AS38.2, (b) yields $\vdash (\exists\underline{x})\underline{N}\Psi(\underline{x})$, so that by (a) and $(28)_2$ we have (c) $\vdash (\exists_1^{\frown}\underline{x})\underline{N}\Psi(\underline{x})$.

Now we assume (d) $(\exists\underline{x})[\Phi(\underline{x})\underline{N}\Psi(\underline{x})]$, whence by rule $\underline{C}$, we obtain (e) $\Phi(\underline{x})$ and (f) $\underline{N}\Psi(\underline{x})$. By (c) and Def. 14.1, from (b) and (f) we deduce (g) $(\iota\underline{x})\Phi(\underline{x}) = {}^{\frown}\underline{x}$. By AS34.4, (g) and (e) yield (h) $\Phi[(\iota\underline{x})\Phi(\underline{x})]$, where $\underline{x}$ does not occur free. Hence by Theor. 33.1 we obtain (d) $\vdash$ (h).

Now we start with (h) $\Phi[(\iota\underline{x})\Phi(\underline{x})]$, whence by (b) we deduce (i) $\Phi[(\iota\underline{x})\Phi(\underline{x})]\underline{N}\Psi[(\iota\underline{x})\underline{N}\Phi(\underline{x})]$. By $(7)_3$ and AS38.2, (b) yields (d). So besides (d) $\vdash$ (h), we have (h) $\vdash$ (d). Then by Theor. 33.2, $(35)_1$ follows.

(II) To prove $(35)_2$ we assume (a) $(\exists\underline{x})\underline{N}\Phi(\underline{x})$ and (b) $\underline{N}(\exists^{(1)}\underline{x})\Phi(\underline{x})$. Then by rule $\underline{C}$ with $\underline{x}$, (a) yields (c) $\underline{N}\Phi(\underline{x})$ and $\underline{N}(\exists^{(1)}\underline{x})\Phi(\underline{x})$. So by $(16)_1$ (d) $\underline{N}\Psi(\underline{x})$ holds. (c) also yields (e) $\Phi(\underline{x})$. By $(7)_3$, from (d) and (e) we deduce (f) $(\exists\underline{x})[\Phi(\underline{x})\underline{N}\Psi(\underline{x})]$, which by $(35)_1$ yields $\Phi[(\iota\underline{x})\Phi(\underline{x})]$. By Theor. 33.2 we conclude that $(35)_2$ holds.

(III) We assume (a) $(\underline{x})[\diamond\Phi(\underline{x}) \equiv \underline{N}\Phi(\underline{x})]$ and (b) $\underline{N}(\exists_1\underline{x})\Phi(\underline{x})$. Thence we easily deduce $(\exists_1\underline{x})\underline{N}\Phi(\underline{x})$, which by Theor. 36.1 yields $(\exists\underline{x})\underline{N}\Phi(\underline{x})$. By $(35)_2$ and $(14)_{1,2,3}$ this and (b) yield $\Phi[(\iota\underline{x})\Phi(\underline{x})]$. We conclude that $(35)_3$ holds. Thence $(35)_4$ follows immediately.

(IV) By $(8')_1$, $(27)_2$, and $(27)_1$, $\vdash (\exists_1^{\frown}\underline{x})\underline{N}\Phi(\underline{x}) \supset \underline{N}(\exists_1\underline{x})\underline{N}\Phi(\underline{x})$. Hence $(35)_4$ yields $(35)_5$. Thence we immediately deduce $(35)_6$.

QED

N39. Some further theorems based on our axioms for descriptions.

Here we consider the analogues for MC^{ν} of some theorems in extensional logic, including the one concerning the definition by cases [Theor. 39.3] and a formal theorem concerning $(\exists_1 \underline{x})$, which apparently is not provable until ASs38.1, 2 are laid down.

We first state the immediate consequence of ASs34.4 and 38.2.

THEOR. 39.1. Let $\underline{x}, \underline{y}, \underline{z}$ be distinct variables of the same type and let Δ be free for $\underline{x}$ and $\underline{z}$ and Δ' free for $\underline{y}$ and $\underline{z}$, in $\Phi(\underline{x}, \underline{y}, \underline{z})$. Then $\vdash \Delta =^{\frown} \Delta' \supset [\Phi(\Delta, \Delta', \Delta) \equiv \Phi(\Delta, \Delta', \Delta')]$.

THEOR. 39.2. Let the variable $\underline{x}$ not occur free in Δ and let $\Psi(\Delta, \Delta)$ result from $\Psi(\underline{x}, \Delta)$ by substituting the term Δ for (the free occurrences) of $\underline{x}$. Furthermore let Δ be free for $\underline{x}$ in $\Psi(\underline{x}, \Delta)$. Then:

(36) $\vdash (\imath \underline{x})(\underline{x} = \underline{y}) = y$, $\vdash (\imath \underline{x})(\underline{x} =^{\frown} \underline{y}) =^{\frown} \underline{y}$—cf. Theor. VIII.2.3 in [32], p. 187.

(37) $\vdash \Psi(\Delta, \Delta) \supset (\exists \underline{x}) \Psi(\underline{x}, \Delta)$, $\vdash (\exists \underline{x}) \underline{x} =^{\frown} (\imath \underline{x}) \underline{p}$—cf. Theor. VIII.2.7 in [32].

Proof. (I) $\vdash (\text{ext } \underline{x}) \underline{x} = \underline{y}$ [Def. 35.1] obviously holds. Hence $(24)_2$ and $(31)_2$ yield $(36)_1$.

(II) By (15) we have $\vdash (\exists_1^{\frown} \underline{x}) \underline{x} =^{\frown} \underline{y}$, which by $(35)_5$ and AS30.9 yields $(36)_2$.

(III) By AS38.2, $\vdash (\underline{x}) \sim \Psi(\underline{x}, \Delta) \supset \sim \Psi(\Delta, \Delta)$, which obviously yields $(37)_1$.

(IV) ASs34.1 and 38.2 yield $\vdash (\imath \underline{x}) \underline{p} =^{\frown} (\imath \underline{x}) \underline{p}$, whence by $(37)_1$ we deduce $(37)_2$.

QED

THEOR. 39.3. Let the variable $\underline{y}$ have no free occurrences in the matrices $\underline{p}_1, \ldots, \underline{p}_{\underline{n}}$ or the terms $\Delta_1, \ldots, \Delta_{\underline{n}}$ ($\underline{n} > 1$). Then, assuming

$$(38)\quad \underline{q} \equiv_D \bigwedge_{\underline{i}=2}^{\underline{n}} \bigwedge_{\underline{j}=1}^{\underline{i}-1} \sim (\underline{p}_{\underline{i}} \underline{p}_{\underline{j}}),\ \underline{r} \equiv_D \bigvee_{\underline{i}=1}^{\underline{n}} \underline{p}_{\underline{i}},\ \Psi(\underline{y}) \equiv_D \bigvee_{\underline{i}=1}^{\underline{n}} (\underline{y} = \Delta_{\underline{i}} \underline{p}_{\underline{i}}) \qquad \text{[Defs. 6.9, 6.10]},$$

we have

$$(39)\quad \vdash \underline{qr} \supset (\exists_1 \underline{y}) \Psi(\underline{y}),\quad \vdash \underline{qp}_{\underline{i}} \supset (\imath \underline{y}) \Psi(\underline{y}) = \Delta_{\underline{i}} \quad (\underline{i} = 1, \ldots, \underline{n}).$$

This theorem is the analogue for MC^{ν} of Theor. VIII.2.9 in [32], p. 190, and since $\Psi(\underline{y})$ is obviously extensional with respect to $\underline{y}$ [$\vdash (\underline{\text{ext}}\ \underline{y}) \Psi(\underline{y})$], our theorem can be proved in the same way as the latter.

Here we prove some properties of extensional matrices in connection with $(\exists_1^{\frown} \underline{x})$, $(\exists_1 \underline{x})$, and $(\imath \underline{x})$. The use of our axioms on descriptions is essential throughout the whole proof.

THEOR. 39.4. (I) $\vdash \underline{N}(\underline{\text{ext}}\ \underline{x})\underline{p} \supset [(\exists_1^{\frown} \underline{x}) \underline{Np} \equiv \underline{N}(\exists_1 \underline{x})\underline{p}]$,

(II) $\vdash \underline{N}(\underline{\text{ext}}\ \underline{x})\underline{p}\,\underline{N}(\exists_1 \underline{x})\underline{p} \supset (\imath \underline{x})\underline{p} =^{\frown} (\imath \underline{x}) \underline{Np}$.

<u>Proof</u>. Let $\underline{p}$ be $\Phi(\underline{x})$ and let $\underline{y}$ not occur in $\Phi(\underline{x})$.

(I) Let us start with (a) $\underline{N}(\underline{\text{ext}}\ \underline{x})\underline{p}$, (b) $(\exists_1^{\frown} \underline{x}) \underline{N} \Phi(\underline{x})$, and, as a hypothesis for reductio ad absurdum, (c) $\sim \underline{N}(\exists_1 \underline{x}) \Phi(\underline{x})$. By $(14)_{1,2,3}$, (b) yields (d) $(\exists \underline{x}) \underline{N} \Phi(\underline{x})$, which by rule $\underline{C}$, yields (d′) $\underline{N} \Phi(\underline{z})$, where $\underline{z}$ is a variable not occurring in $\Phi(\underline{x})\underline{x} = \underline{y}$. From (d) and $(7')_2$ we deduce (d″) $\underline{N}(\exists \underline{x}) \Phi(\underline{x})$.

By Theor. 36.1, from (c) and (d) we obtain (e) $\sim \underline{N}(\exists^{(1)} \underline{x}) \Phi(\underline{x})$, whence by Def. 34.1, $(\exists \underline{x}_1, \underline{x}_2) \diamond [\Phi(\underline{x}_1) \Phi(\underline{x}_2) \underline{x}_1 \neq \underline{x}_2]$. Thence by rule $\underline{C}$ with $\underline{x}_1$ and $\underline{x}_2$ we obtain (e′) $\Phi(\underline{x}_1) \Phi(\underline{x}_2) \underline{x}_1 \neq \underline{x}_2$. We now set—see (4) in N4—

(40) $\Phi_{\underline{i}}(\underline{y}) \equiv_D \sim(\exists_1 \underline{x})\Phi(\underline{x})\Phi(\underline{x}_{\underline{i}})\underline{y} = \underline{x}_{\underline{i}} \vee [(\exists_1 \underline{x})\Phi(\underline{x}) \vee \sim \Phi(\underline{x}_{\underline{i}})]\underline{y} = \underline{z}$

$(\underline{i} = 1, 2)$.

Then by Theor. 39.3, in particular $(39)_2$, we easily obtain

(41) $\vdash \sim(\exists_1 \underline{x})\Phi(\underline{x})\Phi(\underline{x}_{\underline{i}}) \supset^{\frown} (\iota \underline{y})\Phi_{\underline{i}}(\underline{y}) = \underline{x}_{\underline{i}}$,

$\vdash (\exists_1 \underline{x})\Phi(\underline{x}) \vee \sim \Phi(\underline{x}_{\underline{i}}) \supset^{\frown} (\iota \underline{y})\Phi_{\underline{i}}(\underline{y}) = \underline{z}$ $\qquad (\underline{i} = 1, 2)$.

By Theor. 36.1 we have $\vdash (e') \supset^{\frown} \sim(\exists_1 \underline{x})\Phi(\underline{x})$. Hence (e′) yields $\diamond [\sim(\exists_1 \underline{x})\Phi(\underline{x})\Phi(\underline{x}_1)\Phi(\underline{x}_2)\underline{x}_1 \neq \underline{x}_2]$, whence by $(41)_1$ we have $\diamond [(\iota \underline{y})\Phi_1(\underline{y}) = \underline{x}_1 \wedge (\iota \underline{y})\Phi_2(\underline{y}) = \underline{x}_2 \wedge \underline{x}_1 \neq \underline{x}_2]$. This yields (f) $\diamond (\iota \underline{y})\Phi_1(\underline{y}) \neq (\iota \underline{y})\Phi_2(\underline{y})$. On the other hand, by Def. 35.1 and AS38.2, from (a) we deduce (g) $\Phi(\underline{x}_{\underline{i}})\ (\iota \underline{y})\Phi_{\underline{i}}(\underline{y}) = \underline{x}_{\underline{i}} \supset^{\frown} \Phi[(\iota \underline{y})\Phi_{\underline{i}}(\underline{y})]$ $(\underline{i} = 1, 2)$. By $(41)_1$, (g) yields (h) $\sim\overline{(\exists_1 \underline{x})\Phi(\underline{x})\Phi(\underline{x}_{\underline{i}})} \supset^{\frown} \Phi[(\iota \underline{y})\Phi_{\underline{i}}(\underline{y})]$ $(\underline{i} = 1, 2)$. Analogously by $(41)_2$ and Def. 35.1, from (a) and the consequence $\Phi(\underline{z})$ of (d′) we deduce (i) $(\exists_1 \underline{x})\Phi(\underline{x}) \vee \sim \Phi(\underline{x}_{\underline{i}}) \supset^{\frown} \Phi[(\iota \underline{y})\Phi(\underline{y})]$ $(\underline{i} = 1, 2)$. Since obviously $\vdash \underline{N}[\sim(\exists_1 \underline{x})\Phi(\underline{x})\overline{\Phi}(\underline{x}_{\underline{i}}) \vee (\exists_1 \underline{x})\Phi(\underline{x}) \vee \sim \Phi(\underline{x}_{\underline{i}})]$, from (h) and (i) we deduce (l) $\underline{N}\Phi[(\iota \underline{y})\Phi(\underline{y})]$ $(\underline{i} = 1, 2)$.

Hence (a), (b), (c) $\vdash_{\underline{c}}$ (l). Moreover (l) is modally closed and has no free occurrences of $\underline{x}_1$, $\underline{x}_2$, or $\underline{z}$. Then by Theor. 33.1 we have (a), (b), (c) $\vdash$ (l).

The formal theorem $(27)_2$ and (b) yield $\underline{N}(\exists_1^{\frown} \underline{x})\underline{N}\Phi(\underline{x})$, so that by Def. 34.2 and AS38.2, from (l) we deduce $(\iota \underline{y})\Phi_1(\underline{y}) =^{\frown} (\iota \underline{y})\Phi_2(\underline{y})$, which contradicts (f). We conclude that (c) is not compatible with (a) and (b). Hence (a) and (b) yield (m) $\underline{N}(\exists_1 \underline{x})\Phi(\underline{x})$.

Now conversely, we start with (a) $\underline{N}(\underline{\text{ext}}\ \underline{x})\underline{p}$ and (m). Thence by $(31)_2$ we deduce $\underline{N}\Phi[(\iota \underline{y})\Phi(\underline{y})]$, whence by $(37)_1$, (n) $(\exists \underline{x})\underline{N}\Phi(\underline{x})$. By $(28)_2$, from (m) and (n) we deduce $\underline{N}(\exists_1^{\frown} \underline{x})\underline{N}\Phi(\underline{x})$, whence (b) $(\exists_1 \underline{x})\underline{N}\Phi(\underline{x})$.

We conclude that (a), (m) $\vdash$ (b). Since we have also proved that (a), (b) $\vdash$ (m), by three uses of Theor. 33.2 we deduce part (I) of Theor. 39.4.

(II) We first assume (a) $\underline{N}(\underline{ext}\ \underline{x})\underline{p}$ and (m) $\underline{N}(\exists_1 \underline{x})\Phi(\underline{x})$. By part (I), (a) and (m) yield $(\exists_1^{\cap} \underline{x})\underline{N}\Phi(\underline{x})$. Thence by $(8')_1$ and $(27)_1$ we deduce $\underline{N}(\exists_1 \underline{x})\underline{N}\Phi(\underline{x})$. So by AS38.1(I) and 30.7 we easily deduce (p) $(\underline{x})[\underline{N}\Phi(\underline{x}) \supset \underline{x} =^{\cap} (\iota \underline{x})\underline{N}\Phi(\underline{x})]$.

On the other hand, as before, by $(31)_2$ from (a) and (m) we deduce $\underline{N}\Phi[(\iota \underline{x})\Phi(\underline{x})]$. Thence by (p) and AS38.2 we obtain $(\iota \underline{x})\Phi(\underline{x}) =^{\cap} (\iota \underline{x})\underline{N}\Phi(\underline{x})$. Since $\underline{p}$ is $\Phi(\underline{x})$, by Theor. 33.2 we conclude that part (II) of Theor. 39.4 holds.

QED

The following theorem concerns matrices of any kind.

THEOR. 39.5(I) $\vdash \underline{N}(\exists_1 \underline{x})\underline{p} \equiv (\exists_1^{\cap} \underline{x})\underline{N}\underline{p}^{(ex)}$,

(II) $\vdash \underline{N}(\exists_1 \underline{x})\underline{p} \supset (\iota \underline{x})\underline{p} =^{\cap} (\iota \underline{x})\underline{N}\underline{p}^{(ex)}$.

Proof. By $(17)_1$ and Theor. 39.4(I) we obtain $\vdash \underline{N}(\exists_1 \underline{x})\underline{p}^{(ex)} \equiv (\exists_1^{\cap} \underline{x})\underline{N}\underline{p}^{(ex)}$. Thence by $(32)_2$ we obtain part (I) of Theor. 39.5.

Now we assume (a) $\underline{N}(\exists_1 \underline{x})\underline{p}$, which by $(32)_2$ yields (b) $\underline{N}(\exists_1 \underline{x})\underline{p}^{(ex)}$. By Theor. 39.4(II), from (b) and $(17)_1$ we deduce $(\iota \underline{x})\underline{p}^{(ex)} =^{\cap} (\iota \underline{x})\underline{N}\underline{p}^{(ex)}$. Thence by $(33)_1$ we obtain $(\iota \underline{x})\underline{p} =^{\cap} (\iota \underline{x})\underline{N}\underline{p}^{(ex)}$. Hence by Theor. 33.2 part (II) of Theor. 39.5 also holds.

QED

8

On Attributes, Functions, and Natural Numbers

N40. Axioms for attributes and functions. First theorems.

The axioms for attributes and functions laid down in N12—where all our axiom schemes were proved to be L-true—are the intensionality principles ASs12.14, 12.15, the existence principles ASs12.16, 12.17, the new axiom As12.19, and Zermelo's principle AS12.20.

In Memoir 1, fn. 42, we explain why we prefer not to reduce many parts of ASs12.14, 12.15 to a definition of identity, but rather to write them in full as follows:

AS40.1. $(\underline{N})(\forall \underline{x}_1, \ldots, \underline{x}_n)[\underline{F}(\underline{x}_1, \ldots, \underline{x}_n) \equiv \underline{G}(\underline{x}_1, \ldots, \underline{x}_n)] \equiv \underline{F} = \underline{G}$,

AS40.2. $(\underline{N})(\forall \underline{x}_1, \ldots, \underline{x}_n)[\underline{f}(\underline{x}_1, \ldots, \underline{x}_n) = \underline{g}(\underline{x}_1, \ldots, \underline{x}_n)] \equiv \underline{f} = \underline{g}$,

where $\underline{x}_1, \ldots, \underline{x}_n$, $\underline{F}$, $\underline{G}$, $\underline{f}$, and $\underline{g}$ are distinct variables (whose types are, of course, compatible with the condition that ASs40.1, 40.2 should be well formed).

The existence principles are

AS40.3. $(\underline{N})(\exists \underline{F})(\forall \underline{x}_1, \ldots, \underline{x}_n)[\underline{F}(\underline{x}_1, \ldots, \underline{x}_n) \equiv \underline{p}]$,

AS40.4. $(\underline{N})(\exists \underline{f})(\forall \underline{x}_1, \ldots, \underline{x}_n)[\underline{f}(\underline{x}_1, \ldots, \underline{x}_n) = \Delta]$,

where $\underline{x}_1, \ldots, \underline{x}_n$, $\underline{F}$, and $\underline{f}$ are distinct variables, and where $\underline{F}$ and $\underline{f}$ do not occur free in $\underline{p}$ and Δ respectively.

As we hinted in N12, the strong axiom AS12.19 enables us to characterize our Γ-cases within MC^{ν} by a certain absolute concept El. In Memoir 3 it is shown that this in turn allows us, for example, (1) to prove a certain theorem of relative completeness for MC^{ν}, (2) to define within MC^{ν} an "intensional" descriptor, $\iota_{\underline{u}}$—in contrast with our descriptor $\imath$, which by $(30)_1$ and $(31)_1$ has a rather extensional character—where the parameter $\underline{u}$ on which $\iota_{\underline{u}}$ depends is meant to range on $\underline{El}$, and (3) to define an intensional descriptor, ι_{ρ} (independent of any parameter), for the real case $\gamma_{\underline{\mathfrak{R}}}$ [N11] after having introduced a suitable constant ρ governed by the axiom $\rho \in \underline{El}$.

Since a large part of our theory on attributes and functions can be constructed without using AS12.19 and the Zermelo principle AS12.20, here we will dispense with these axioms.

THEOR. 40.1. Let $\underline{x}_1, \ldots, \underline{x}_n$, $\underline{F}$ and $\underline{f}$ be distinct variables by which ASs40.1, 2 are well formed, and let $\underline{F}$ and $\underline{f}$ not occur free in the matrix $\underline{p}$ and the term Δ respectively. Then

(42) $\vdash (\exists_1 \underline{F}) \Phi(\underline{F})$ where

$$\Phi(\underline{F}) \equiv_D (\forall \underline{x}_1, \ldots, \underline{x}_n) [\underline{F}(\underline{x}_1, \ldots, \underline{x}_n) \equiv \underline{p}],$$

(43) $\vdash (\exists_1 \underline{f}) \Psi(\underline{f})$ where

$$\Psi(\underline{f}) \equiv_D (\forall \underline{x}_1, \ldots, \underline{x}_n) [\underline{f}(\underline{x}_1, \ldots, \underline{x}_n) = \Delta],$$

(44) $\vdash (\underline{\text{ext}}\ \underline{F}) \Phi(\underline{F})$ where $(42)_2$ holds,

$\vdash (\underline{\text{ext}}\ \underline{f}) \Psi(\underline{f})$ where $(43)_2$ holds,

(45) $\vdash (\lambda \underline{x}_1, \ldots, \underline{x}_n)(\underline{p})(\underline{x}_1, \ldots, \underline{x}_n) \equiv \underline{p}$, [Def. 4.4]

$\vdash (\lambda \underline{x}_1, \ldots, \underline{x}_n)(\Delta)(\underline{x}_1, \ldots, \underline{x}_n) = \Delta$ [Def. 4.5]

(46) $\vdash (\exists \underline{F})(\forall \underline{x}_1, \ldots, \underline{x}_n)[\underline{F}(\underline{x}_1, \ldots, \underline{x}_n) \equiv {}^\frown \underline{p}]$,

$\vdash (\exists \underline{f})(\forall \underline{x}_1, \ldots, \underline{x}_n)[\underline{f}(\underline{x}_1, \ldots, \underline{x}_n) = {}^\frown \Delta]$.

Note that the theorems $(46)_1$ and $(46)_2$ strengthen ASs40.3, 40.4 respectively; and (45) assures that the basic properties of Church's lambda operator hold in MC^{ν}—cf. Memoir 1 (35).

Proof of Theor. 40.1. (I) As to (42) and (43), let $\underline{G}$ and $\underline{g}$ be variables having the same type as $\underline{F}$ and $\underline{f}$ respectively and not occurring in $\Phi(\underline{F})$ and $\Psi(\underline{f})$ respectively.

To prove $(42)_1$ we start with (a) $\Phi(\underline{F})\Phi(\underline{G})$. Then by $(42)_2$ we obtain $(\forall \underline{x}_1, \ldots, \underline{x}_n)[\underline{F}(\underline{x}_1, \ldots, \underline{x}_n) \equiv \underline{G}(\underline{x}_1, \ldots, \underline{x}_n)]$, which by (the first part of) AS40.1 yields (b) $\underline{F} = \underline{G}$. Then by Theor. 33.2 we have $\vdash$ (a) $\supset$ (b). Thence by Theor. 32.2 we obtain $\vdash (\forall \underline{F}, \underline{G})$ $[\Phi(\underline{F})\Phi(\underline{G}) \supset \underline{F} = \underline{G}]$, which by Def. 34.1 is (equivalent to) (c) $\vdash (\exists^{(1)}\underline{F})\Phi(\underline{F})$. By Theor. 36.1 and $(42)_2$, from (c) and AS40.3 we deduce $(42)_1$.

To prove (43) we start with (a′) $\Psi(\underline{f})\Psi(\underline{g})$. Thence by $(43)_2$ we deduce (b) $(\forall \underline{x}_1, \ldots, \underline{x}_n)\underline{f}(\underline{x}_1, \ldots, \underline{x}_n) = \underline{g}(\underline{x}_1, \ldots, \underline{x}_n)$, which by (the first part of) AS40.2 yields (b′) $\underline{f} = \underline{g}$. Then by Theor. 33.2 we obtain $\vdash$ (a′) $\supset$ (b′). Thence by Theor. 32.2 we obtain $\vdash (\forall \underline{f}, \underline{g})$ $[\Psi(\underline{f})\Psi(\underline{g}) \supset \underline{f} = \underline{g}]$, which by Def. 34.1 is equivalent to (c′) $\vdash (\exists^{(1)}\underline{f})$ $\Psi(\underline{f})$. By $(43)_2$ and AS40.4 we have $\vdash (\exists \underline{f})\Psi(\underline{f})$, so that from Theor. 36.1 and (c′) we deduce $(43)_1$.

(II) As to (44) we first remark that by (the second part of) AS40.1 the matrix (a) $\underline{F} = \underline{G}$ yields $(\forall \underline{x}_1, \ldots, \underline{x}_n)[\underline{F}(\underline{x}_1, \ldots, \underline{x}_n) \equiv \underline{G}(\underline{x}_1, \ldots, \underline{x}_n)]$, so that by $(42)_2$, (a) and $\Phi(\underline{F})$ yield $\Phi(\underline{G})$. By Theor. 33.2 we conclude that $\vdash \underline{F} = \underline{G}\,\Phi(\underline{F}) \supset \Phi(\underline{G})$, whence by Theor. 32.2 and Def. 35.1 we obtain $(44)_1$.

Now we remark that by (the second part of) AS40.2 the matrix (a′) $\underline{f} = \underline{g}$ yields $(\forall \underline{x}_1, \ldots, \underline{x}_n)\underline{f}(\underline{x}_1, \ldots, \underline{x}_n) = \underline{g}(\underline{x}_1, \ldots, \underline{x}_n)$, so that by $(43)_2$, the matrices (a′) and $\Psi(\underline{f})$ yield $\Psi(\underline{g})$. By Theor. 33.2 we conclude that $\vdash \underline{f} = \underline{g}\,\Psi(\underline{f}) \supset \Psi(\underline{g})$, whence by Theor. 32.2 and Def. 35.1 we obtain $(44)_2$.

(III) As to $(45)_1$ and $(45)_2$ we remark that by $(31)_2$, from $(42)_1$ and $(44)_1$ we deduce (a) $\vdash \Phi[(\imath\underline{F})\Phi(\underline{F})]$, and from $(43)_1$ and $(44)_2$ we deduce (b) $\vdash \Psi[(\imath\underline{f})\Psi(\underline{f})]$. By $(42)_2$ and Def. 4.4, $(\imath\underline{F})\Phi(\underline{F})$ is

$(\lambda \underline{x}_1, \ldots, \underline{x}_n)\underline{p}$, and by $(43)_2$ and Def. 4.5, $(\imath \underline{f})\Psi(\underline{f})$ is $(\lambda \underline{x}_1, \ldots, \underline{x}_n)\Delta$. Hence by $(42)_2$ and $(43)_2$, (a) and (b) yield (with $\underline{n}$ uses of AS30.6) $(45)_1$ and $(45)_2$ respectively.

(IV) Now observe that by assertions (a) and (b) in (III) and by Theor. 32.2 we obtain $\vdash \underline{N}\Phi[(\imath \underline{F})\Phi(\underline{F})]$ and $\vdash \underline{N}\Psi[(\imath \underline{f})\Psi(\underline{f})]$. So by $(37)_1$ we have $\vdash (\exists \underline{F})\underline{N}(\underline{F})$ and $\vdash (\exists \underline{f})\underline{N}\Psi(\underline{f})$, i.e. (46).

QED

N41. Some properties of the attributes MConst, MSep, and Abs.

Some of the properties of the concepts MConst, MSep, and Abs proved in this section will be useful in Memoir 3.

THEOR. 41.1. Let $\underline{F}$ and Φ be attributes of the respective types $\underline{t}$ and $(\underline{t})$ $(\underline{t} \in \tau^{\nu})$. Then

(I) $\vdash \underline{F} \in \underline{\text{MConst}}_{\underline{t}} \equiv \underline{F}^{\cup} = \underline{F}^{\cap}$,

(II) $\vdash \underline{\text{MConst}}_{\underline{t}} \in \underline{\text{MConst}}_{(\underline{t})}$,

(III) $\vdash \underline{\text{MConst}}_{\underline{t}} \in \underline{\text{MSep}}_{(\underline{t})}$,

(IV) $\vdash \underline{\text{MConst}}_{\underline{t}} \in \underline{\text{Abs}}_{(\underline{t})}$,

(V) $\vdash \underline{F} \subset \underline{G}\, \underline{F} \in \underline{\text{MConst}}\ \underline{G} \in \underline{\text{Abs}} \supset \underline{F} \in \underline{\text{Abs}}$,

(VI) $\vdash \Phi \subset \underline{\text{MConst}}_{\underline{t}}\ \Phi \in \underline{\text{MConst}}_{(\underline{t})} \supset \Phi \in \underline{\text{Abs}}_{(\underline{t})}$,

(VII) $\vdash \underline{\text{Abs}}_{\underline{t}} \in \underline{\text{MConst}}_{(\underline{t})}$,

(VIII) $\vdash \underline{\text{Abs}}_{\underline{t}} \in \underline{\text{Abs}}_{(\underline{t})}$,

(IX) $\vdash \underline{\text{MSep}}_{\underline{t}} \in \underline{\text{Ext}}_{(\underline{t})}$.

Proof. We deduce (I) from Defs. 13.2, 6.6, 6.8.

Since $\vdash \underline{F}^{\cup} = {}^{\cup}\underline{F}^{\cap} \supset \underline{F}^{\cup} = {}^{\cap}\underline{F}^{\cap}$ obviously holds, by Def. 13.2 (I) yields (II).

To prove (III) we start with (a) $\underline{G}, \underline{H} \in \underline{MConst}$ and (b) $\underline{G} =^{\cup} \underline{H}$, where $\underline{G}$ and $\underline{H}$ are distinct variables of the attribute type $\underline{t}$. By AS40.1, (b) yields $\Diamond (\forall \underline{x}_1, \ldots, \underline{x}_n) [\underline{G}(\underline{x}_1, \ldots, \underline{x}_n) \equiv \underline{H}(\underline{x}_1, \ldots, \underline{x}_n)]$ (where $\underline{x}_1, \ldots, \underline{x}_n$ are distinct variables). This yields (c) $(\forall \underline{x}_1, \ldots, \underline{x}_n) [\underline{G}(\underline{x}_1, \ldots, \underline{x}_n) \equiv^{\cup} \underline{H}(\underline{x}_1, \ldots, \underline{x}_n)]$, whence by (a) and Def. 13.2 we easily deduce $(\forall \underline{x}_1, \ldots, \underline{x}_n) [\underline{G}(\underline{x}_1, \ldots, \underline{x}_n) \equiv^{\cap} \underline{H}(\underline{x}_1, \ldots, \underline{x}_n)]$, which by AS40.1 yields $\underline{G} =^{\cap} \underline{H}$.

By Theor. 33.2 and rule $\underline{G}$ we conclude that $\vdash (\forall \underline{G}, \underline{H}) (\underline{G}, \underline{H} \in \underline{MConst} \wedge \underline{G} =^{\cup} \underline{H} \supset \underline{G} =^{\cap} \underline{H})$, whence, by Def. 18.7, (III) follows.

By Def. 18.8 we deduce (IV) from (II) and (III).

Part (V) is an immediate consequence of Defs. 13.2, 18.7, and 18.8.

Since the attribute type $\underline{t} \in \tau^{\nu}$ is arbitrary, from (V) we obtain $\vdash \Phi \subset \underline{MConst}_{\underline{t}}\ \Phi \in \underline{MConst}_{(\underline{t})}\ \underline{MConst}_{\underline{t}} \in \underline{Abs}_{(\underline{t})} \supset \Phi \in \underline{Abs}_{(\underline{t})}$, which by (IV) yields (VI).

To prove (VII) we start with (d) $\underline{F} \in^{\cup} \underline{Abs}$, whence by Def. 18.8, we have (e) $\underline{F} \in^{\cup} \underline{MConst}$ and (f) $\underline{F} \in^{\cup} \underline{MSep}$. For the sake of simplicity we assume that $\underline{F}$ is a property variable. Then by (e), (II), and Def. 13.2 we obtain (g) $\underline{F} \in^{\cap} \underline{MConst}$ and $(\underline{x}, \underline{y} \in^{\cup} \underline{F} \equiv \underline{x}, \underline{y} \in^{\cap} \underline{F})$, whence $\vdash \Diamond (\underline{x}, \underline{y} \in \underline{F}\, \underline{x} =^{\cup} \underline{y} \supset \underline{x} =^{\cap} \underline{y}) \equiv \underline{N}(\underline{x}, \underline{y} \in \underline{F}\, \underline{x} =^{\cup} \underline{y} \supset \underline{x} =^{\cap} \underline{y})$. Thence by Theor. 32.2 and Def. 18.7, we have $\underline{F} \in^{\cup} \underline{MSep} \equiv \underline{F} \in^{\cap} \underline{MSep}$, which by (f) yields $\underline{F} \in^{\cap} \underline{MSep}$. This and (g) yield $\underline{F} \in^{\cap} \underline{Abs}$ by Def. 18.8.

By Theors. 33.2 and 32.2 we conclude that $\vdash (\underline{F}) [\underline{F} \in^{\cup} \underline{Abs} \supset \underline{F} \in^{\cap} \mathrm{Abs}]$. The same conclusion, and hence (VII), holds in case $\underline{F}$ is an arbitrary attribute variable.

To prove (VIII), we deduce from (V) that

$$\vdash \underline{Abs}_{\underline{t}} \subset \underline{MConst}_{\underline{t}} \wedge \underline{Abs}_{\underline{t}} \in \underline{MConst}_{(\underline{t})} \wedge \underline{MConst}_{\underline{t}} \in \underline{Abs}_{(\underline{t})} \supset \underline{Abs}_{\underline{t}} \in \underline{Abs}_{(\underline{t})},$$

and from Def. 18.8 that $\vdash \underline{\text{Abs}}_{\underline{t}} \subseteq \underline{\text{MConst}}_{\underline{t}}$. So by (VII) and (IV) we have (VIII).

Briefly, the matrices $\underline{F} \in \underline{\text{MSep}}$ and $\underline{G} = \underline{F}$ yield $\underline{G} \in \underline{\text{MSep}}$ by Def. 18.7. So by Theor. 33.2 and Def. 6.11 we easily see that (IX) holds.

QED

Through the following theorem we present some simplification properties of modally separated and absolute attributes in MC^{ν} that are connected with the use of $\exists^{(1)\frown}$ and $\exists_1^{\frown}$ and with the possibility of using descriptions like those in ordinary speech; we also exhibit some theorems in MC^{ν} which make it easier to apply in practice these simplification properties.

THEOR. 41.2. Let $\underline{p}$ be a matrix and $\underline{x}$ a variable in MC^{ν}. Then

(I) $\vdash (\lambda \underline{x})\underline{p} \in \underline{\text{MSep}} \wedge (\exists^{(1)}\underline{x})\underline{p} \supset (\exists^{(1)\frown}\underline{x})\underline{p}$,

(II) $\vdash (\lambda \underline{x})\underline{p} \in \underline{\text{MSep}} \wedge (\exists_1 \underline{x})\underline{p} \supset (\exists_1^{\frown} \underline{x})\underline{p}$,

(III) $\vdash (\lambda \underline{x})\Phi(\underline{x}) \in \underline{\text{Abs}} \wedge (\exists_1 \underline{x})\Phi(\underline{x}) \supset \Phi[(\imath \underline{x})\Phi(\underline{x})]$—cf. conventions (a) and (b) in N30.

Furthermore, setting $\underline{F} =_D (\lambda \underline{x}_1)\underline{p}$ and $\underline{R} =_D (\lambda \underline{x}_1, \ldots, \underline{x}_n)\underline{p}$ ($\underline{n} > 1$), we have

(IV) $\vdash \underline{R} \in \underline{\text{MConst}} \supset \underline{F} \in \underline{\text{MConst}}$,

(V) $\vdash \underline{R} \in \underline{\text{MSep}} \supset \underline{F} \in \underline{\text{MSep}}$,

(VI) $\vdash \underline{R} \in \underline{\text{Abs}} \supset \underline{F} \in \underline{\text{Abs}}$.

Proof. (I) follows from Defs. 18.7 and 34.1, 34.2; (II) follows from (I), Theor. 36.1, and $(14)_{1,2,3}$.

To prove (III) we start with (a) $(\lambda \underline{x})\Phi(\underline{x}) \in \underline{\text{Abs}} \wedge (\exists_1 \underline{x})\Phi(\underline{x})$. Then by (II) we deduce (b) $(\exists_1^{\frown} \underline{x})\Phi(\underline{x})$. By (a) and Defs. 18.8, 13.2, we easily obtain $(\underline{x})[\Diamond \Phi(\underline{x}) \equiv \underline{N}\Phi(\underline{x})]$, which by (b) and $(35)_6$ yields $\Phi[(\imath \underline{x})\Phi(\underline{x})]$. By Theor. 33.2 we conclude that (III) holds.

Parts (IV) to (VI) are easy consequences of Defs. 13.2 and 18.7, 18.8 for $\underline{n} = 1$ and $\underline{n} > 1$.

QED

N42. General theorems on equivalence and substitution.

We first extend AS34.4—cf. Theor. IX.2.5 in [30], p. 216—and (20) to terms (perhaps not consisting of a variable):

THEOR. 42.1. Let $\underline{x}, \underline{y}, \underline{z}$ be distinct variables and let $\underline{x}, \underline{y}$, and the terms Δ and Δ' be free for $\underline{z}$ in the term $\underline{A}(\underline{x}, \underline{y}, \underline{z})$. Then

(47) $\vdash \underline{x} =^{\frown} \underline{y} \supset \underline{A}(\underline{x}, \underline{y}, \underline{x}) =^{\frown} \underline{A}(\underline{x}, \underline{y}, \underline{y})$, $\vdash (\text{ext } \underline{z})\underline{A}(\underline{x}, \underline{y}, \underline{z})\underline{x} = \underline{y} \supset \underline{A}(\underline{x}, \underline{y}, \underline{z}) = \underline{A}(\underline{x}, \underline{y}, \underline{y})$,

(47′) $\vdash \Delta =^{\frown} \Delta' \supset \underline{A}(\Delta, \Delta', \Delta) =^{\frown} \underline{A}(\Delta, \Delta', \Delta')$, $\vdash (\text{ext } \underline{z})\underline{A}(\Delta, \Delta', \underline{z})\Delta = \Delta' \supset \underline{A}(\Delta, \Delta', \Delta) = \underline{A}(\Delta, \Delta', \Delta')$.

Proof. Let $\Phi(\underline{x}, \underline{y}, \underline{z})$ be $\underline{A}(\underline{x}, \underline{y}, \underline{x}) =^{\frown} \underline{A}(\underline{x}, \underline{y}, \underline{z})$. Then by AS34.4 we deduce (a) $\vdash \underline{x} =^{\frown} \underline{y} \supset [\Phi(\underline{x}, \underline{y}, \underline{x}) \equiv \Phi(\underline{x}, \underline{y}, \underline{y})]$. Since ASs34.1 and 38.2 yield (b) $\vdash \Phi(\underline{x}, \underline{y}, \underline{x})$, from (a) we obtain $\vdash \underline{x} =^{\frown} \underline{y} \supset \Phi(\underline{x}, \underline{y}, \underline{y})$, i.e. $(47)_1$.

Now we start with (c) $(\text{ext } \underline{z})\underline{A}(\underline{x}, \underline{y}, \underline{z})$ and (d) $\underline{x} = \underline{y}$. By Def. 35.2, (c) yields (e) $(\forall \underline{u}, \underline{v})\, [\underline{u} = \underline{v} \supset \underline{A}(\underline{x}, \underline{y}, \underline{u}) = \underline{A}(\underline{x}, \underline{y}, \underline{v})]$, where $\underline{u}, \underline{v}$ are distinct variables not occurring in $\underline{A}(\underline{x}, \underline{y}, \underline{z})$. (d) and (e) yield (f) $\underline{A}(\underline{x}, \underline{y}, \underline{x}) = \underline{A}(\underline{x}, \underline{y}, \underline{y})$. Hence (c), (d) $\vdash$ (f), so that by Theor. 33.2, $(47)_2$ also holds.

$(47')_1$ and $(47')_2$ follow from $(47)_1$ and $(47)_2$ respectively by AS38.2.

QED

Now we want to extend to MC^{ν} the general substitution theorem in extensional logic, precisely Theors. IX.2.6 and IX.2.9 in [32], pp. 217–18 for matrices and terms respectively. We do so by a single theorem. Therefore we first set a single hypothesis corresponding to Rosser's hypotheses $\underline{H}_2$ and $\underline{H}_3$.

HYP. 42.1. $\underline{p}_1, \ldots, \underline{p}_n, \underline{q}, \underline{r}$ are matrices, $\Delta_1, \ldots, \Delta_{\underline{m}}$ are terms, $\underline{x}_1, \ldots, \underline{x}_{\underline{a}}$ are variables, and $\underline{y}$ is a variable distinct from $\underline{x}_{\underline{i}}$ ($\underline{i} = 1, \ldots, \underline{a}$) and not occurring in any of $\underline{p}_1, \ldots, \underline{p}_{\underline{n}}$, $\underline{q}, \underline{r}, \Delta_1, \ldots, \Delta_{\underline{m}}$. Δ is a designator (wff) in ML^ν built up out of some or all of the wffs $\underline{y} = \underline{y}, \underline{p}_1, \ldots, \underline{p}_{\underline{n}}, \underline{q}, \underline{r}, \Delta_1, \ldots, \Delta_{\underline{m}}$ by means of parentheses, commas, $\sim$, $\wedge$, $\underline{N}$, and using the formation rules ($\underline{f}_2$) to ($\underline{f}_9$) in N3 (each part listed may be used as often as desired). $\Delta^{\underline{q}}$ and $\Delta^{\underline{r}}$ are the results of replacing all occurrences of $\underline{y} = \underline{y}$ in Δ by $\underline{q}$ and $\underline{r}$ respectively.

The symbol $\asymp$ stands for $\equiv$ or $=$ depending on whether Δ is a matrix or a term.

THEOR. 42.2. Assume Hyp. 42.1 and let $\underline{y}_1, \ldots, \underline{y}_{\underline{b}}$ be variables such that $\underline{x}_1, \ldots, \underline{x}_{\underline{a}}$ do not occur free in $(\forall \underline{y}_1, \ldots, \underline{y}_{\underline{b}})$ $(\underline{q} \equiv^\cap \underline{r})$. Then

$$(48) \quad \vdash \underline{X} \supset \Delta^{\underline{q}} \asymp^\cap \Delta^{\underline{r}} \quad \text{where} \quad \underline{X} \equiv_D (\forall \underline{y}_1, \ldots, \underline{y}_{\underline{b}})(\underline{q} \equiv^\cap \underline{r}).$$

Proof. We use induction on the number ν of symbols in Δ, and for every subdesignator Δ_0 of Δ we denote by $\Delta_0^{\underline{q}}$ and $\Delta_0^{\underline{r}}$ the analogues for Δ_0 of $\Delta^{\underline{q}}$ and $\Delta^{\underline{r}}$ respectively—see Hyp. 42.1.

If $\nu = 0$, Δ is not a designator, hence Theor. 42.2 holds.

Now let $\underline{n}$ be any integer and let Theor. 42.2 hold for every $\nu \leq \underline{n}$; furthermore let $\nu = \underline{n} + 1$. The following five cases, namely:

1. Δ is $\underline{y} = \underline{y}$,
2. Δ does not contain $\underline{y} = \underline{y}$,
3. Δ is $\sim\underline{Y}$,
4. Δ is $\underline{Y} \wedge \underline{Z}$,
5. Δ is $(\underline{x})\underline{Y}$,

are handled just like the cases in the proof of Theor. VI.5.4 in [32], p. 109.

There remain the following five cases:

Case 6. Δ is $\underline{NY}$. Then by the hypothesis of the induction, (a) $\vdash \underline{X} \supset (\underline{Y}^{\underline{q}} \equiv^{\cap} \underline{Y}^{\underline{r}})$. Furthermore $\underline{X}$ is modally closed—cf. $(48)_2$. Hence by ASs30.7, 30.8, from (a) we have $\vdash \underline{X} \supset (\underline{NY}^{\underline{q}} \equiv^{\cap} \underline{NY}^{\underline{r}})$. This is $(48)_1$.

Case 7. Δ is $\underline{A} = \underline{B}$. Then by the hypothesis of induction, $\vdash \underline{X} \supset \underline{A}^{\underline{q}} =^{\cap} \underline{A}^{\underline{r}}$ and $\vdash \underline{X} \supset \underline{B}^{\underline{q}} =^{\cap} \underline{B}^{\underline{r}}$, whence $\vdash \underline{X} \supset \underline{A}^{\underline{q}} = \underline{B}^{\underline{q}} \equiv^{\cap} \underline{A}^{\underline{r}} = \underline{B}^{\underline{r}}$. This is $(48)_1$.

Case 8. Δ is $(\iota\underline{x})\underline{Y}$. Then by the hypothesis of the induction, $\vdash \underline{X} \supset (\underline{Y}^{\underline{q}} \equiv^{\cap} \underline{Y}^{\underline{r}})$. Thence, since $\underline{X}$ is closed with respect to $\underline{x}$, by AS30.5 we have $\vdash \underline{X} \supset (\underline{x})\,(\underline{Y}^{\underline{q}} \equiv^{\cap} \underline{Y}^{\underline{r}})$. So by $(30)_1$ we have $\vdash \underline{X} \supset [(\iota\underline{x})\underline{Y}^{\underline{q}} =^{\cap} (\iota\underline{x})\underline{Y}^{\underline{r}}]$, which is $(48)_1$.

Case 9. Δ is the matrix $\Phi(\underline{A}_1, \ldots, \underline{A}_{\underline{c}})$. Then by the hypothesis of the induction, we have (a) $\vdash \underline{X} \supset \Phi^{\underline{q}} =^{\cap} \Phi^{\underline{r}}$ and (b) $\vdash \underline{X} \supset \underline{A}_{\underline{i}}^{\underline{q}} =^{\cap} \underline{A}_{\underline{i}}^{\underline{r}}$ $(\underline{i} = 1, \ldots, \underline{c})$. By $\underline{c}$ uses of Theor. 39.1, from (b) we obtain (c) $\vdash \underline{X} \supset [\Phi^{\underline{q}}(\underline{A}_1^{\underline{q}}, \ldots, \underline{A}_{\underline{c}}^{\underline{q}}) \equiv^{\cap} \Phi^{\underline{q}}(\underline{A}_1^{\underline{r}}, \ldots, \underline{A}_{\underline{c}}^{\underline{r}})]$.

By Theor. 39.1, from (a) and (c) we obtain $\vdash \underline{X} \supset [\Phi^{\underline{q}}(\underline{A}_1^{\underline{q}}, \ldots, \underline{A}_{\underline{c}}^{\underline{q}}) \equiv^{\cap} \Phi^{\underline{r}}(\underline{A}_1^{\underline{r}}, \ldots, \underline{A}_{\underline{c}}^{\underline{r}})]$. This is $(48)_1$.

Case 10. Δ is the term $\phi(\underline{A}_1, \ldots, \underline{A}_{\underline{c}})$. Then by the hypothesis of the induction, we have (a) $\vdash \underline{X} \supset \phi^{\underline{q}} =^{\cap} \phi^{\underline{r}}$ and (b) $\vdash \underline{X} \supset \underline{A}_{\underline{i}}^{\underline{q}} =^{\cap} \underline{A}_{\underline{i}}^{\underline{r}}$ $(\underline{i} = 1, \ldots, \underline{c})$.

By $\underline{c}$ uses of $(47')_1$, from (b) we obtain (c) $\vdash \underline{X} \supset \phi^{\underline{q}}(\underline{A}_1^{\underline{q}}, \ldots, \underline{A}_{\underline{c}}^{\underline{q}}) =^{\cap} \phi^{\underline{q}}(\underline{A}_1^{\underline{r}}, \ldots, \underline{A}_{\underline{c}}^{\underline{r}})$.

By Theor. 39.1, from (a) and (c) we obtain $\vdash \underline{X} \supset [\phi^{\underline{q}}(\underline{A}_1^{\underline{q}}, \ldots, \underline{A}_{\underline{c}}^{\underline{q}}) =^{\cap} \phi^{\underline{r}}(\underline{A}_1^{\underline{r}}, \ldots, \underline{A}_{\underline{c}}^{\underline{r}})]$. This is $(48)_1$.

QED

The next theorem is a weaker form of the equivalence theorem [Theor. 42.2], whereas Theor. 42.4 is the corresponding substitution theorem (for matrices of any kind). We may regard Theor. 42.3 as the total analogue for MC^{ν} of Theor. IX.2.7 and IX.2.10 in [32], p. 218 taken together, and Theor. 42.4 as the total analogue of Theor. IX.2.8 in [32].

THEOR. 42.3. If Hyp. 42.1 and $\vdash \underline{q} \equiv \underline{r}$ hold, then $\vdash \Delta^{\underline{q}} \asymp \Delta^{\underline{r}}$.

Proof. By Theor. 32.2, $\vdash \underline{q} \equiv \underline{r}$ implies $(\forall \underline{y}_1, \ldots, \underline{y}_{\underline{b}})(\underline{q} \equiv^{\frown} \underline{r})$, whence by Theor. 42.2, $\vdash \Delta^{\underline{q}} \asymp \Delta^{\underline{r}}$ follows.

QED

THEOR. 42.4. If Hyp. 42.1, $\vdash \underline{q} \equiv \underline{r}$, and $\vdash \Delta^{\underline{q}}$ hold, then $\vdash \Delta^{\underline{r}}$.

N43. General theorems on extensionalization. An extensional form of the equivalence theorem in MC^{ν}.

THEOR. 43.1. We assume that $\underline{x}, \underline{z}_1, \ldots, \underline{z}_{\underline{n}}$ are distinct (and not necessarily individual) variables, that $0 \leq \underline{r} \leq \underline{r}' \leq \underline{s}$ holds, and that $\Delta_1, \ldots, \Delta_{\underline{s}}$ are terms, of which $\Delta_1, \ldots, \Delta_{\underline{r}'}$ have a nonindividual type. In addition let Δ be a designator built up out of the matrices $\underline{p}_1, \ldots, \underline{p}_{\underline{m}}$ and the terms $\Delta_1, \ldots, \Delta_{\underline{s}}$ by means of operations of the following kind:

(i) to form $\sim \underline{p}$, $\underline{p} \wedge \underline{q}$, $\Delta' = \Delta''$, $(\underline{z}_{\underline{i}})\underline{p}$, or $(\iota \underline{z}_{\underline{i}})\underline{p}$ out of the designators $\underline{p}$, $\underline{q}$, $\underline{z}_{\underline{i}}$, Δ', and Δ'';

(ii) to form the term or matrix $\Delta_{\underline{l}}(\Delta'_1, \ldots, \Delta'_{\underline{a}})$ out of the designator $\Delta_{\underline{l}}$ with $\underline{l} \leq \underline{r}$ (which will be assumed to be extensional—cf. $(4\overline{9})_2$) and out of some designators $\Delta'_1, \ldots, \Delta'_{\underline{a}}$;

(iii) to form the term or matrix $\Delta_{\underline{l}}(\xi_1, \ldots, \xi_{\underline{a}})$ out of $\Delta_{\underline{l}}$ with $\underline{r} < \underline{l} \leq \underline{r}'$ and out of some terms $\xi_1, \ldots, \xi_{\underline{a}}$ where $\underline{x}$ has no free occurrences.

Then—cf. Defs. 6.12, 35.1, 35.2—

(49) $$\vdash \underline{X} \supset (\forall \underline{z}_1, \ldots, \underline{z}_{\underline{n}}) (\underline{ext}\ \underline{x})\Delta,$$

where

(49′) $$\underline{X} \equiv_D (\forall \underline{z}_1, \ldots, \underline{z}_{\underline{n}}) [\bigwedge_{\underline{i}=1}^{\underline{m}} (\underline{ext}\ \underline{x})\underline{p}_{\underline{i}} \bigwedge_{\underline{j}=1}^{\underline{s}} (\underline{ext}\ \underline{x})\Delta_{\underline{j}} \bigwedge_{\underline{l}=1}^{\underline{r}} \underline{Ext}\,(\Delta_{\underline{l}})].$$

Let us remark that by (49′) and the validity of (49) for $\Delta = \Delta'_1$, $\ldots$, $\Delta = \Delta'_{\underline{a}}$ as an inductive hypothesis, the matrix $\underline{X}$ implies that the designator $\Delta_{\underline{l}}(\Delta'_1, \ldots, \Delta'_{\underline{a}})$, with $1 \leq \underline{l} \leq \underline{r}$, considered in (ii), is formed with an extensional attribute or function $\Delta_{\underline{l}}$, whose expression, like those of $\Delta'_1, \ldots, \Delta'_{\underline{a}}$, is extensional with respect to $\underline{x}$ [$(\underline{ext}\ \underline{x})\Delta_{\underline{l}}$, $(\underline{ext}\ \underline{x})\Delta_{\underline{j}}$ for $\underline{j} = 1, \ldots, \underline{a}$].

For $\underline{r} < \underline{l} \leq \underline{r}'$ the matrix $\underline{X}$ does not imply that the attribute or function $\Delta_{\underline{l}}$ should be extensional, and in spite of the truth of $(\underline{ext}\ \underline{x})\Delta_{\underline{l}}$ it is possible that the terms $\xi_1, \ldots, \xi_{\underline{n}}$ are extensional with respect to $\underline{x}$ while the designator $\Delta_{\underline{l}}(\xi_1, \ldots, \xi_{\underline{n}})$ is not. Therefore in (iii), $\xi_1, \ldots, \xi_{\underline{n}}$ are assumed not to have free occurrences of $\underline{x}$.

Proof of Theor. 43.1. We use induction on the number ν of operations by which Δ is built up out of $\underline{p}_1, \ldots, \underline{p}_{\underline{m}}, \Delta_1, \ldots, \Delta_{\underline{s}}$.

For $\nu = 0$, (49) obviously holds.

Now suppose that $\nu' > 0$ and that our theorem holds for every $\nu < \nu'$; let $\nu = \nu'$ hold.

Case 1. Δ is one of $\underline{p}_1, \ldots, \underline{p}_{\underline{m}}, \Delta_1, \ldots, \Delta_{\underline{s}}$. Then (49) holds obviously.

Case 2. Δ is $\sim\Phi(\underline{x})$, and by the hypothesis of the induction, $\underline{X}$ yields (a) $(\underline{ext}\ \underline{x})\Phi(\underline{x})$, which by Def. 35.1 is equivalent to $(\forall \underline{x}, \underline{y}) [\underline{x} = \underline{y}\, \Phi(\underline{x}) \supset \Phi(\underline{y})]$ where, as in the sequel, $\underline{y}$ does not occur in Δ. Thence we easily deduce $(\forall \underline{x}, \underline{y})\, [\underline{x} = \underline{y} \sim \Phi(\underline{y}) \supset \sim\Phi(\underline{x})]$, which by Def. 35.1 yields $(\underline{ext}\ \underline{x}) \sim \Phi(\underline{x})$, i.e. $(\underline{ext}\ \underline{x})\, \Delta$. Hence since by (49′) $\underline{X}$ is closed with respect to $\underline{z}_1, \ldots, \underline{z}_n$, by Theors. 33.2 and 32.2, (49) holds in case 2.

Case 3. Δ is $\Phi_1(\underline{x})\, \Phi_2(\underline{x})$. By the hypothesis of induction, $\underline{X}$ yields $(\underline{ext}\ \underline{x})\, \Phi_i(\underline{x})$ $(\underline{i} = 1, 2)$. Thence by Def. 35.1 we easily deduce $(\forall \underline{x}, \underline{y})\, [\underline{x} = \underline{y}\, \Phi_1(\underline{x})\, \Phi_2(\underline{x}) \supset \Phi_1(\underline{y})\, \Phi_2(\underline{y})]$. Hence by Def. 35.1, $(\underline{ext}\ \underline{x})\, \Delta$, so that on the basis of Theors. 33.2 and 32.2 we may conclude, as before, that Theor. 43.1 holds in case 3.

Case 4. Δ is $\phi_1(\underline{x}) = \phi_2(\underline{x})$. By the hypothesis of induction and $(20)_2$, $\underline{X}$ yields $(\forall \underline{x}, \underline{y})\, [\underline{x} = \underline{y} \supset \phi_i(\underline{x}) = \phi_i(\underline{y})]$ $(\underline{i} = 1, 2)$. Thence by ASs34.2, 34.3, and 38.2 we deduce $(\forall \underline{x}, \underline{y})\, [\underline{x} = \underline{y}\, \phi_1(\underline{x}) = \phi_2(\underline{x}) \supset \phi_1(\underline{y}) = \phi_2(\underline{y})]$, which by Def. 35.1 is $(\underline{ext}\ \underline{x})\, \Delta$. Hence by Theors. 33.2 and 32.2, Theor. 43.1 holds in case 4.

Case 5. Δ is $(\underline{z}_i)\, \Phi(\underline{x}, \underline{z}_i)$. By the hypothesis of the induction, $\underline{X}$ yields $(\forall \underline{z}_i, \underline{x}, \underline{y})\, [\underline{x} = \underline{y}\, \Phi(\underline{x}, \underline{z}_i) \supset \Phi(\underline{y}, \underline{z}_i)]$. Thence by AS30.4 we easily deduce $(\forall \underline{x}, \underline{y})\, [\underline{x} = \underline{y}(\underline{z}_i)\, \Phi(\underline{x}, \underline{z}_i) \supset (\underline{z}_i)\, \Phi(\underline{y}, \underline{z}_i)]$. Hence by Def. 35.1 we have $(\underline{ext}\ \underline{x})\, \Delta$, so that by Theors. 33.2 and 32.2, Theor. 43.1 holds in case 5.

Case 6. Δ is $(\iota \underline{z}_i)\, \Phi(\underline{x}, \underline{z}_i)$. By the hypothesis of the induction and AS30.6, $\underline{X}$ yields $(\underline{z}_i)\, (\underline{ext}\ \underline{x})\, \Phi(\underline{x}, \underline{z}_i)$. By $(20)_1$ this easily yields in turn $(\forall \underline{x}, \underline{y})\, \{\underline{x} = \underline{y} \supset [(\underline{z}_i)\, \Phi(\underline{x}, \underline{z}_i) \equiv (\underline{z}_i)\, \Phi(\underline{y}, \underline{z}_i)]\}$, so that by $(30)_1$ we deduce $(\forall \underline{x}, \underline{y})\, [\underline{x} = \underline{y} \supset (\iota \underline{z}_i)\, \Phi(\underline{x}, \underline{z}_i) = (\iota \underline{z}_i)\, \Phi(\underline{y}, \underline{z}_i)]$. Thence by $(20)_2$ we conclude $(\underline{ext}\ \underline{x})\, \Delta$, so that by Theors. 33.2 and 32.2, Theor. 43.1 holds in case 6.

Case 7. Δ is $\Delta_{\underline{1}}[\phi_1(\underline{x}), \ldots, \phi_{\underline{a}}(\underline{x})]$ where $1 \le \underline{1} \le \underline{r}$. Let $\phi_0(\underline{x})$ be $\Delta_{\underline{1}}$, let the variable $\underline{y}$ not occur in $\underline{x} = \underline{x} \wedge \Delta \wedge \underline{X}$, and denote $\phi_0(\underline{y})$—see conventions (b) and (c) in N30—by $\Delta^{\underline{y}}_{\underline{1}}$; furthermore let $\underline{X}$ hold. Then by the hypothesis of the induction and by $(20)_2$, we have (a)

$(\forall \underline{x}, \underline{y})\ [\underline{x} = \underline{y} \supset \phi_{\underline{j}}(\underline{x}) = \phi_{\underline{j}}(\underline{y})]\ (\underline{j} = 0, \ldots, \underline{a})$. By (49′) $\underline{X}$ yields $\underline{\mathrm{Ext}}(\Delta_{\underline{1}})$, whence by either (22) or (23) we have (b) $(\forall \underline{x}_1, \underline{y}_1, \ldots, \underline{x}_{\underline{a}}, \underline{y}_{\underline{a}})\ \{\underline{x}_1 = \underline{x}_1 \wedge \ldots \wedge \underline{x}_{\underline{a}} = \underline{y}_{\underline{a}} \supset [\Delta_{\underline{1}}(\underline{x}_1, \ldots, \underline{x}_{\underline{a}}) \asymp (\underline{y}_1, \ldots, \underline{y}_{\underline{a}})]\}$, where $\asymp$ stands for $\equiv$ or $=$ according to whether Δ is a matrix or a term, and where $\underline{x}_1, \underline{y}_1, \ldots, \underline{x}_{\underline{a}}, \underline{y}_{\underline{a}}$ are $2\underline{a}$ distinct variables.

Now we assume (c) $\underline{x} = \underline{y}$. Thence by AS38.2, by (a) (for $\underline{j} = 1, \ldots, \underline{a}$) and (b) we obtain (d) $\Delta_{\underline{1}}[\phi_1(\underline{x}), \ldots, \phi_{\underline{a}}(\underline{x})] \asymp \Delta_{\underline{1}}[\phi_1(\underline{y}), \ldots, \phi_{\underline{a}}(\underline{y})]$.

By (a) (for $\underline{j} = 0$), (c) yields (e) $\Delta_{\underline{1}} = \Delta_{\underline{1}}^{\underline{y}}$. By ASs40.1 or 40.2 and by AS38.2, from (d) and (e) we easily deduce (f) $\Delta_{\underline{1}}[\phi_1(\underline{x}), \ldots, \phi_{\underline{a}}(\underline{x})] \asymp \Delta_{\underline{1}}^{\underline{y}}[\phi_1(\underline{y}), \ldots, \phi_{\underline{a}}(\underline{y})]$.

Let us denote Δ by $\Phi(\underline{x})$. Then (f) is $\Phi(\underline{x}) \asymp \Phi(\underline{y})$. By Theor. 33.2 $\underline{X}$ yields (c) $\supset$ (f). Since $\underline{x}$ and $\underline{y}$ do not occur free in $\underline{X}$, by rule $\underline{G}$, from (c) $\supset$ (f) we obtain $(\forall \underline{x}, \underline{y})\ \{\underline{x} = \underline{y} \supset [\Phi(\underline{x}) \asymp \Phi(\underline{y})]\}$. Thence by either $(20)_1$ or $(20)_2$ we deduce $(\underline{\mathrm{ext}}\ \underline{x})\, \Phi(\underline{x})$. Hence by Theors. 33.3 and 32.2, Theor. 43.1 holds in case 7.

Case 8. Δ is $\Delta_{\underline{1}}(\xi_1, \ldots, \xi_{\underline{a}})$ where $\xi_1, \ldots, \xi_{\underline{a}}$ are any terms containing no free occurrences of $\underline{x}$ and where $\underline{r} < \underline{1} \leq r'$.

Let $\Delta_{\underline{1}}^{\underline{y}}$ be as in case 7, and assume (a) $\underline{x} = \underline{y}$. By (49′), $\underline{X}$ yields $(\underline{\mathrm{ext}}\ \underline{x})\ \Delta_{\underline{1}}$, so that by Def. 35.2 and (a) we deduce (b) $\Delta_{\underline{1}} = \Delta_{\underline{1}}^{\underline{y}}$. Thence by ASs40.1 or 40.2 we obtain (c) $\Delta_{\underline{1}}(\xi_1, \ldots, \xi_{\underline{a}}) \asymp \Delta_{\underline{1}}^{\underline{y}}(\xi_1, \ldots, \xi_{\underline{a}})$. Thus by Theor. 33.2 we conclude that $\underline{X}$ yields (a) $\supset$ (c). By rule $\underline{G}$, (a) $\supset$ (c) yields (d) $(\forall \underline{x}, \underline{y})\ \{\underline{x} = \underline{y} \supset [\Delta_{\underline{1}}(\xi_1, \ldots, \xi_{\underline{a}}) \asymp \Delta_{\underline{1}}^{\underline{y}}(\xi_1, \ldots, \xi_{\underline{a}})]\}$. Since $\underline{x}$ does not occur free in $\xi_1, \ldots, \xi_{\underline{a}}$, by (20), (d) yields $(\underline{\mathrm{ext}}\ \underline{x})\ \Delta$. Thence by Theor. 33.3 and rule $\underline{G}$ we easily obtain (49).

QED

The following analogue of (16) also holds:

THEOR. 43.2. If α is an $\underline{n}$-ary attribute, then

(50) $\vdash \alpha \subseteq \alpha^{(\underline{e})}, \qquad \vdash \underline{\text{Ext}}(\alpha) \equiv \alpha = \alpha^{(\underline{e})}.$

Proof. $(50)_1$ follows immediately from Defs. 18.4, 18.9. As to $(50)_2$ let us remark that $(21)_1$ and $(19)_2$ yield (a) $\vdash \underline{\text{Ext}}(\alpha) \equiv \bigwedge_{\underline{i}=1}^{\underline{n}} (\underline{\text{ext}}\ \underline{x}_{\underline{i}})\ \alpha(\underline{x}_1, \ldots, \underline{x}_{\underline{n}})$. Then by $\underline{n}$ uses of $(16)_2$ we easily obtain (b) $\vdash \underline{\text{Ext}}(\alpha) \equiv (\forall \underline{x}_1, \ldots, \underline{x}_{\underline{n}}) [\alpha(\underline{x}_1, \ldots, \underline{x}_{\underline{n}}) \equiv \alpha(\underline{x}_1, \ldots, \underline{x}_{\underline{n}})^{(\underline{\text{ex}}_1) \ldots (\underline{\text{ex}}_{\underline{n}})}]$. Thence by $(18)_2$ and AS40.1 we easily obtain $(50)_2$.

QED

In N42 we proved a general theorem on equivalence [Theor. 42.2] which can be considered to be the total analogue (i.e. the ordinary one together with the modal analogue) for MC^{ν} of Theors. IX.2.6 and IX.2.10 in [32] taken together. Now we prove, so to speak, an ordinary (extensional) counterpart for MC^{ν} of the latter theorems taken together. To this end we first consider the following hypothesis—somewhat similar to Hyp. 42.1:

HYP. 43.1. $\underline{p}_1, \ldots, \underline{p}_{\underline{m}}, \underline{q}, \underline{r}$ are matrices, $\Delta_1, \ldots, \Delta_{\underline{b}}$ are terms, $\Delta_1, \ldots, \Delta_{\underline{a}}$ with $0 \leq \underline{a} < \underline{b}$ have a nonindividual type, $\underline{y}$ is a variable distinct from the variables $\underline{x}_1, \ldots, \underline{x}_{\underline{n}}$, and $\underline{y}$ does not occur in $\underline{p}_1, \ldots, \underline{p}_{\underline{m}}, \underline{q}, \underline{r}, \Delta_1, \ldots, \Delta_{\underline{b}}$. Furthermore Δ is a designator built up out of some or all of $\underline{y} = \underline{y}$, $\underline{p}_1, \ldots, \underline{p}_{\underline{m}}, \underline{q}, \underline{r}, \Delta_1, \ldots, \Delta_{\underline{b}}$ by means of operations of the following kinds—cf. Theor. 43.1:

(i) to form $\sim \underline{p}'$, $\underline{p}' \wedge \underline{q}'$, $\Delta' = \Delta''$, $(\underline{x}_{\underline{i}})\underline{p}'$, and $(\imath\, \underline{x}_{\underline{i}})\underline{p}'$ out of the designators $\underline{p}'$, $\underline{q}'$, Δ', and Δ'', and

(ii) to form the term or matrix $\Delta_{\underline{l}}(\underline{B}_1, \ldots, \underline{B}_{\underline{s}})$ from the designators $\Delta_{\underline{l}}, \underline{B}_1, \ldots, \underline{B}_{\underline{s}}$, where $1 \leq \underline{l} \leq \underline{a}$.

In addition, for every subdesignator Δ_0 of Δ (Δ_0 may be Δ), $\Delta_0^{\underline{q}}$ and $\Delta_0^{\underline{r}}$ are the results of replacing all occurrences of $\underline{y} = \underline{y}$ in Δ_0 by $\underline{q}$ and $\underline{r}$ respectively.

Lastly $\asymp$ stands for $\equiv$ or $=$ as in Hyp. 42.1.

THEOR. 43.3. Let Hyp. 43.1 hold and let $\underline{y}_1, \ldots, \underline{y}_{\underline{c}}$ be variables such that $\underline{x}_1, \ldots, \underline{x}_{\underline{n}}$ do not occur free in $\underline{X}$, where $\underline{X}$ is $(\forall \underline{y}_1, \ldots, \underline{y}_{\underline{c}})\ (\underline{q} \equiv \underline{r})$. Then

(51) $\qquad \vdash \underline{A}\,\underline{X} \supset \Delta^{\underline{q}} \asymp \Delta^{\underline{r}} \qquad$ where $\qquad A \equiv_D \bigwedge_{\underline{l}=1}^{\underline{a}} \underline{\text{Ext}}\ (\Delta_{\underline{l}}^{\underline{q}})$.

Proof. We use induction on the number ν of symbols in Δ.

Our theorem obviously holds for $\nu = 0$. Let $\nu' > 0$ hold, assume Theor. 43.3 for every $\nu < \nu'$, and let $\nu = \nu'$; and start with $\underline{A}$—see $(51)_2$. Then cases 1–5, 7, and 8, considered or hinted at in the proof of Theor. 41.1, are to be handled here just as they were there. There remains the following case (corresponding to cases 9 and 10 in the proof of Theor. 41.2 taken together):

Case 9. Δ is $\Delta_{\underline{l}}(\underline{B}_1, \ldots, \underline{B}_{\underline{s}})$ where $1 \leq \underline{l} < \underline{a}$. Then by the hypothesis of the induction, from $\underline{A}\,\underline{X}$ we deduce (a) $\Delta_{\underline{l}}^{\underline{q}} = \Delta_{\underline{l}}^{\underline{r}}$ and (b) $\underline{B}_{\underline{i}}^{\underline{q}} = \underline{B}_{\underline{i}}^{\underline{r}}$ ($\underline{i} = 1, \ldots, \underline{s}$).

$\underline{A}$ yields $\underline{\text{Ext}}\ (\Delta_{\underline{l}}^{\underline{q}})$. So by Defs. 6.11 or 6.12, by (22) or (23), and by AS38.2, from (b) we deduce (c) $\Delta_{\underline{l}}^{\underline{q}}(\underline{B}_1^{\underline{q}}, \ldots, \underline{B}_{\underline{s}}^{\underline{q}}) \asymp \Delta_{\underline{l}}^{\underline{q}}(\underline{B}_1^{\underline{r}}, \ldots, \underline{B}_{\underline{s}}^{\underline{r}})$. By either ASs40.1, or 40.2, from (a) and (c) we deduce $\Delta_{\underline{l}}^{\underline{q}}(\underline{B}_1^{\underline{q}}, \ldots, \underline{B}_{\underline{s}}^{\underline{q}}) \asymp \Delta_{\underline{l}}^{\underline{r}}(\underline{B}_1^{\underline{r}}, \ldots, \underline{B}_{\underline{s}}^{\underline{r}})$. This is $\Delta^{\underline{q}} \asymp \Delta^{\underline{r}}$. By Theor. 33.2 $(51)_1$ holds in case 9.

QED

N44. On the concept of closure in MC^{ν}.

In Memoir 1 we introduced the concept of closure in MC^{ν} [Def. 26.4] and listed several semantical theorems on this concept, theorems of the form $\Vdash \underline{p}$—see e.g. Memoir 1, (68). Such theorems are the semantical analogues for ML^ν of some extensional syntactical theorems in [32]; and they hold in that the modal calculus MC^ν is valid in ML^ν and the analogues for MC^ν of the syntactical theorems in [32] hold and can be proved (in MC^ν) in substantially the same way as in [32]. So for each of the above theorems belonging to extensional logic we have a (semantical) analogue, $\Vdash \underline{p}$, in ML^ν, and also a syntactical analogue in MC^ν, i.e. $\vdash \underline{p}$.

We shall call $\vdash \underline{p}$ the syntactical analogue of $\Vdash \underline{p}$ and, for instance, when $\Vdash \underline{p}$ is the formula (65) in Memoir 1 [N26], we shall refer to $\vdash \underline{p}$ briefly as SA(65). On the basis of the aforementioned provability of certain syntactical analogues we need not prove explicitly these syntactical analogues for MC^ν.

In N26 we also considered some semantical theorems having no extensional analogues. Each of them has a syntactical analogue that can easily be proved in a corresponding way. Therefore we will not prove the syntactical analogues of the semantical theorems proved in N26. The same considerations hold for N27, where natural numbers are dealt with.

As a lemma we now prove the following theorem:

THEOR. 44.1. $\vdash (\underline{x})\,[\alpha(\underline{x}) \supset \{\underline{x}\} \subseteq \beta] \equiv \alpha^{(\underline{e})} \subseteq \beta$ [Defs. 18.4, 18.9, 18.12].

Proof. We start with (a) $(\underline{x})\,[\alpha(\underline{x}) \supset \{\underline{x}\} \subseteq \beta]$ and (b) $\underline{x} \in \alpha^{(\underline{e})}$. By Def. 18.9 and rule $\underline{C}$ with $\underline{y}$, (b) yields (c) $\underline{y} = \underline{x}\,\alpha(\underline{y})$, whence by (a) we deduce (d) $\{\underline{y}\} \subseteq \beta$. By Def. 18.12, from (c) we deduce (e) $\underline{x} \in \{\underline{y}\}$. By Def. 18.4, (d) and (e) yield (f) $\underline{x} \in \beta$. Hence (a), (b) $\vdash_{\underline{C}}$ (f).

By Theor. 33.3, Def. 18.9, and rule $\underline{G}$ we conclude that (a) yields $(\underline{x})\,(\underline{x} \in \alpha^{(\underline{e})} \supset \underline{x} \in \beta)$, hence [Def. 18.4] (g) $\alpha^{(\underline{e})} \subseteq \beta$, using rule $\underline{C}$. Then by Theor. 33.1 we have (a) $\vdash$ (g).

Now conversely, let us start with (g) $\alpha^{\underline{(e)}} \subseteq \beta$, (h) $\alpha(\underline{x})$, and (i) $\underline{y} \in \{\underline{x}\}$. By Def. 18.12 (i) yields (l) $\underline{y} = \underline{x}$, whence by (h) and Def. 18.9 we deduce $\underline{y} \in \alpha^{\underline{(e)}}$. Then by (g) and Def. 18.4 we have (m) $\underline{y} \in \beta$. So (g), (h), (i) $\vdash$ (m). Then by Theor. 33.2 and Def. 18.4 we conclude that (g), (h) $\vdash \{\underline{x}\} \subseteq \beta$, so that by Theor. 33.2 and rule $\underline{G}$ we obtain (g) $\vdash_{\underline{c}}$ (a), whence by Theor. 33.1, (g) $\vdash$ (a).

Since also (a) $\vdash$ (g), by Theor. 33.2 our theorem holds.

QED

By Def. 18.4, $\vdash \alpha^{\underline{(e)}} \subseteq \alpha^{\underline{(e)}}$, so that taking β to be $\alpha^{\underline{(e)}}$ in Theor. 44.1,

(52) $$\vdash (\underline{x})\, [\alpha(\underline{x}) \supset \{\underline{x}\} \subseteq \alpha^{\underline{(e)}}].$$

Here we prove a syntactical theorem on closure. One part of it—namely (54)—is SA(71), i.e. the syntactical analogue of (71) in Memoir 1, and another part—(55)—is SA(69).

In N26 we gave sufficient hints for proving (69). In spite of this we will now also prove (55) because it follows easily from (53) and other parts of Theor. 44.2 that are, I think, of interest in themselves.

THEOR. 44.2. Let α and β be predicates of type $\underline{t}$ and let $\underline{R}$ and $\underline{S}$ be $(\underline{n}+1)$-ary predicates of type $(\underline{t}, \ldots, \underline{t})$. Then

(53) $\vdash \alpha \subseteq \beta\ \underline{R} \subseteq \underline{S} \supset \underline{\text{clos}}\,(\alpha, \underline{R}) \subseteq \underline{\text{clos}}\,(\beta, \underline{S})$, $\quad \vdash \underline{\text{Ext}}\,(\underline{\text{clos}})$,

(54) $\vdash [\underline{\text{clos}}\,(\alpha, \underline{R})]^{\underline{(e)}} \subseteq \underline{\text{clos}}\,(\alpha^{\underline{(e)}}, \underline{R}^{\underline{(e)}})$ [Defs. 18.4, 18.9, 26.4],

(55) $\vdash \underline{\text{Ext}}_{(\underline{t})}\,(\alpha)\ \underline{\text{Ext}}_{(\underline{t}, \ldots, \underline{t})}\,(\underline{R}) \supset \underline{\text{Ext}}_{(\underline{t})}\,[\underline{\text{clos}}\,(\alpha, \underline{R})]$.

Proof. (I) $(53)_1$ can be proved as in extensional logic, using $SA(65)_2$, SA(66), and SA(67)—see N26.

(II) By Defs. 26.4 and 26.3 and AS40.1, $\vdash \alpha = \beta\ \underline{R} = \underline{S} \supset \underline{clos}(\alpha, \underline{R}) = \underline{clos}(\beta, \underline{S})$. Thence by Def. 6.12 we obtain $(53)_2$.

(III) As to (54), note that by AS40.3 and rule $\underline{C}$ with β we obtain:

(56) $\vdash_{\underline{c}} (\underline{x}) [\beta(\underline{x}) \equiv \underline{x} \in \underline{clos}(\alpha, \underline{R}) \wedge \{\underline{x}\} \subseteq \underline{clos}(\alpha^{(\underline{e})}, \underline{R}^{(\underline{e})})]$ [Defs. 18.4, 18.12].

By SA$(65)_2$ and Def. 18.9, $\vdash \alpha^{(\underline{e})} \subseteq \underline{clos}(\alpha^{(\underline{e})}, \underline{R}^{(\underline{e})})$. Thence by Theor. 44.1 we obtain (a) $\vdash (\underline{x}) [\alpha(\underline{x}) \supset \{\underline{x}\} \subseteq \underline{clos}(\alpha^{(\underline{e})}, \underline{R}^{(\underline{e})})]$. Furthermore, by Def. 18.4, SA$(65)_2$ also yields (b) $\vdash (\underline{x}) [\alpha(\underline{x}) \supset \underline{x} \in \underline{clos}(\alpha, \underline{R})]$. By (56) and Def. 18.4, from (a) and (b) we obtain (c) $\vdash_{\underline{c}} \alpha \subseteq \beta$.

Now we assume (d) $\underline{x}_1, \ldots, \underline{x}_n \in \beta$, (e) $\underline{x}_1, \ldots, \underline{x}_n \in \underline{clos}(\alpha, \underline{R})$, and (f) $\underline{R}(\underline{x}_1, \ldots, \underline{x}_n, \underline{z})$. By $(50)_1$ we have $\vdash \alpha \subseteq \alpha^{(\underline{e})}$ and $\vdash \underline{R} \subseteq \underline{R}^{(\underline{e})}$. As a consequence, by $(53)_1$ from (e) we obtain (g) $\underline{x}_1, \ldots, \underline{x}_n \in \underline{clos}(\alpha^{(\underline{e})}, \underline{R}^{(\underline{e})})$.

By (56) and SA (66), from (e) and (f) we deduce (h) $\underline{z} \in \underline{clos}(\alpha, \underline{R})$. Hence (d), (e), (f) $\vdash$ (g) $\wedge$ (h).

Now we also assume (i) $\underline{u} \in \{\underline{z}\}$, where $\underline{u}$ is a new variable. By Def. 18.12, (i) yields (l) $\underline{u} = \underline{z}$. Thence by (f) and Def. 18.9 we deduce (m) $\underline{R}^{(\underline{e})}(\underline{x}_1, \ldots, \underline{x}_n, \underline{u})$.

By SA(66), from (g) and (m) we deduce (n) $\underline{u} \in \underline{clos}(\alpha^{(\underline{e})}, \underline{R}^{(\underline{e})})$.

We conclude that (d), (e), (f), (i) $\vdash_{\underline{c}}$ (n). Then by Theor. 33.3 we obtain (d), (e), (f) $\vdash_{\underline{c}}$ (i) $\supset$ (n). By rule $\underline{G}$ and Def. 18.4, from (i) $\supset$ (n) we deduce (p) $\{\underline{z}\} \subseteq \underline{clos}(\alpha^{(\underline{e})}, \underline{R}^{(\underline{e})})$. Hence (d), (e), (f) $\vdash_{\underline{c}}$ (p).

By (56), from (h) and (p) we deduce (q) $\beta(\underline{z})$. Hence (d), (e), (f) $\vdash_{\underline{c}}$ (q). Then by Theor. 33.3 and rule $\underline{G}$ we have (r) $\vdash_{\underline{c}} (\forall \underline{x}_1, \ldots, \underline{x}_n, \underline{z}) [\underline{x}_1, \ldots, \underline{x}_n \in \beta \cap \underline{clos}(\alpha, \underline{R}) \wedge \underline{R}(\underline{x}_1, \ldots, \underline{x}_n, \underline{z}) \supset \beta(\underline{z})]$.

By SA(67), from (c) and (q) we obtain (s) $\vdash_{\overline{c}} \underline{\text{clos}}(\alpha, \underline{R}) \subseteq \beta$. Furthermore (56) yields (s′) $\vdash_{\overline{c}} \beta \subseteq \underline{\text{clos}}(\alpha, \underline{R})$. From (s) and (s′), by Def. 18.4 we obtain $\vdash_{\overline{c}} (\underline{x}) [\beta(\underline{x}) \equiv \underline{x} \in \underline{\text{clos}}(\alpha, \underline{R})]$. So by (56) $\vdash_{\overline{c}} (\underline{x}) [\underline{x} \in \underline{\text{clos}}(\alpha, \underline{R}) \supset \{\underline{x}\} \subseteq \underline{\text{clos}}(\alpha^{\underline{(e)}}, \underline{R}^{\underline{(e)}})]$, whence by Theor. 44.1, $\vdash_{\overline{c}} [\underline{\text{clos}}(\alpha, \underline{R})]^{\underline{(e)}} \subseteq \underline{\text{clos}}(\alpha^{\underline{(e)}}, \underline{R}^{\underline{(e)}})$. Then by Theor. 33.1 (54) holds.

(IV) Now to prove (55) we assume (a) $\underline{\text{Ext}}(\alpha)$ and (b) $\underline{\text{Ext}}(\underline{R})$. Thence by $(50)_2$ we deduce (c) $\alpha = \alpha^{\underline{(e)}}$ and (d) $\underline{R} = \underline{R}^{\underline{(e)}}$.

By Def. 6.12 and $(53)_2$, from (c) and (d) we deduce $\underline{\text{clos}}(\alpha, \underline{R}) = \underline{\text{clos}}(\alpha^{\underline{(e)}}, \underline{R}^{\underline{(e)}})$.

Thence by $(54)_1$, Def. 18.4, and AS40.1, we obtain $[\underline{\text{clos}}(\alpha, \underline{R})]^{\underline{(e)}} \subseteq \underline{\text{clos}}(\alpha, \underline{R})$. So by $(50)_1$ (and Theor. 33.2) we deduce $\underline{\text{clos}}(\alpha, \underline{R}) = [\underline{\text{clos}}(\alpha, \underline{R})]^{\underline{(e)}}$, which by $(50)_2$ yields $\underline{\text{Ext}}[\underline{\text{clos}}(\alpha, \underline{R})]$. Then by Theor. 33.2, (55) holds.

QED

N45. On natural numbers in MC^ν.

In N27 the natural absolute concept $\underline{\text{Nn}}$ of natural number was defined in ML^ν [Def. 27.5]; also some basic properties of $\underline{\text{Nn}}$ (and $\underline{\text{Nn}}^{\underline{(e)}}$) were considered from the semantical point of view.

In particular we assumed that each of our domains of individuals $\underline{D}_1, \ldots, \underline{D}_\nu$ [N6] has infinitely many elements. Then we showed—referring to [32]—that Peano's axioms hold in ML^ν both when the usual concepts of natural number, predecessor, and identity are taken to be, in order, $\underline{\text{Nn}}$, $\underline{\text{Pred}}^\frown$ [Defs. 27.4, 27.5], and $=^\frown$, and when they are taken to be, in order, $\underline{\text{Nn}}^{\underline{(e)}}$, $\underline{\text{Pred}}$, and $=$.

One of Peano's axioms cannot be a logical axiom. That is, from the semantical point of view it holds only if the domains of individuals $\underline{D}_1, \ldots, \underline{D}_\nu$ fulfill a certain condition (that of being infinite).

Such an axiom is e.g. axiom 13 in [32], p. 279. It has two semantical analogues for ML^ν, namely $(78)_1$ and $(78)_2$ in N27. The first refers to $\underline{Nn}$ and $=^\frown$, the other to $\underline{Nn}^{(e)}$ and $=$. The theorem $\Vdash \underline{Nn} \in \underline{MConst}$ [Def. 13.2]—see $(79)_1$ in N27—was proved in Memoir 1 without referring to these two semantical analogues and in a way that shows clearly how to prove

$$(57) \qquad \vdash \underline{Nn} \in \underline{MConst} \qquad [\text{Def. } 13.2]$$

from our preceding logical axioms.

Now we lay down an axiom similar to $SA(78)_1$ and $SA(78)_2$. From it we infer two analogues of Peano's nonlogical axiom, $SA(78)_1$ and $SA(78)_2$. Moreover we prove in MC^ν the syntactical analogues of those semantical theorems on natural numbers considered in N27 whose proofs were neither explicitly given there nor clearly suggested by any proof in [32].

AS45.1. $(\underline{N})\ \underline{m}, \underline{n} \in \underline{Nn}\ \underline{m}+1 = \underline{n}+1 \supset \underline{m} = \underline{n}$ [Defs. 27.3, 5].

Then by (57) and Defs. 13.2, 6.3, 6.7 and 18.9, we easily obtain $SA(78)_1$ and $SA(78)_2$, i.e.

$$(58) \quad \vdash \underline{m}, \underline{n} \in \underline{Nn} \wedge \underline{m}+1 =^\frown \underline{n}+1 \supset \underline{m} =^\frown \underline{n}, \quad \vdash \underline{m}, \underline{n} \in \underline{Nn}^{(e)} \wedge \underline{m}+1 = \underline{n}+1 \supset \underline{m} = \underline{n}.$$

THEOR. 45.1(I) $\underline{Nn} \in \underline{MSep}$ [Def. 18.7],

(II) $\underline{Nn} \in \underline{Abs}$ [Def. 18.8].

Proof. This proof is based on the proof of $(79)_{2,3}$ and in particular on SA(75). By AS40.1 and rule $\underline{C}$ with the variable β we obtain—cf. (80) in Memoir 1—

$$(59) \quad \vdash (\underline{x})\ \{\beta(\underline{x}) \equiv \underline{x} \in \underline{Nn}\ (\underline{y})\ [\underline{y} \in \underline{Nn}\ \underline{y} =^\smile \underline{x} \supset \underline{y} =^\frown \underline{x}]\},$$

which says that β is the modally separated part of $\underline{Nn}$.

Now we assume (a) $\beta(0)$, (b) $\underline{x} \in \beta \cap \underline{Nn}$, and—as a hypothesis for reductio ad absurdum—(c) $\sim \beta(\underline{x}+1)$. The matrix (b) and SA(74)$_3$ yield $\underline{x}+1 \in \underline{Nn}$, so that by (59) and (c) we have $(\exists \underline{y})\, [\underline{y} \in \underline{Nn} \wedge \underline{y} =^{\cup} \underline{x}+1 \wedge \sim \underline{y} =^{\cap} \underline{x}+1]$. So by rule $\underline{C}$ we obtain (d) $\underline{y} \in \underline{Nn}$, (e) $\underline{y} =^{\cup} \underline{x}+1$, and (e') $\sim \underline{y} =^{\cap} \underline{x}+1$.

By (d), (e), SA(74)$_1$, SA(76), and rule $\underline{C}$ with $\underline{m}$, we have (f) $\underline{m} \in \underline{Nn}\ \underline{y} =^{\cap} \underline{m}+1$. By AS45.1, from (d), (e), and (f) we deduce (g) $\underline{m} \in \underline{Nn}\ \underline{m} =^{\cup} \underline{x}$.

Furthermore, on the one hand (e') and (f) yield $\sim \underline{y} =^{\cap} \underline{x}+1 \wedge \underline{y} =^{\cap} \underline{m}+1$, whence by AS34.4 we deduce (h) $\sim \underline{m} =^{\cap} \underline{x}$. On the other hand by (59) and (b) we have $\underline{m} \in \underline{Nn}\ \underline{m} =^{\cup} \underline{x} \supset \underline{m} =^{\cap} \underline{x}$, so that by (g) we obtain $\underline{m} =^{\cap} \underline{x}$, which contradicts (h).

We conclude that (c) is not compatible with (b). Hence by Theor. 33.3 and rule $\underline{G}$, we obtain (i) $\vdash_{\underline{c}} (\underline{x})\, [\underline{x} \in \beta \cap \underline{Nn} \supset \beta(\underline{x}+1)]$.

By SA(75) from (a)—i.e. $\beta(0)$—and (i) we deduce $\underline{Nn} \subset \beta$, whence by (59) $\underline{Nn} = \beta$. Then by (59) and Def. 18.7 we have (1) $\vdash_{\underline{c}} \underline{Nn} \in \underline{MSep}$.

Since β, $\underline{y}$, and $\underline{m}$ do not occur free in (1), by Theor. 33.1, Theor. 45.1(I) holds.

By Def. 18.8, (57), and (I), Theor. 45.1(II) also holds.

QED

THEOR. 45.2(I) $\vdash (\lambda\, \alpha, \beta)\, (\alpha + \beta) \in \underline{Ext}$ [Def. 27.3],

(II) $\vdash \underline{Pred} \in \underline{Ext}$ [Def. 27.4],

(III) $\vdash \underline{clos}\, (\{0\}, \underline{Pred}) \in \underline{Ext}$,

(IV) $\underline{Nn}^{(e)} = \underline{clos}(\{0\}, \underline{Pred})$.

Proof. By Defs. 27.3, 35.2, 4.5, and Theor. 43.1, we obtain $\vdash (\underline{ext}\ \alpha, \beta)\, (\alpha + \beta)$, whence we easily deduce (I) by Defs. 35.2 and 6.12. Thence by Def. 27.4 we obtain (II).

By SA $(42)_1$—or by Defs. 6.11 and 18.12—we have $\vdash \{0\} \in \underline{\text{Ext}}$. So by (II) and (55), (III) holds.

To prove (IV) we observe that by Def. 27.5, $\underline{\text{Nn}}$ is $\underline{\text{clos}}(\{0\}^\frown$, $\underline{\text{Pred}}^\frown)$, so that from (54) we obtain (a) $\vdash \underline{\text{Nn}}^{(\underline{e})} \subseteq \underline{\text{clos}}(\{0\}^{\frown(\underline{e})}$, $\underline{\text{Pred}}^{\frown(\underline{e})})$. By Defs. 6.6, 18.12, 27.1, and 27.4, we obtain both[20] $\vdash \{0\}^{\frown(\underline{e})} =^\frown \{0\}$ and $\underline{\text{Pred}}^{\frown(\underline{e})} =^\frown \underline{\text{Pred}}$, so that by AS34.4, from (a) we obtain (b) $\vdash \underline{\text{Nn}}^{(\underline{e})} \subseteq \underline{\text{clos}}(\{0\}, \underline{\text{Pred}})$.

Now to prove the converse of (b) we want to use SA(67). So we first remark that by AS40.1 and rule $\underline{\text{C}}$ with β we have

$$(60) \qquad \vdash_{\underline{c}} (\underline{x})\,[\beta(\underline{x}) \equiv \underline{x} \in \underline{\text{clos}}(\{0\}, \underline{\text{Pred}})\, \underline{x} \in \underline{\text{Nn}}^{(\underline{e})}].$$

By SA$(74)_2$ $\vdash 0 \in \underline{\text{Nn}}$. Thence by $(50)_1$ and Def. 18.4 we obtain (c) $\vdash 0 \in \underline{\text{Nn}}^{(\underline{e})}$. Furthermore by SA$(65)_2$, (d) $\vdash \{0\} \subseteq \underline{\text{clos}}(\{0\}$, $\underline{\text{Pred}})$. By (60), from (c) and (d) we deduce (e) $\{0\} \subseteq \beta$.

Now we assume (f) $\underline{x} \in \underline{\text{clos}}(\{0\}, \underline{\text{Pred}}) \cap \beta$ and (g) $\underline{z} = \underline{x}+1$. By SA(66) and Def. 27.4, from (f) and (g) we deduce (h) $\underline{z} \in \underline{\text{clos}}(\{0\}$, $\underline{\text{Pred}})$.

On the other hand (f) yields $\beta(\underline{x})$, whence by (60), $\underline{x} \in \underline{\text{Nn}}^{(\underline{e})}$. Thence by Def. 18.9 and rule $\underline{\text{C}}$ with $\underline{y}$ we obtain (i) $\underline{y} = \underline{x}\,\underline{y} \in \underline{\text{Nn}}$, whence by SA$(74)_3$, (l) $\underline{y}+1 \in \underline{\text{Nn}}$. (i) yields $\underline{x} = \underline{y}$, whence we have $\underline{x}+1 = \underline{y}+1$. Thence by Def. 18.9, (g), and (l), we obtain $\underline{z} \in \underline{\text{Nn}}^{(\underline{e})}$. Thence by (h) and (60) we deduce $\beta(\underline{z})$.

We conclude that (f) and (g) yield $\beta(\underline{z})$. So by Theor. 33.3, Def. 27.4, and rule $\underline{\text{G}}$ we have (m) $(\forall \underline{x}, \underline{z})\,[\underline{x} \in \underline{\text{clos}}(\{0\}, \underline{\text{Pred}}) \cap \beta \wedge$ $\underline{\text{Pred}}(\underline{x}, \underline{z}) \supset \beta(\underline{z})]$.

By SA(67), from (e) and (m) we deduce $\underline{\text{clos}}(\{0\}, \underline{\text{Pred}}) \subseteq \beta$, which by (60) allows us to conclude with (n) $\vdash_{\underline{c}} \underline{\text{clos}}(\{0\}, \underline{\text{Pred}})$ $\subseteq \underline{\text{Nn}}^{(\underline{e})}$. By Theor. 33.1, (n) and (b) yield (IV).

QED

Notes to Memoir 2

1. Substantially, EC^{ν} can be obtained immediately by transferring Rosser's theory [32] of extensional logic to EL^{ν}.

2. We denote by $\overline{EL}^{\nu+1}$ that part of $EL^{\nu+1}$ which consists of the extensional translations of the designators in ML^{ν}. By the extensional translation $(MC^{\nu})^{\eta}$ of MC^{ν} we mean that calculus based on $\overline{EL}^{\nu+1}$ whose axioms are the extensional translations of the axioms of MC^{ν}.

3. The deduction and generalization theorems have already been proved in modal logic—cf. [3], [26], and [14], chap. 8—but for different systems and in a different way, i.e. directly. The main reason that we prove e.g. the deduction theorem for MC^{ν} is in some respects more general than the system considered by Barcan in [3], but the usefulness of proving all metatheorems for MC^{ν} by the same (new) procedure. Cf. fn. 4.

4. The ordinary and modal analogues for MC^{ν} of some metatheorems in extensional logic—e.g. those of Rosser's theorem VI.7.2 in [32] on rules $\underline{G}$ and $\underline{C}$—are stated in a single theorem [Theor. 33.1] to be called the total analogue of the former. As far as I know no such total analogue of Rosser's theorem has been proved in any preceding theory of modal logic. The same holds for the deduction theorem in connection with rules $\underline{C}$ and $\underline{G}$ [Theor. 33.3].

5. E.g. the well-known theorem on prenex forms has an ordinary analogue for MC^{ν} [Theor. 32.9] but not a modal one [N32].

6. From Theor. 38.1 it is apparent that the iota operator, used to construct descriptions, behaves in a strongly extensional way.

7. The semantical analogues of some of the (syntactical) theorems on $=$, $\neg$, $(\exists \underline{x})$, $(\exists_1 \underline{x})$ [Def. 11.3], and $(\exists_1^{\cap} \underline{x})$ [Def. 14.1] were enunciated in N14 but not proved. Since the calculus MC^{ν} is valid in ML^{ν} [Theor. 29.1], the proof in MC^{ν}—given explicitly here—for the syntactical theorems mentioned above immediately yields proofs for their semantical analogues. Some fallacies concerning $=$, $\neg$, $(\exists \underline{x})$, $(\exists_1 \underline{x})$, and $(\exists_1^{\cap} \underline{x})$ were shown in N14.

8. As in [11] and [12], strictly speaking no existence and uniqueness conditions are required for laying down formal definitions in ML^{ν} or EL^{ν}. However such conditions are needed to assure a common use of the definienda.

9. Def. 29.1 remains substantially unchanged if we replace rule (2) by the following rule (2′): $\underline{p}_i$ is a premise, i.e. one of $\underline{p}_1, \ldots, \underline{p}_m$.

10. Observe that ASs30.1–4 are ASs12.1–4; AS30.5 is AS12.6; AS30.6 is a special case of AS12.8; and for $\underline{r} = 1, 2, 3$, AS6 + $\underline{r}$ is AS12.3 + 2$\underline{r}$.

11. The following theorems in MC^ν correspond to theorems in [32]: $(4)_1$ to Theor. VI.5.1, p. 107; $(4)_{2,3}$ to Theor. VI.6.2.I, II, p. 115; $(5)_1$–$(6)_3$ to Theors. VI.6.5.I and VI.6.6.I, III, V, VII, IX, p. 119; $(7)_1$ to Theor. VI.7.5, p. 136; $(7)_2$ to Theor. VI.6.8.I, p. 121; $(7)_3$ to Theor. VI.7.3 and $(8)_1$ to its corollary, p. 134; $(8)_2$ to Theor. VI.7.4.I, p. 134; $(8)_3$ to Theor. VI.6.7.I, p. 120; $(9)_1$ to Theor. VI.7.6, p. 136; $(9)_2$ to Theor. VI.7.7, p. 136; and $(9)_3$ to Theor. VI.6.8.II, p. 121.

12. Theorems $(5')_{1,2,3}$, $(6')_{1,2,3}$, $(8')_{1,2,3}$, and $(9')_{1,2}$ are the modal analogues of $(5)_{1,2,3}$, $(6)_{1,2,3}$, $(8)_{1,2,3}$, and $(9)_{1,2}$. Note that $(7')_{\underline{i}}$ is, so to speak, the $\underline{i}$-th modal analogue of $(7)_1$ ($\underline{i} = 1,2,3$) and that, on the basis of the remarks preceding Theor. 31.4, the modal analogues of $(7)_2$ and $(7)_3$ need not be considered. $(9')_3$ follows easily from $(5')_2$.

13. The theory $\mathscr{T}$ may have some axioms that are not modally closed.

14. In connection with requirement (d), in N53 we shall make certain remarks about ordinary sentences, e.g. of astronomy, which are usually schematized within modal logic into sentences that are not modally closed.

15. The $\underline{i}$-th definition $\underline{p}_i$ in $\mathscr{T}$ is a totally closed matrix having the form $(\underline{N})\underline{c}_{\underline{i}}(\underline{x}_1, \ldots, \underline{x}_{\underline{n}}) \equiv \Delta_{\underline{i}}$ or the form $(\underline{N})\underline{c}_{\underline{i}}(\underline{x}_1, \ldots, \underline{x}_{\underline{n}}) = \Delta_{\underline{i}}$ [Def. 29.1]. Then $\underline{p}_{\underline{i}}^{\eta}$ is closed and in the first case has the form $(\ldots)\,\gamma_{\underline{i}}(\underline{y}_1, \ldots, \underline{y}_{\underline{n}}) \equiv \Delta_{\underline{i}}^{\eta}$, so that it can be accepted as a definition; in the second case $\underline{p}_{\underline{i}}^{\eta}$ has the form $(\ldots)(\kappa)\gamma_{\underline{i}}(\underline{y}_1, \ldots, \underline{y}_{\underline{n}}) = \frac{\underline{t}}{\kappa}\Delta_{\underline{i}}^{\eta}$; thus it cannot be a definition according to Def. 29.1. However by (2) [N30] $\underline{p}_{\underline{i}}^{\eta}$ is equivalent to a matrix of the form $(\ldots)$ $\gamma_{\underline{i}}(\underline{y}_1, \ldots, \underline{y}_{\underline{n}}) = \overline{\Delta}_{\underline{i}}^{\eta}$, which can constitute a definition in $EL^{\nu+1}$.

16. AS34.4 can be given a weaker form by assuming, in addi-

tion to our hypotheses, that $\underline{x}$, $\underline{y}$, $\underline{z}$ are distinct variables and $\underline{x}$ and $\underline{y}$ have no bound occurrences in $\Phi(\underline{x}, \underline{y}, \underline{z})$. In this second alternative form, AS34.4 is quite similar to axiom scheme 7A in [32], p. 163. By the preceding axiom schemes the first form of AS34.4 follows from the second. For the sake of brevity we accept the first form.

17. The proof of the syntactical analogue of theorem $(26)_1$ in N14 requires the axioms for descriptions [N38]. Therefore it will be considered later—see Theor. 39.5(II).

18. In part (III) of the proof no matrix has yet been denoted by assumption (c). In this case, by writing "(c) $\vdash (\forall \underline{x}, \underline{y}) \ldots$" we mean to denote the theorem $\vdash (\forall \underline{x}, \underline{y}) \ldots$ itself. If (c) had already denoted a matrix, then "(c) $\vdash (\forall \underline{x}, \underline{y}) \ldots$" would mean "the matrix (c) yields $(\forall \underline{x}, \underline{y}) \ldots$."

19. The matrix $(\exists \underline{x})\,[\Phi(\underline{x})\,\underline{N}\,\Psi(\underline{x})]$ in $(35)_1$ may hold even if $(\exists_1 \underline{x})\,\Phi(\underline{x})$ does not, but then $\Phi(\underline{a}^*)$ [Def. 11.2] must hold.

20. One can prove, in general, that

$$(*)\quad \vdash \underline{N}\,\underline{Ext}\,(\underline{F})\,(\exists \underline{x}_1, \ldots, \underline{x}_n)\,\underline{N}\,\underline{F}\,(\underline{x}_1, \ldots, \underline{x}_n) \supset \underline{F}^{\frown(\underline{e})} = \underline{F}.$$

This result can be strengthened on the basis of the theorem

$$(**)\quad \vdash \underline{N}\,\underline{Ext}\,(\underline{F})\,\underline{N}(\exists \underline{x}_1, \ldots, \underline{x}_n)\,\underline{F}\,(\underline{x}_1, \ldots, \underline{x}_n) \supset (\exists \underline{x}_1, \ldots, \underline{x}_n)\,\underline{N}\,\underline{F}\,(\underline{x}_1, \ldots, \underline{x}_n).$$

We can prove (**) using Zermelo's principle AS12.20 and a certain consequence of the strong axiom AS12.19, namely (90) in Theor. 49.2, presented in Memoir 3.

MEMOIR 3

Memoir 3: Elementary Possible Cases, Intensional Descriptions, and Completeness in the Modal Calculus MC^{ν}

N46. Introduction.

On the basis of the strong axiom AS12.19 we define [chaps. 9, 12] within MC^{ν} an analogue of the Γ-cases [N6]—i.e. the elementary possible cases—and an analogue of the semantical system constructed for ML^{ν} in an ordinary extensional metalanguage [chap. 2].

We define [chap. 10] an intensional description operator, $\iota_{\underline{u}}$, depending on a parameter (variable), $\underline{u}$, and a combination, $\imath_{\underline{u}}$, of $\iota_{\underline{u}}$ with the primitive description operator $\imath$ in MC^{ν}, which has an extensional character; furthermore we deal with a constant representing the real possible elementary case.

In chapter 12 we prove a theorem [Theor. 63.1] of relative completeness of MC^{ν}.

To describe the contents of this memoir in more detail, let us remember that in chapters 1 and 2 the ν-sorted modal language ML^{ν}—including attributes, functions, and descriptions of any type having a finite level—was defined, and extensional semantical rules were considered for it [Ns8, 11]. Furthermore the $\underline{L}$-truths [Def. 9.5] of the axioms of the ν-sorted modal calculus MC^{ν} based on ML^{ν} were proved. This calculus is studied in chapters 6 to 8 from the syntactical point of view, without however taking into account the strong axiom AS12.19, which in effect asserts that for every attribute $\underline{F}$ there is a modally constant attribute $\underline{G}$ [Def. 13.2] having the same extensionalization as $\underline{F}$ ($\underline{G}^{(\underline{e})} = \underline{F}^{(\underline{e})}$) [Def. 18.9].

In this memoir, we first define [Defs. 48.1, 48.2] on the basis of AS12.19 the absolute property $\underline{El}$ [Def. 18.8] and the matrix $|_{\underline{u}}$ (where $\underline{u}$ is a variable that occurs free in $|_{\underline{u}}$). By these definitions $\underline{u} \in \underline{El}$ and $|_{\underline{u}}$ in effect express that $\underline{u}$ is an elementary possible case and that the elementary possible case $\underline{u}$ takes place, respectively. This can be asserted on the basis of either semantical considerations or, independently, the syntactical results $(83)_{3,4}$ and $(\overline{89})$, i.e.

(a) $\vdash \underline{N}(\exists_1 \underline{u})\, |_{\underline{u}}$ [Def. 11.3], $\vdash (\underline{u})(\underline{u} \in \underline{El} \supset \diamond\, |_{\underline{u}})$,

(b) $\vdash \diamond\, (|_{\underline{u}} \underline{p}) \equiv \underline{u} \in \underline{El}\, \underline{N}(|_{\underline{u}} \supset^{\frown} \underline{p})$ for every matrix p.

In N52 we consider the calculus MC^{ν}_{ρ} obtained from MC^{ν} by adding the simple axiom $\rho \in \underline{El}$ [A52.1], where ρ is a constant denoting the "real elementary possible case." The sentence $|_{\rho}$ appears to coincide with the sentence $\underline{n}$, introduced as a primitive by Meredith and Prior in [27] in order to deal with counterfactual conditionals. So the strong axiom AS12.19 allows us to generalize the propositional constant $\underline{n}$ into the matrix $|_{\underline{u}}$ connected with any elementary possible case $\underline{u}$, and we do this without assuming anything primitive or stating any additional axiom.

The description operator $\imath$ in MC^{ν} has a rather extensional character—cf. $(30)_1$ and $(31)_1$ in N38. By an essential use of the matrix $|_{\underline{u}}$ we introduce, by means of a metalinguistic definition [Def. 50.1], an intensional description operator, $\iota_{\underline{u}}$, depending on the parameter (variable) $\underline{u}$.

The dependence of $|_{\underline{u}}$ on $\underline{u}$ causes $|_{\underline{u}}$ to appear among the conditions under which the description $(\iota_{\underline{u}} \underline{x})\underline{p}$ has the usual replacement property. This appears natural when we consider (1) certain ambiguities of ordinary language [N53], (2) the trouble caused by an intensional description operator that is independent of any parameter and that is used—unlike $|_\rho$—in connection with more than one elementary possible case [N54], and (3) the efficiency of MC^ν_ρ in affording various solutions of certain philosophical puzzles considered by Thomason and Stalnaker in [34] and [35], including some solutions sketched by these authors [N55]. Some comments will be made on the semantical theories used in [34] and [35] to solve those puzzles, some of which are not completely formal nor formalizable within rigorous theories of the same authors.

The theory of MC^ν developed up to N49 is sufficient for proving, along the lines of a suggestion by Lemmon—cf. [22]—and from a general point of view, that various systems of modal logic—precisely Fey's system $\underline{T}$, the Brouwerian system $\underline{B}$, S4, and S5—are not rivals but can coexist—see chapter 13. This proof is based in part on some analogues in MC^ν of Kripke's semantical theories for $\underline{T}$, $\underline{B}$, S4, and S5—cf. [22]. In this regard the Barcan formula and its converse are discussed in N69.

The combination $\text{ɿ}_{\underline{u}}$ of $\iota_{\underline{u}}$ and ɿ [N51] has the usual replacement property of descriptions whenever either $\iota_{\underline{u}}$ or ɿ has this property.

In chapter 12, using certain slightly modified versions ($|'_{\underline{u}}$ and $\underline{El}'$) of the matrix $|_{\underline{u}}$ and property $\underline{El}$, we first set up, within MC^ν itself, an analogue of the extensional semantical system constructed for ML^ν and based on the sets $\underline{D}_1, \ldots, \underline{D}_\nu, \underline{D}_{\nu+1}$ (= Γ)—cf. Ns6, 7. The analogues—defined within MC^ν—of the QIs defined for ML^ν in chapter 2 are completely formed with absolute concepts and allow us to consider a certain translation ($\Delta \to \Delta^*$), called the <u>star translation</u>, of the extensional languages EL^ν and $EL^{\nu+1}$ into ML^ν. Incidentally this translation is different from Carnap's translations of extensional languages into modal languages—see [13], p. 894—in that the star translation of a predicate $\underline{F}$ (considered in an extensional language) is an absolute predicate, whereas Carnap's translation of $\underline{F}$ would be an extensional predicate.

The entailment relation is invariant for the star translation, and every totally closed matrix $\underline{p}$ in MC^{ν} is strictly equivalent to a matrix constructed with the absolute concepts mentioned above, more precisely, with the star translation $\underline{p}^{\eta *}$ of the extensional translation $\underline{p}^{\eta}$ of $\underline{p}$ into $EL^{\nu+1}$ [Ns15, 62].

This result enables us to generalize the invariance of the entailment relation under the extensional translation ($\Delta \rightarrow \Delta^{\eta}$) of MC^{ν} into $EC^{\nu+1}$ proved in N31. On the one hand this invariance consists of the fact that $\underline{p}_1, \ldots, \underline{p}_{\underline{m}}$ entail $\underline{p}_0$ in MC^{ν} if and only if $\underline{p}_1^{\eta}, \ldots, \underline{p}_{\underline{m}}^{\eta}$ entail $\underline{p}_0^{\eta}$ in $(MC^{\nu})^{\overline{\eta}}$—and $(MC^{\nu})^{\eta}$ includes, according to its definition considered in Memoir 2, fn. 2, the lower predicate calculus of $EC^{\nu+1}$; however, a priori, it is not clear whether every theorem in $EC^{\nu+1}$ consisting of a matrix in $(MC^{\nu})^{\eta}$ is also a theorem in $(MC^{\nu})^{\eta}$. On the other hand the aforementioned result concerning the star translation enables us to prove [Theor. 63.1] that $\underline{p}_1, \ldots, \underline{p}_{\underline{m}}$ entail $\underline{p}_0$ in MC^{ν} if and only if $\underline{p}_1^{\eta}, \ldots, \underline{p}_{\underline{m}}^{\eta}$ entail $\underline{p}_0^{\eta}$ in $EC^{\nu+1}$.

Since the extensional translation $\Delta \rightarrow \Delta^{\eta}$ of ML^{ν} into $EL^{\nu+1}$ characterizes our semantical system for ML^{ν} [chap. 2], Theor. 63.1 substantially expresses the relative completeness of MC^{ν} with respect to this semantical system [N63]. More precisely Theor. 63.1 in effect implies that $\underline{p}_1, \ldots, \underline{p}_{\underline{m}}$ entail $\underline{p}_0$ in MC^{ν} if and only if we can prove in our (extensional) semantical metalanguage that $\underline{p}_1, \ldots, \underline{p}_{\underline{m}}$ $\underline{L}$-imply $\underline{p}_0$ in our semantical system for ML^{ν} (which, as above, is understood not to be more specified than it is in chap. 2—cf. N64).

So the axioms $\mathscr{A}$ of MC^{ν} which are quite different from the axioms of common extensional and modal calculi and are to be added to the other axioms of MC^{ν}, say $\mathscr{A}'$, in order to obtain relative completeness with respect to our semantical system for ML^{ν} [Ns6, 7, 11]—are, at most, the strong axiom AS12.19 and AS25.1, which asserts that the number of individuals is the same in all possible cases. More, it is obvious [N64] that the axioms $\mathscr{A}$ must be equivalent with the last two axioms.

Let us add that in our opinion we can prove the analogue of Theor. 63.1 (of relative completeness) for the semantical system briefly considered for ML^{ν} in N25. In this system AS12.19 is valid and AS25.1 is not. As a consequence, AS12.19 is the only axiom scheme that is (obviously) independent of the axioms $\mathscr{A}'$, whose validity in the semantical system for ML^{ν} considered in N25 can be proved in the common extensional metalanguage.

9

Definition in MC^{ν} of Some Analogues of the Γ-Cases and Their Occurrence

N47. Definition in ML^{ν} of the first analogue ElR of the concept of elementary possible case.

In chapter 12 we show that a matrix is provable in MC^{ν} if and only if its extensional translation [N15] can be proved in $EC^{\nu+1}$; i.e. in effect we replace $(MC^{\nu})^{\eta}$ with $EC^{\nu+1}$ in Theor. 31.1. In this chapter we lay the groundwork; in particular we want to construct within ML^{ν} itself some preliminary analogues of the Γ-cases and quasi intensions that were considered within our semantical metalanguage for ML^{ν} in N7. Some of these preliminaries are interesting in themselves, as is the definition in MC^{ν} of some analogues of the Γ-cases [Ns47, 48], especially in that the properties of these analogues can be proved through an essential use of the strong axiom AS12.19 and independently of AS25.1. The same holds for our translation $\Delta \rightarrow \Delta^*$ of the extensional language EL^{ν} into the modal language ML^{ν}. In particular this translation may be interesting in that to carry it out we use absolute properties and not extensional ones, as suggested by Carnap—cf. [13]. The introduction of our intensional description operators in chapter 10 is also essentially based on the present chapter.

Now, to deal with the subject mentioned in the section heading, we consider an arbitrary matrix $\underline{p}$ in ML^{ν}. The natural number concept

$$\xi =_D (\imath \underline{x})(\underline{px} = 1 \vee \sim \underline{px} = 0), \tag{61}$$

where $\underline{x}$ is the first variable that has the same type as the term 1 and that does not occur free in $\underline{p}$, characterizes the range of $\underline{p}$. It is natural to say that the range of $\underline{p}$ is <u>proper</u> if it is neither Γ nor Λ, i.e. if both $\underline{p}$ and $\sim\underline{p}$ are possible. For the sake of brevity we call ξ the range of $\underline{p}$ and we define <u>proper range</u> (<u>PR</u>) in ML^{ν} as follows:

DEF. 47.1. $\underline{PR}(\underline{x}) \equiv_D \underline{N}(\underline{x} = 0 \vee \underline{x} = 1)\underline{x} =^{\smile} 0 \underline{x} =^{\smile} 1.$

Ordinarily the possible case where $\underline{p}$ holds is not an elementary possible case in that it has possible (proper) subcases, i.e. a matrix $\underline{q}$ can be found such that both $\underline{pq}$ and $\underline{p}\sim\underline{q}$ are possible. In terms of Γ-cases the range of $\underline{p}$ is not elementary in that it has more than one element.

Incidentally, if $\underline{N}(\underline{x} = 0 \vee \underline{x} = 1)$ is true, $\underline{x}$ could briefly be called a range.

In accordance with these considerations we now introduce in ML the concept of (part of proper range, or briefly) <u>subrange</u> (<u>SubR</u>)

and the concept of elementary range (ElR), using natural number concepts like those in (61) and Def. 47.1:

DEF. 47.2. $\underline{\text{SubR}}(\underline{x}, \underline{y}) \equiv_D \underline{x}, \underline{y} \in \underline{\text{PR}} \wedge \underline{x} \leq^{\cap} \underline{y}$,

DEF. 47.3. $\underline{\text{ElR}}(\underline{x}) \equiv_D (\underline{x} \in \underline{\text{PR}})(\underline{y})[\underline{\text{SubR}}(\underline{y}, \underline{x}) \supset^{\cap} \underline{x} =^{\cap} \underline{y}]$.

From Defs. 47.1–3 and 13.2 we obviously have

(62) $\vdash \underline{\text{PR}} \in \underline{\text{MConst}}$, $\vdash \underline{\text{SubR}} \in \underline{\text{MConst}}$, $\vdash \underline{\text{ElR}} \in \underline{\text{MConst}}$.

We want to prove that there exists exactly one elementary possible case that actually holds—cf. $(63)_{4,5}$ below. In the proof of this theorem we use the (new) strong axiom AS12.19, the version AS12.23 of the axiom (very common in modal logic) that asserts the existence of contingent propositions, and not axiom AS25.1 on the numbers of individuals in different Γ-cases. We conjecture that the use of AS12.19 is essential in the proof of Theor. 47.1 for proving the existence of an elementary case that actually holds.

THEOR. 47.1. The following holds in MC^{ν}:

$$(63) \quad \begin{array}{l} \vdash (\exists \underline{x})(\underline{x} \in \underline{\text{ElR}}\ \underline{x} = 1), \quad \vdash (\exists^{(1)^{\cap}} \underline{x})(\underline{x} \in \underline{\text{ElR}}\ \underline{x} = 1) \\ \vdash \underline{x} \in \underline{\text{ElR}} \supset \Diamond\ \underline{x} = 1, \quad \vdash (\exists_1^{\cap} \underline{x})(\underline{x} \in \underline{\text{ElR}}\ \underline{x} = 1), \\ \vdash (\exists_1 \underline{x})(\underline{x} \in \underline{\text{ElR}}\, \underline{x} = 1). \end{array}$$

Proof. (I) By AS12.19 and rule C with F we obtain

(64) $\vdash_{\underline{c}} (\underline{x})[\underline{F}(\underline{x}) \equiv \underline{x} = 1\, \underline{N}(\underline{x} = 1 \vee \underline{x} = 0 \mid] \wedge \underline{F}^{\cup} = \underline{F}^{\cap}$ [Defs. 6.6, 6:8].

Speaking intuitively, we now construct the number ξ_1 representing the actual elementary possible case, say the one holding in the Γ-case γ, as follows. We consider that number ξ_1 such that (1) in the Γ-case γ ξ_1 equals 1 and (2) in every other Γ-case γ' ξ_1 equals the smallest number $\underline{x}$ that represents a proper range and equals 1 in γ. Taking (64) into account, we set

$$\xi_1 =_D (\iota \underline{y}) \Phi(\underline{y}) \text{ where}$$
$$(65) \qquad \Phi(\underline{y}) \equiv_D (\exists \underline{x})(\underline{x} \in \underline{F}\underline{x} = 0)\underline{y} = 0 \vee \sim (\exists \underline{x})(\underline{x} \in \underline{F}\underline{x} = 0)\underline{y} = 1.$$

Obviously $\vdash \underline{N}(\exists_1 \underline{y})\Phi(\underline{y})$ and $\vdash \underline{N}(\underline{\text{ext}}\ \underline{y})\Phi(\underline{y})$, so that by $(31)_2$,

$$(66) \quad \vdash \underline{N}[(\exists \underline{x})(\underline{x} \in \underline{F}\underline{x} = 0)\xi_1 = 0 \vee \sim(\exists \underline{x})(\underline{x} \in \underline{F}\underline{x} = 0)\ \xi_1 = 1].$$

From As12.23, by rule $\underline{C}$ with $\underline{v}_{11}$ and $\underline{v}_{(1)1}$ we obtain

$$(67) \qquad \vdash_{\underline{c}} \diamond \overline{\underline{p}} \diamond \sim\overline{\underline{p}} \quad \text{where} \quad \overline{\underline{p}} \equiv_D \underline{v}_{11} \in \underline{v}_{(1)1}.$$

By (64), $\vdash_{\underline{c}} \sim(\exists \underline{x})(\underline{x} \in \underline{F}\underline{x} = 0)$, so that by (66) we have (a) $\vdash_{\underline{c}} \xi_1 = 1$, whence (b) $\vdash_{\underline{c}} \xi_1 =^{\cup} 1$. Furthermore (66) implies (c) $\vdash \underline{N}(\xi_1 = 0 \vee \xi_1 = 1)$.

Let ξ_2 and ξ_3 be what ξ—cf. (61)—becomes for the respective instances $\overline{\underline{p}}$ and $\sim\overline{\underline{p}}$ of $\underline{p}$. Then by (61), $(67)_1$, and $\vdash \overline{\underline{p}} \vee \sim\overline{\underline{p}}$, we have

$$(68) \quad \vdash \bigvee_{\underline{r}=2}^{3} \xi_{\underline{r}} = 1 \wedge \xi_{\underline{r}} =^{\cup} 0 \wedge \underline{N}(\xi_{\underline{r}} = 1 \vee \xi_{\underline{r}} = 0)] \qquad \text{[Def. 6.10]},$$

whence by (64) we obtain $\vdash_{\underline{c}} \underline{F}(\xi_2) \vee \underline{F}(\xi_3)$ and (d) $\vdash_{\underline{c}} \underline{F}^{\cap}(\xi_2) \vee \underline{F}^{\cap}(\xi_3)$.

Since (67) and the definitions of ξ_2 and ξ_3 imply $\vdash \xi_2 =^{\cup} 0 \wedge \xi_3 =^{\cup} 0$, by (d) we have $\vdash_{\underline{c}} \diamond (\exists \underline{x})(\underline{x} \in \underline{F}\underline{x} = 0)$—cf. (64). By (66) this implies (e) $\vdash_{\underline{c}} \xi_1 =^{\cup} 0$. From (b), (c), (e), and Def. 47.1, we have (f) $\vdash_{\underline{c}} \underline{PR}(\xi_1)$. Thus by (a),

$$(69) \qquad \vdash_{\underline{c}} \xi_1 = 1 \wedge \underline{PR}(\xi_1).$$

Now we note that for the arbitrary matrix $\underline{p}$, (61) implies

$$(70) \qquad \vdash \underline{p} \supset^{\cap} \xi = 1, \qquad \vdash \sim\underline{p} \supset^{\cap} \xi = 0, \qquad \vdash \underline{N}(\xi = 1 \vee \xi = 0),$$

whence by (64) we obtain $\underline{p} \vdash_{\underline{c}} \underline{F}^{\cap}(\xi)$. So by $(70)_2$, $\underline{p} \vdash_{\underline{c}} \sim\underline{p} \supset^{\cap} \underline{F}(\xi)\ \xi = 0$. Then $\underline{p} \vdash_{\underline{c}} \sim\underline{p} \supset^{\cap} (\exists \underline{x})(\underline{x} \in \underline{F}\underline{x} = 0)$, whence by (66) we conclude that

(71) $\underline{p} \vdash_{\underline{c}} \sim\underline{p} \supset^{\cap} \xi_1 = 0$ for every matrix $\underline{p}$.

Taking $\underline{p}$ to be $\underline{x} = 1$ or $\underline{x} = 0$, from (71) we obtain, respectively [Theor. 33.3],

(72) $\vdash_{\underline{c}} \underline{x} = 1 \supset (\underline{x} = 0 \supset^{\cap} \xi_1 = 0)$, $\vdash_{\underline{c}} \underline{x} = 0 \supset (\underline{x} = 1 \supset^{\cap} \xi_1 = 0)$.

We now assume

(73) $\underline{\text{SubR}}\,(\underline{x}, \xi_1)$, whence by Def. 47.2, $\underline{x} \in \underline{\text{PR}}$ and $\underline{x} \leq^{\cap} \xi_1$.

From $(73)_2$ and Def. 47.1 we have $\underline{x} =^{\cup} 1$, which by $(72)_2$ yields $\underline{x} = 0 \supset \diamond (\underline{x} = 1 \wedge \xi_1 = 0)$, whence $\underline{x} = 0 \supset \sim\underline{x} \leq^{\cap} \xi_1$. So $(73)_3$ implies $\sim\underline{x} = 0$, whence by $(73)_2$ and Def. 47.1 we obtain (g) $\underline{x} = 1$.

From (g) and $(72)_1$ we obtain (h) $\underline{x} = 0 \supset^{\cap} \xi_1 = 0$. By Defs. 47.1 and 47.2, $(73)_1$ yields $\underline{N}(\underline{x} = 1 \vee \underline{x} = 0)$ and $\underline{N}(\xi_1 = 1 \vee \xi_1 = 0)$. So by (h) and $(73)_3$ we obtain $\underline{x} =^{\cap} \xi_1$.

By the deduction theorem Theor. 33.3, (ordinary) rule $\underline{G}$ [Defs. 33.1(4), 33.2], and $(62)_2$, we conclude that

(74) $\vdash_{\underline{c}} (\underline{x})\,[\underline{\text{SubR}}\,(\underline{x}, \xi_1) \supset^{\cap} \underline{x} =^{\cap} \xi_1]$.

From (69), (74), and Def. 47.3 we obtain

(75) $\vdash_{\underline{c}} \xi_1 = 1\ \underline{\text{ElR}}\,(\xi_1)$,

whence $\vdash_{\underline{c}} (\exists\underline{x})\,[\underline{x} = 1\ \underline{\text{ElR}}\,(\underline{x})]$, which implies $(63)_1$ by Theor. 33.1.

(II) We assume

(76) $\underline{x} = 1$, $\underline{\text{ElR}}\,(\underline{x})$,

whence by Def. 47.3 we have

(77) $(\underline{y})\,[\underline{\text{SubR}}\,(\underline{y}, \underline{x}) \supset \underline{y} =^{\cap} \underline{x}]$, $\underline{x} \in \underline{\text{PR}}$.

By $(72)_1$, $(76)_1$ yields $\underline{x} = 0 \supset^{\cap} \xi_1 = 0$. Furthermore, by $(77)_2$, (69), and Def. 47.1 we have $\underline{N}(\underline{x} = 1 \vee \underline{x} = 0)$ and $\underline{N}(\xi_1 = 1 \vee \xi_1 = 0)$. So we have $\xi_1 \leq \underline{x}$. By (69), $(77)_2$, and Def. 47.2 this implies

$\underline{\text{SubR}}(\xi_1, \underline{x})$, whence by $(77)_1$ we have $\underline{x} = {}^\frown\xi_1$. By Theors. 33.3 and rule $\underline{G}$ we conclude that $\vdash_{\underline{c}} (\underline{x}) [\underline{x} = 1\ \underline{\text{ElR}}(\underline{x}) \supset \underline{x} = {}^\frown\xi_1]$, whence we easily obtain $\vdash_{\underline{c}} (\forall\underline{x}, \underline{y}) [\underline{x}, \underline{y} \in \underline{\text{ElR}}\,\underline{x} = 1 = \underline{y} \supset \underline{x} = {}^\frown\underline{y}]$.

Then by Theor. 33.1 and Def. 34.2 we obtain $(63)_2$.

(III) From Defs. 47.1 and 47.3 we deduce $(63)_3$.

(IV) From $(63)_{1,2}$ and $(14)_{1,2,3}$ we obtain $(63)_4$. Thence $(63)_5$ follows by $(27)_1$.

N48. The second analogue El of the concept of elementary possible case.

Now we want to prove in MC^ν that the class $\underline{\text{ElR}}$ has only two elements, so that it is expedient to represent the elementary possible cases (or ranges) by means of another class, $\underline{\text{El}}$, closely related to $\underline{\text{ElR}}$.

By (75), $(63)_2$, and Def. 34.2, $\underline{\text{ElR}}(\underline{x})\underline{x} = 1$ yields $\underline{x} = {}^\frown\xi_1$. Then by (71) we have $\underline{p} \wedge \underline{x} \in \underline{\text{ElR}} \wedge \underline{x} = 1 \vdash_{\underline{c}} {\sim}\underline{p} \supset {}^\frown\underline{x} = 0$, whence by Theors. 33.1 and 33.3,

$$(78) \qquad \vdash \underline{p}\underline{x} \in \underline{\text{ElR}}\,\underline{x} = 1 \supset ({\sim}\underline{p} \supset {}^\frown\underline{x} = 0) \quad \text{for every matrix } \underline{p}.$$

Thence by the modal generalization theorem included in Theor. 32.2, and by $(9')_1$ and $(5')_2^*$, we deduce

$$(79) \qquad \vdash \diamond (\underline{p}\underline{x} \in \underline{\text{ElR}}\,\underline{x} = 1) \supset ({\sim}\underline{p} \supset {}^\frown\underline{x} = 0) \text{ for every matrix } \underline{p}.$$

By $(67)_1$ we have (a) $\diamond\,\overline{\underline{p}}$ and (b) $\diamond {\sim}\overline{\underline{p}}$. Furthermore $(63)_1$ implies $\vdash \underline{N}(\exists\underline{x}) (\underline{x} \in \underline{\text{ElR}}\,\underline{x} = 1)$, so that from (a) and (b) we have, respectively,

$$(80) \quad \diamond\, [\overline{\underline{p}}(\exists\underline{x}) (\underline{x} \in \underline{\text{ElR}}\,\underline{x} = 1)], \qquad \diamond\, [{\sim}\overline{\underline{p}}(\exists\underline{x}) (\underline{x} \in \underline{\text{ElR}}\,\underline{x} = 1)].$$

We now assume (c) $\overline{\underline{p}}$, and from $(63)_1$ we deduce (d) $\vdash_{\underline{c}} \underline{z} \in \underline{\text{ElR}}\,\underline{z} = 1$. Since $\underline{x}$ cannot occur in $\underline{p}$—cf. $(67)_2$—$(80)_2$ yields $(\exists\underline{x})\ \diamond [{\sim}\overline{\underline{p}}(\underline{x}$

$\in \underline{\mathrm{ElR}})\underline{x} = 1]$, whence by rule $\underline{C}$ with $\underline{x}$ we obtain (e) $\diamond(\sim\overline{\underline{p}}\ \underline{x} \in \underline{\mathrm{ElR}}\,\underline{x} = 1)$. Thence by (79) with $\underline{p}$ identified with $\sim\overline{\underline{p}}$, we have $\overline{\underline{p}} \supset^{\frown} \underline{x} = 0$, which by (c) implies (f) $\underline{x} = 0$. From $(62)_3$ and (e) we also have (g) $\underline{x} \in \underline{\mathrm{ElR}}$.

The results (d), (f), and (g) yield $\underline{x}, \underline{z} \in \underline{\mathrm{ElR}}\,\underline{x} \neq \underline{z}$. Then by Defs. 47.3 and 47.1 we also have $(\underline{y})\ [\underline{y} \in \underline{\mathrm{ElR}} \supset (\underline{y} = \underline{x} \vee \underline{y} = \underline{z})]$. So we easily deduce $\underline{\mathrm{ElR}} \in 2$.

The same result is reached beginning with the assumption $\sim\overline{\underline{p}}$ instead of (c) and using $(80)_1$ instead of $(80)_2$. Thus we conclude that

$$\vdash \underline{\mathrm{ElR}} \in 2. \tag{81}$$

Of course, intuitively, the elementary ranges—or the elementary possible cases—are usually much greater than 2; they are as many as the Γ-cases. In order to comply with this fact and to represent the Γ-cases in ML^{ν} by means of a concept that, unlike $\underline{\mathrm{ElR}}$, is absolute, we introduce within ML^{ν} the class $\underline{\mathrm{El}}$ of <u>absolute elementary ranges</u>:

DEF. 48.1. $\underline{u} \in \underline{\mathrm{El}} \equiv_D (\exists \underline{x})\ (\underline{x} \in \underline{\mathrm{ElR}}\ \underline{u} =^{\frown} \{\underline{x}\}^{(\underline{i})})$ —cf. Def. 18.13—

where the variable $\underline{x}$ has the index 1.

From Def. 48.1 $(62)_3$, and Def. 13.2, the first of the following theorems is easily deduced:

$$\vdash \underline{\mathrm{El}} \in \underline{\mathrm{MConst}}, \quad \vdash \underline{\mathrm{El}} \in \underline{\mathrm{MSep}}, \quad \vdash \underline{\mathrm{El}} \in \underline{\mathrm{Abs}}. \tag{82}$$

Now we assume (a) $\underline{u}, \underline{v} \in \underline{\mathrm{El}}$ and (b) $\underline{u} =^{\smile} \underline{v}$. By (a), Def. 48.1, and rule $\underline{C}$ with $\underline{x}$ and $\underline{y}$, we obtain (c) $\underline{u} =^{\frown} \{\underline{x}\}^{(\underline{i})}\ \underline{v} =^{\frown} \{\underline{y}\}^{(\underline{i})}$ and (d) $\underline{x}, \underline{y} \in \underline{\mathrm{ElR}}$.

Def. 18.13, ASs12.11, 12.12, (b), and (c) yield $\diamond\ \{\underline{x}\}^{(\underline{i})} = \{\underline{y}\}^{(\underline{i})}$, whence by Def. 18.13, $\diamond\ \underline{x} =^{\frown} \underline{y}$. Thence we have $\underline{x} =^{\frown} \underline{y}$, and by (c) we deduce (d) $\underline{u} =^{\frown} \underline{v}$. We conclude that (a) $\wedge$ (b) $\vdash_{\underline{C}}$ (d), so that by

Theors. 33.3 and 33.1 we obtain $\vdash \underline{u}, \underline{v} \in \underline{El}\,\underline{u} = {}^{\cup}\underline{v} \supset \underline{u} = {}^{\cap}\underline{v}$. Thence $(82)_2$ follows by Theor. 32.2 and Def. 18.7.

By Def. 18.8, $(82)_{1,2}$ yield $(82)_3$.

The following abbreviating definition is useful:

DEF. 48.2. $|_{\underline{u}} \equiv_D \underline{u} \in \underline{El} \wedge (\imath \underline{x})\, \underline{u}\, (\underline{x}) = 1.$

Now let us remark that by Def. 48.1, $(82)_2$, $(63)_{1,4,5,3}$, and Theor. 41.2(I), (II), we easily obtain

$$(83)\quad \begin{array}{l} \vdash (\exists \underline{u})\, |_{\underline{u}}, \quad \vdash (\exists^{(1)} \underline{u})\, |_{\underline{u}}, \quad \vdash (\exists_1 \underline{u})\, |_{\underline{u}}, \quad \vdash \underline{u} \in \underline{El} \supset \diamond\, |_{\underline{u}}, \\ \vdash (\exists^{(1)} {}^{\cap}\underline{u})\, |_{\underline{u}}, \quad \vdash (\exists_1^{\cap} \underline{u})\, |_{\underline{u}}. \end{array}$$

Intuitively, by Def. 48.2 and $(83)_2$ $|_{\underline{u}}$ is a translation into MC^{ν} of "γ is the elementary possible case that actually holds."

N49. Characterizations of Np and $\diamond$ p using ElR and El. A property of El.

Let us assume $\underline{x} \in \underline{ElR}$. Then both $\diamond (\underline{x} = 1\, \underline{p})$ and $\underline{x} = 1 \supset {}^{\cap}\underline{p}$ express in ML^{ν} that the (arbitrary) matrix $\underline{p}$ holds in the Γ-case characterized by $\underline{x}$—cf. (85) below. So $\underline{Np}$ and $\diamond\, \underline{p}$ can be characterized in ML^{ν} in a way similar to their characterizations in our semantical metalanguage [N8]. More precisely we prove the following theorem:

THEOR. 49.1. For every matrix $\underline{p}$ and variable $\underline{x}$,

$$(84)\qquad \vdash \underline{x} \in \underline{ElR} \supset [\diamond (\underline{x} = 1\, \underline{p}) \equiv \underline{x} = 1 \supset {}^{\cap}\underline{p}],$$

and if $\underline{x}$ does not occur free in $\underline{p}$,

$$(85)\quad \vdash (\underline{x})\, [\underline{x} \in \underline{ElR} \supset \diamond (\underline{x} = 1\, \underline{p})] \equiv \underline{Np} \equiv (\underline{x})\, [\underline{x} \in \underline{ElR} \supset (\underline{x} = 1 \supset {}^{\cap}\underline{p})],$$

$$(86)\quad \vdash (\exists \underline{x})\, [\underline{x} \in \mathrm{ElR}\, (\underline{x} = 1 \supset {}^{\cap}\underline{p})] \equiv \diamond\, \underline{p} \equiv (\exists \underline{x})\, [\underline{x} \in \underline{ElR}\, \diamond (\underline{x} = 1\, \underline{p})].$$

Proof. (I) We assume (a) $\underline{x} \in \underline{ElR}$, (b) $\diamond\,(\underline{x} = 1\,\underline{p})$, and (c) $\underline{x} = 1$. From (a), (b), and $(62)_3$ we obtain $\diamond\,(\underline{x} \in \underline{ElR}\,\underline{u} = 1\,\underline{p})$, which by (79) yields (d) $\sim\underline{p} \supset^{\cap} \underline{x} = 0$.

Furthermore (c) yields $\sim\underline{x} = 0$, whence by (d) we obtain $\underline{p}$.

We conclude [Theor. 33.2] that (a), (b) $\vdash \underline{x} = 1 \supset \underline{p}$. Then by $(62)_3$ and Theor. 32.2 we have

$$(87) \qquad \underline{x} \in \underline{ElR},\ \diamond\,(\underline{x} = 1\,\underline{p}) \vdash \underline{x} = 1 \supset^{\cap} \underline{p}.$$

Now we conversely assume (a) $\underline{x} \in \underline{ElR}$ and (e) $\underline{x} = 1 \supset^{\cap} \underline{p}$. From (a) and $(63)_3$ we deduce $\underline{x} =^{\cup} 1$, whence by (e), $\diamond\,(\underline{x} = 1\,\underline{p})$. We conclude that $\underline{x} \in \underline{ElR}$, $\underline{x} = 1 \supset^{\cap} \underline{p} \vdash \diamond\,(\underline{x} = 1\,\underline{p})$.

Then by (87) we easily deduce (84) [Theor. 33.2].

(II) We assume (a) $\underline{Np}$ and (b) $\underline{x} \in \underline{ElR}$ where the variable $\underline{x}$ does not occur free in $\underline{p}$. By $(63)_3$, (b) yields $\underline{x} =^{\cup} 1$, so that by (a) we have (c) $\diamond\,(\underline{p}\,\underline{x} = 1)$. By Theor. 33.2 $\underline{Np} \vdash$ (b) $\supset$ (c), whence by Theor. 32.2, $\underline{Np} \vdash (\underline{x})$ [(b) $\supset$ (c)] (for $\underline{x}$ does not occur free in $\underline{p}$). By Theor. 33.2 this yields (d) $\vdash \underline{Np} \supset (\underline{x})\,[\underline{x} \in \underline{ElR} \supset \diamond\,(\underline{p}\,\underline{x} = 1)]$.

Now we conversely assume (e) $\diamond \sim\underline{p}$. By $(63)_1$, the modal generalization theorem included in Theor. 33.2, and (e), we have $\diamond\,[\sim\underline{p}\,(\exists\underline{x})\,(\underline{x} \in \underline{ElR}\,\underline{x} = 1)]$, whence since $\underline{x}$ does not occur free in $\underline{p}$, we deduce $(\exists\underline{x})\,\diamond\,[\sim\underline{p}\,\underline{x} \in \underline{ElR}\,\underline{x} = 1]$, which by rule $\underline{C}$ yields (f) $\diamond\,(\sim\underline{p}\,\underline{x} \in \underline{ElR}\,\underline{x} = 1)$.

Thence by (79), we deduce $\underline{p} \supset^{\cap} \underline{x} = 0$, so that (g) $\sim \diamond\,(\underline{p}\,\underline{x} = 1)$ holds.

Furthermore (f) and $(62)_3$ yield $\underline{x} \in \underline{ElR}$. This and (g) yield (h) $(\exists\underline{x})\,[\underline{x} \in \underline{ElR} \sim \diamond\,(\underline{p}\,\underline{x} = 1)]$.

We conclude that $\vdash$ (e) $\supset$ (h), whence $\vdash \sim$(h) $\supset \sim$(e), which can be written $\vdash (\underline{x})\,[\underline{x} \in \underline{ElR} \supset \diamond\,(\underline{p}\,\underline{x} = 1)] \supset \underline{Np}$.

At this point, remembering (d), we deduce $(85)_1$.

By (84), $(85)_2$ easily follows from $(85)_1$.

(III) By $(85)_1$, $\vdash \sim \underline{N} \sim \underline{p} \equiv (\exists \underline{x}) [\underline{x} \in \underline{ElR} \sim \diamond (\underline{x} = 1 \sim \underline{p})]$, whence $(86)_1$. By (84), $(86)_2$ easily follows from $(86)_1$.

QED

The forthcoming theorem is the analogue of Theor. 49.1 for $\underline{El}$.

As a preliminary we remark that by Defs. 18.13 and 14.1 we have

(88) $\vdash (\exists_1^{\frown} \underline{x}) \{\underline{x}\}^{(i)} = \underline{u}$ whence $\vdash \underline{u} = {}^{\frown}\{\underline{x}\}^{(i)} \supset (\iota \underline{x})\, \underline{u}\, (\underline{x}) = {}^{\frown}\underline{x}$ — cf. ASs40.1, 38.1(I).

THEOR. 49.2. For every matrix $\underline{p}$ in ML^{ν} we have [Def. 48.2]:

(89) $\vdash \underline{u} \in \underline{El} \supset [\diamond (\mid_{\underline{u}} \underline{p}) \equiv {}^{\frown} (\mid_{\underline{u}} \supset {}^{\frown} \underline{p})]$,

$(\overline{89})$ $\vdash \diamond (\mid_{\underline{u}} \underline{p}) \equiv {}^{\frown}\, \underline{u} \in \underline{El}\, (\mid_{\underline{u}} \supset {}^{\frown} \underline{p})$.

Furthermore let the variable $\underline{u}$ not occur free in $\underline{p}$. Then

(90) $\vdash (\underline{u}) [\underline{u} \in \underline{El} \supset \diamond (\mid_{\underline{u}} \underline{p})] \equiv \underline{Np} \equiv (\underline{u}) (\mid_{\underline{u}} \supset {}^{\frown} \underline{p})$,

(91) $\vdash (\exists \underline{u}) [\underline{u} \in \underline{El} \wedge (\mid_{\underline{u}} \supset {}^{\frown} \underline{p})] \equiv \diamond \underline{p} \equiv (\exists \underline{u}) \diamond (\mid_{\underline{u}} \underline{p})$.

Proof. We assume (a) $\underline{u} \in \underline{El}$ and either (b) $\diamond [(\iota \underline{x})\, \underline{u}\, (\underline{x}) = 1 \underline{p}]$ or (b′) $(\iota \underline{x})\, \underline{u}\, (\underline{x}) = 1 \supset^{\frown} \underline{p}$. By Def. 48.1 and rule $\underline{C}$ with $\underline{x}$, (a) yields (c) $\underline{x} \in \underline{ElR}$ and (d) $\underline{u} = {}^{\frown} \{\underline{x}\}^{(i)}$, whence by $(88)_2$ we deduce (e) $(\iota \underline{x})\, \underline{u}\, (\underline{x}) = {}^{\frown}\underline{x}$.

Thus (b) and (b′) yield (f) $\diamond (\underline{x} = 1 \underline{p})$ and (f′) $\underline{x} = 1 \supset^{\frown} \underline{p}$ respectively. In addition, by (c) and (84) the matrices (f) and (f′) yield one another. By (e) from (f′) and (f) we deduce (b′) and (b) respectively.

By Theor. 33.3, $(82)_1$, and modal rule $\underline{G}$ [Defs. 33.1(4′), 33.2], we conclude that (a) $\vdash_{\underline{c}}$ (b) $\equiv {}^{\frown}$(b′). So by Def. 48.2 and Theors. 33.1 and 33.2, (89) holds.

To prove $(\overline{89})$ we assume $\Diamond (\mid_{\underline{u}} \underline{p})$, whence $\Diamond \mid_{\underline{u}}$, which by Def. 48.2 yields $\Diamond \underline{u} \in \underline{El}$. Thence by $(\overline{82})_1$ we deduce $\underline{u} \in \underline{El}$. We conclude that $\vdash \Diamond (\mid_{\underline{u}} \underline{p}) \supset \underline{u} \in \underline{El}$. Then (89) implies $(\overline{89})$.

Now we want to prove $(90)_1$. To this end, instead of (a) and (b) we assume (g) $(\underline{u}) \{\underline{u} \in \underline{El} \supset \Diamond [(\iota \underline{x}) \, \underline{u} \, (\underline{x}) = 1 \, \underline{p}]\}$, where the variables $\underline{x}$ and $\underline{u}$ do not occur free in $\underline{p}$.

We also assume (g′) $\underline{x} \in \underline{ElR}$. Furthermore, by rule $\underline{C}$ with $\underline{u}$ from $\vdash \{\underline{x}\}^{(\underline{i})} = {}^\cap\{\underline{x}\}^{(\underline{i})}$ we have $\underline{u} = {}^\cap\{\underline{x}\}^{(\underline{i})}$. Hence $\underline{u} \in \underline{El}$ [Def. 48.1]. This and (g) yield $\Diamond [(\iota \underline{x}) \underline{u} (\underline{x}) = 1 \underline{p}]$, hence (g″) $\Diamond (\underline{x} = 1 \underline{p})$. By Theor. 33.2 (g′) $\supset$ (g″), i.e. (h) $\underline{x} \in \underline{ElR} \supset \Diamond (\underline{x} = 1 \underline{p})$ whence, by rule $\underline{G}$, (h′) $(\underline{x}) [\underline{x} \in \underline{ElR} \supset \Diamond (\underline{x} = 1 \underline{p})]$. By $(85)_1$ this yields $\underline{Np}$. Thus (g) $\vdash \underline{Np}$.

Conversely we now assume (i) $\underline{Np}$ and (a) $\underline{u} \in \underline{El}$. From (i) and $(85)_1$ we have (h′), and from (a), by Def. 48.1 and rule $\underline{C}$ with $\underline{x}$, we deduce (c) $\underline{x} \in \underline{ElR}$ and (d) $\underline{u} = {}^\cap\{\underline{x}\}^{(\underline{i})}$, which by $(88)_2$ yields (e) $(\iota \underline{x}) \, \underline{u} \, (\underline{x}) = {}^\cap \underline{x}$.

From (h′) we have (h), which together with (c) $\underline{x} \in \underline{ElR}$ yields (f) $\Diamond (\underline{x} = 1 \underline{p})$. Thence by (e) we obtain (b) $\Diamond [(\iota \underline{x}) \, \underline{u} \, (\underline{x}) = 1 \underline{p}]$. So [Theor. 33.1] (i), (a) $\vdash$ (b), whence (i) $\vdash$ (a) $\supset$ (b) [Theor. 33.2] and (i) $\vdash (\underline{u}) [(a) \supset (b)]$, i.e. $\underline{Np} \vdash$ (g).

Remembering that (g) $\vdash \underline{Np}$ also holds, by Theor. 33.2 we have $\vdash$ (g) $\equiv \underline{Np}$.

Since, in addition, by Def. 48.2 we have $\vdash$ (g) $\equiv (\underline{u}) [\underline{u} \in \underline{El} \supset \Diamond (\mid_{\underline{u}} \underline{p})]$, $(90)_1$ holds.

To prove $(90)_2$ we first start with $\underline{Np}$ and $\mid_{\underline{u}}$. By Def. 48.2 $\mid_{\underline{u}}$ yields $\underline{u} \in \underline{El}$, so that from $\underline{Np}$ and $(90)_1$ we deduce $\Diamond (\mid_{\underline{u}} \underline{p})$, which by $(\overline{89})$ yields $\mid_{\underline{u}} \supset {}^\cap \underline{p}$. This matrix and $\mid_{\underline{u}}$ yield $\underline{p}$. Since $\underline{u}$ does not occur free in $\underline{p}$, by Theors. 33.2 and 32.2 and AS30.7 we easily conclude that (m) $\vdash \underline{Np} \supset (\underline{u}) (\mid_{\underline{u}} \supset {}^\cap \underline{p})$.

Now we start with (n) $(\underline{u})(|_{\underline{u}} \supset^{\frown} \underline{p})$ and (o) $\underline{u} \in \underline{El}$, whence by $(\overline{89})$ we deduce $\diamond(|_{\underline{u}}\underline{p})$. By Theors. 33.2 and 32.2 we conclude that $\vdash (n) \supset (\underline{u})[\underline{u} \in \underline{El} \supset \diamond(|_{\underline{u}}\underline{p})]$, whence by $(90)_1$ we have $\vdash (n) \supset \underline{Np}$. This result and (m) yield $(\overline{90})_2$.

Lastly from $(90)_{1,2}$ we can easily obtain $(91)_{1,2}$ respectively by deductions like that of $(86)_1$ from $(85)_1$—see (III) in the proof of Theor. 49.1.

QED

THEOR. 49.3. For all matrices $\underline{p}_1, \ldots, \underline{p}_m$ let $\phi(\underline{p}_1, \ldots, \underline{p}_m)$ be a matrix built up from $\underline{p}_1, \ldots, \underline{p}_m$, (,), $\sim$, $\wedge$, and any of the quantifiers $(\underline{x}_1), \ldots, (\underline{x}_n)$, with $\underline{x}_1, \ldots, \underline{x}_n$ different from the variable $\underline{u}$. Then

$$\text{(92)} \quad \underline{u} \in \underline{El} \vdash |_{\underline{u}} \supset^{\frown} \phi(\underline{p}_1, \ldots, \underline{p}_m) \equiv \phi(|_{\underline{u}} \supset^{\frown} \underline{p}_1, \ldots, |_{\underline{u}} \supset^{\frown} \underline{p}_m)$$

and

$$(\overline{92}) \quad \vdash \diamond[|_{\underline{u}} \phi(\underline{p}_1, \ldots, \underline{p}_m)] \equiv \underline{u} \in \underline{El}\, \phi[\diamond(|_{\underline{u}}\underline{p}_1), \ldots, \diamond(|_{\underline{u}}\underline{p}_m)].$$

Proof. (I) We prove (92) by induction. If $\phi(\underline{p}_1, \ldots, \underline{p}_n)$ is $\underline{p}_i$, then (92) holds trivially. Otherwise $\phi(\underline{p}_1, \ldots, \underline{p}_m)$ has the form $\sim\underline{p}$ or $\underline{p} \wedge \underline{q}$ or else $(\underline{x}_i)\underline{p}$ $(1 \leq \underline{i} \leq \underline{n})$, where the matrices $\underline{p}$ and $\underline{q}$ are built up beginning from $\underline{p}_1, \ldots, \underline{p}_m$, (,), $\sim$, $\wedge$, and quantifiers $(\underline{x}_1), \ldots, (\underline{x}_n)$, with $\underline{x}_1, \ldots, \underline{x}_n$ different from $\underline{u}$. We assume that (92) holds for $\underline{p}$ and $\underline{q}$ as our hypothesis of induction. We consider the following cases.

Case 1. $\phi(\underline{p}_1, \ldots, \underline{p}_m)$ is $\sim\underline{p}$. We start with (a) $\underline{u} \in \underline{El}$. Since $|_{\underline{u}} \supset^{\frown} \sim\underline{p}$ can be replaced with $\sim\diamond(|_{\underline{u}}\underline{p})$, by (a) and (89) it is strictly equivalent to $\sim(|_{\underline{u}} \supset^{\frown} \underline{p})$. We conclude by the hypothesis of induction that (92) holds.

Case 2. $\phi(\underline{p}_1, \ldots, \underline{p}_m)$ is $\underline{p} \wedge \underline{q}$. We start with (a) $\underline{u} \in \underline{El}$. Then $|_{\underline{u}} \supset^\frown \underline{pq}$ is equivalent to $(|_{\underline{u}} \supset^\frown \underline{p})\, (|_{\underline{u}} \supset \underline{q})$. So (92) obviously follows from the inductive hypothesis.

Case 3. $\phi(\underline{p}_1, \ldots, \underline{p}_m)$ is $(\underline{x})\underline{p}$ and $\underline{x}$ is different from $\underline{u}$. Then by Def. 48.2 and the inductive hypothesis the validity of (92) is obvious.

(II) In order to prove $(\overline{92})$ we first start with (a) $\underline{u} \in \underline{El}$, whence by (89) we deduce (b) $\diamond (|_{\underline{u}} \underline{p}) \equiv^\frown (|_{\underline{u}} \supset^\frown \underline{p})$. So by (92) and the equivalence theorem [Theor. 42.2], from (a) we easily deduce (c) $\diamond [|_{\underline{u}}\ \phi(\underline{p}_1, \ldots, \underline{p}_m)] \equiv \phi[\diamond (|_{\underline{u}} \underline{p}_1), \ldots, \diamond (|_{\underline{u}} \underline{p}_m)]$. We conclude that $\vdash \underline{u} \in \underline{El} \supset$ (c).

Furthermore $\vdash \diamond |_{\underline{u}} \supset \underline{u} \in \underline{El}$—cf. $(\overline{89})$ for $\underline{p} =_D |_{\underline{u}}$. Now it is easy to see that $(\overline{92})$ holds.

QED

THEOR. 49.4. Let the variable $\underline{u}$—for which $|_{\underline{u}}$ is a wff—not occur free in $\underline{p}$. Then

(I) $\vdash \diamond [|_{\underline{u}} (\exists^{(1)\frown} \underline{x})\underline{p}] \equiv \underline{u} \in \underline{El}\ (\exists^{(1)\frown} \underline{x}) \diamond (|_{\underline{u}} \underline{p})$

and

(II) $\vdash \diamond [|_{\underline{u}} (\exists_1^\frown \underline{x})\underline{p}] \equiv \underline{u} \in \underline{El}\ (\exists_1^\frown \underline{x}) \diamond (|_{\underline{u}} \underline{p})$.

Proof. By Defs. 34.2 and 14.1 the matrix $(\exists^{(1)\frown} \underline{x})\underline{p}$ $[(\exists_1^\frown \underline{x})\underline{p}]$ can be identified with a suitable choice of the matrix $\phi(\underline{p}_1, \ldots, \underline{p}_m)$ considered in Theor. 49.3, for $\underline{m} = 3$ and $\underline{p}_1, \underline{p}_2, \underline{p}_3$ identical with $\Phi(\underline{x})$, $\Phi(\underline{y})$, and $\underline{x} =^\frown \underline{y}$ respectively, where $\Phi(\underline{x})$ is $\underline{p}$ and the variable $\underline{y}$ is different from $\underline{x}$ and does not occur free in $\underline{p}$. Then $(\overline{89})$ becomes (I) [(II)].

QED

10

A Definition in MC^{ν} of Some New Description Operators; Formal Treatment of the Real Elementary Case by Means of a Primitive Constant

N50. Definition in MC^{ν} of the intensional description operator $\iota_{\underline{u}}$.

The description operator $\imath$—which is a primitive symbol in ML^{ν}—can be considered to be extensional on the basis of $(30)_1$, $(31)_1$, and $(33)_1$. Using the class El of elementary cases—whose definition in a satisfactory form was possible on the basis of AS12.19—we now want to introduce, by means of a metalinguistic definition (where the use of $\imath$ is essential), an intensional description operator, $\iota_{\underline{u}}$, depending on the parameter $\underline{u}$ which ranges over El.

In N51 we shall introduce a combination, $\imath_{\underline{u}}$, of $\imath$ and $\iota_{\underline{u}}$. This combination can be used in the usual way—i.e. it has the well-known replacement property of descriptions—whenever either $\imath$ or $\iota_{\underline{u}}$ has this property. So $\imath_{\underline{u}}$ appears more useful than $\imath$ and $\iota_{\underline{u}}$.

In connection with the dependence of $\iota_{\underline{u}}$ and $\imath_{\underline{u}}$ on $\underline{u}$, the matrix $|_{\underline{u}}$ occurs among the conditions permitting the replacement property of $\iota_{\underline{u}}$ or $\imath_{\underline{u}}$ mentioned above. This has no analogue in ordinary language. However by $(83)_1$ the additional condition $|_{\underline{u}}$ is not substantially restrictive. Moreover the same condition helps avoid some ambiguities of ordinary language—cf. N53. The dependence of $\iota_{\underline{u}}$ on $\underline{u}$ will disappear when we identify $\underline{u}$ with a constant, ρ, representing the real elementary case [N52].

For every matrix $\underline{p}$ and variable $\underline{x}$ we set

DEF. 50.1. $(\iota_{\underline{u}}\underline{x})\underline{p} =_D (\imath\underline{x}) \diamond [\,|_{\underline{u}}\underline{p}\,(\exists^{(1)\frown}\underline{x})\underline{p}]$,

where $\underline{u}$ is a variable of such a type that $|_{\underline{u}}$ is well formed. We shall be interested in $(\iota_{\underline{u}}\underline{x})\underline{p}$ only when $\underline{u}$ has no free occurrences in $\underline{p}$.

The next theorem presents the main properties of $\iota_{\underline{u}}$. In particular parts (III) and (IV) constitute the analogue for $\iota_{\underline{u}}$ of AS38.1.

THEOR. 50.1. Let the variable $\underline{u}$, free in the matrix $|_{\underline{u}}$ [Def. 48.2], be distinct from $\underline{x}$ and not occur free in $\Phi(\underline{x})$. Then

(I) $\vdash \Diamond [|_{\underline{u}} (\exists_1^{\cap}\underline{x}) \Phi(\underline{x})] \supset \Diamond \{|_{\underline{u}} \Phi[(\iota_{\underline{u}}\underline{x}) \Phi(\underline{x})]\}$

(II) $\vdash |_{\underline{u}} (\exists_1^{\cap}\underline{x}) \Phi(\underline{x}) \supset^{\cap} \Phi[(\iota_{\underline{u}}\underline{x}) \Phi(\underline{x})]$,

(III) $\vdash \Diamond [|_{\underline{u}} \underline{p} (\exists^{(1)\cap}\underline{x})\underline{p}] \supset \underline{x} =^{\cap} (\iota_{\underline{u}}\underline{x})\underline{p}$,

(IV) $\vdash \sim \Diamond [|_{\underline{u}} (\exists_1^{\cap}\underline{x})\underline{p}] \supset (\iota_{\underline{u}}\underline{x})\underline{p} =^{\cap} \underline{a}^*$,

(V) $\vdash (\iota_{\underline{u}}\underline{x}) \Phi(\underline{x}) = (\iota_{\underline{u}}\underline{y}) \Phi(\underline{y})$ where condition (c) in N30 holds—cf. $(30)_2$.

Proof. (I) We start with (a) $\Diamond [|_{\underline{u}} (\exists_1^{\cap}\underline{x}) \Phi(\underline{x})]$, whence by Theor. 49.4(II) we deduce (b) $(\exists_1^{\cap}\underline{x}) \Diamond [|_{\underline{u}} \Phi(\underline{x})]$.

By $(\overline{89})$ (a) yields $|_{\underline{u}} \supset^{\cap} (\exists_1^{\cap}\underline{x}) \overline{\Phi}(\underline{x})$, whence by $(14)_{1,2,3}$ $|_{\underline{u}} \supset (\exists^{(1)\cap}\underline{x}) \Phi(\underline{x})$. Thence we easily deduce (c) $(\underline{x}) \{\Diamond [|_{\underline{u}} \Phi(\underline{x})] \equiv^{\cap} \Diamond [|_{\underline{u}} \Phi(\underline{x}) (\exists^{(1)\cap}\underline{x}) \Phi(\underline{x})]\}$. By $(30)_1$ and Def. 50.1, (c) yields (d) $(\iota\underline{x}) \Diamond [|_{\underline{u}} \Phi(\underline{x})] =^{\cap} (\iota_{\underline{u}}\underline{x}) \Phi(\underline{x})$.

By Def. 48.2 and the hypothesis, $\underline{x}$ does not occur in $|_{\underline{u}}$. Then by $(35)_6$ and (b) we easily have $\Diamond [|_{\underline{u}} \Phi\{(\iota\underline{x}) \Diamond [|_{\underline{u}} \Phi(\underline{x})]\}]$, whence by (d) and $(47')_1$ we deduce $\Diamond \{|_{\underline{u}} \Phi[(\iota_{\underline{u}}\underline{x}) \Phi(\underline{x})]\}$. We conclude [Theor. 33.2] that (I) holds.

(II) By $(\overline{89})$ $\vdash \Diamond \{|_{\underline{u}} \Phi[(\iota_{\underline{u}}\underline{x}) \Phi(\underline{x})]\} \supset \{|_{\underline{u}} \supset^{\cap} \Phi[(\iota_{\underline{u}}\underline{x}) \Phi(\underline{x})]\}$. So from $(8')_1$, (I), and Theor. 32.2, we easily deduce (II).

(III) We start with (a) $\Diamond [|_{\underline{u}} \underline{p} (\exists^{(1)\cap}\underline{x})\underline{p}]$, whence by Theor. 49.4(I) we easily deduce $(\exists^{(1)\cap}\underline{x}) \Diamond (|_{\underline{u}}\underline{p})$, which by Def. 34.2 yields (b) $(\forall\underline{x}, \underline{y}) \{\Diamond [|_{\underline{u}} \Phi(\underline{x})] \Diamond [|_{\underline{u}} \Phi(\underline{y})] \supset \underline{x} =^{\cap} \underline{y}\}$, where $\Phi(\underline{x})$ is $\underline{p}$ and the variable $\underline{y}$ does not occur in $\underline{x} = \underline{x} \wedge |_{\underline{u}} \wedge \Phi(\underline{x})$—cf. conventions (a) and (b) in N30. Furthermore (a) yields $\diamond [|_{\underline{u}} \Phi(\underline{x})]$, whence by (b) we easily obtain (c) $(\underline{y}) \diamond [|_{\underline{u}} \Phi(\underline{y})] \supset \underline{x} =^{\cap} \underline{y}$.

By $(14)_{1,2,3}$, from (a) we easily deduce $\diamond [|_{\underline{u}} (\exists_1^{\cap} \underline{x}) \Phi(\underline{x})]$, which by (I) yields (d) $\diamond \{|_{\underline{u}} \Phi[(\iota_{\underline{u}} \underline{x}) \Phi(\underline{x})]\}$. From (c) and (d) we deduce $\underline{x} =^{\cap} (\iota_{\underline{u}} \underline{x}) \Phi(\underline{x})$. Now we may conclude [Theor. 33.2] that (III) holds.

(IV) We start with (a) $\sim \diamond [|_{\underline{u}} (\exists_1^{\cap} \underline{x}) \underline{p}]$. From $(14)_{1,2,3}$ we easily deduce $\vdash |_{\underline{u}} \underline{p} (\exists^{(1)\cap} \underline{x}) \underline{p} \supset^{\cap} |_{\underline{u}} (\exists_1^{\cap} \underline{x}) \underline{p}$. So (a) yields $\sim \diamond [|_{\underline{u}} \underline{p} (\exists^{(1)\cap} \underline{x}) \underline{p}]$, whence by Def. 50.1 and AS38.1(II) we easily infer $(\iota_{\underline{u}} \underline{x}) \underline{p} =^{\cap} \underline{a}^*$. By Theor. 33.2 we conclude that (IV) holds.

(V) This is an easy consequence of Def. 50.1 and $(30)_2$.

QED

N51. Definition in MC^{ν} of the combination $\mathit{ɿ}_{\underline{u}}$ of $\iota_{\underline{u}}$ and $\mathit{ɿ}$.

We now introduce the combination $\mathit{ɿ}_{\underline{u}}$ of $\iota_{\underline{u}}$ and $\mathit{ɿ}$ by means of the following metalinguistic definition:

DEF. 51.1. $(\mathit{ɿ}_{\underline{u}} \underline{x}) \underline{p} =_D (\mathit{ɿ} \underline{x}) \{\diamond [|_{\underline{u}} (\exists_1^{\cap} \underline{x}) \underline{p}]\ \underline{x} = (\iota_{\underline{u}} \underline{x}) \underline{p} \vee \sim \diamond [|_{\underline{u}} (\exists_1^{\cap} \underline{x}) \underline{p}]\ \underline{x} = (\mathit{ɿ} \underline{x}) \underline{p}\}$.

The usefulness of $\mathit{ɿ}_{\underline{u}}$ appears from the following theorem, especially from its fourth part, which says in effect that for the description $(\mathit{ɿ}_{\underline{u}} \underline{x}) \underline{p}$ the ordinary replacement property of descriptions holds whenever it does for either $(\iota_{\underline{u}} \underline{x}) \underline{p}$ or $(\mathit{ɿ} \underline{x}) \underline{p}$.

THEOR. 51.1. We assume that $\Phi(\underline{x})$ is the matrix $\underline{p}$ and that $\Phi(\underline{x})$ and $\Phi(\underline{y})$ fulfill condition (c) in N30—cf. conventions (a) and (b) in N30. Furthermore let the variable $\underline{u}$, by whose logical type $|_{\underline{u}}$ is well formed, not occur free in $\underline{p}$ or $\Phi(\underline{x})$. Then

(I) $\vdash \diamond [|_{\underline{u}} (\exists_1^{\cap} \underline{x}) \underline{p}] \supset (\mathit{ɿ}_{\underline{u}} \underline{x}) \underline{p} =^{\cap} (\iota_{\underline{u}} \underline{x}) \underline{p}$,

(II) $\vdash \sim \Diamond [\,|_{\underline{u}}\, (\exists^{\frown}_1 \underline{x})\underline{p}] \supset (℩_{\underline{u}}\underline{x})\underline{p} =^{\frown} (℩\underline{x})\underline{p}$,

(II′) $\vdash |_{\underline{u}} \sim (\exists^{\frown}_1 \underline{x})\underline{p} \supset (℩_{\underline{u}}\underline{x})\underline{p} =^{\frown} (℩\underline{x})\underline{p}$,

(III) $\vdash |_{\underline{u}} (\exists^{\frown}_1 \underline{x}) \Phi(\underline{x}) \supset \Phi[(℩_{\underline{u}}\underline{x})\Phi(\underline{x})]$—cf. $(35)_{5,6}$,

(IV) $\vdash |_{\underline{u}} \sim \Phi(\underline{a}^*)$ $\vdash \Phi[(℩\underline{x})\Phi(\underline{x})] \vee \Phi[(\iota_{\underline{u}}\underline{x})\Phi(\underline{x})] \supset \Phi[(℩_{\underline{u}}\underline{x})\Phi(\underline{x})]$,

(V) $\vdash |_{\underline{u}} \underline{p}\, (\exists_1 \underline{x})\underline{p} \supset \underline{x} = (℩_{\underline{u}}\underline{x})\underline{p}$,

(V′) $\vdash |_{\underline{u}} \underline{p}\, (\exists_1 \underline{x})\underline{p} \supset (℩\underline{x})\underline{p} = (℩_{\underline{u}}\underline{x})\underline{p}$,

(VI) $\vdash (℩_{\underline{u}}\underline{x})\Phi(\underline{x}) = (℩_{\underline{u}}\underline{y})\Phi(\underline{y})$—cf. $(30)_2$,

(VII) $\vdash |_{\underline{u}} (\exists_1 \underline{x})\underline{p} \supset [(℩_{\underline{u}}\underline{x})\underline{p} = \underline{x} \equiv \underline{p}^{(\underline{ex})}]$—cf. $(31)_1$,

(VIII) $\vdash |_{\underline{u}} (\underline{\mathrm{ext}}\ \underline{x})\Phi(\underline{x})\, (\exists_1 \underline{x})\Phi(\underline{x}) \supset \Phi[(℩_{\underline{u}}\underline{x})\Phi(\underline{x})]$—cf. $(31)_2$,

(IX) $\vdash |_{\underline{u}} (\exists\underline{x})\, \underline{N}\Phi(\underline{x})\, \underline{N}(\exists_1 \underline{x})\Phi(\underline{x}) \supset \Phi[(℩_{\underline{u}}\underline{x})\Phi(\underline{x})]$—cf. $(35)_2$,

(X) $\vdash |_{\underline{u}} \sim (\exists_1 x)\underline{p} \supset (℩_{\underline{u}}\underline{x})\underline{p} = \underline{a}^*$.

(V) is the analogue for $℩_{\underline{u}}$ of AS38.1(I). (VI) through (IX) are the analogues for $℩_{\underline{u}}$ of some useful properties of $℩$. Let us observe that from $(\overline{\mathrm{III}})$ we obtain $(35)_5$ by replacing $℩$ with $℩_{\underline{u}}$ and assuming that $\Phi(\underline{x})$ has the form $\underline{Np}$, which constitutes an advantage of $℩_{\underline{u}}$ with respect to $℩$.

The properties $(30)_1$ and (33) of $℩$—which are not replacement properties of $(\underline{m})\underline{p}$—do not hold for $℩_{\underline{u}}$.

Proof of Theor. 51.1. (I) and (II) follow easily from Def. 51.1 and well-known properties of modally closed matrices.

To prove (II′) we start with $|_{\underline{u}} \sim (\exists^{\frown}_1 \underline{x})\underline{p}$, which by $(8')_1$ and $(\overline{92})$ yields $\sim \Diamond [\,|_{\underline{u}} (\exists^{\frown}_1 \underline{x})\underline{p}]$. Thence by (II) we deduce $(℩_{\underline{u}}\underline{x})\underline{p} =^{\frown} (℩\underline{x})\underline{p}$. So by Theor. 33.2 we conclude that (II′) holds.

(III) follows from $(8')_1$, (I), and Theor. 50.1(II) (and AS38.2).

To prove (IV) we start with $|_{\underline{u}}$, (a) $\sim\Phi(\underline{a}^*)$, and (b) $(\exists^{\frown}_1 \underline{x})\Phi(\underline{x})$. By $|_{\underline{u}}$ and (b) we have $\Diamond [\,|_{\underline{u}} (\exists^{\frown}_1 \underline{x})\Phi(\underline{x})]$. Thence on the basis of

(III), we easily obtain (c) $\diamond \{ \mid_{\underline{u}} \Phi[(\imath_{\underline{u}} \underline{x}) \Phi(\underline{x})]\}$. By $(\overline{89})$, from (c) and $\mid_{\underline{u}}$ we deduce (d) $\Phi[(\imath_{\underline{u}} \underline{x}) \Phi(\underline{x})]$. Then $\mid_{\underline{u}}$, (a), and (b) yield the R.H.S. of $\vdash$ in (IV).

Now we start with $\mid_{\underline{u}}$, (a) and $\sim$(b) (instead of (b)). From $\mid_{\underline{u}}$, $\sim$(b), and (II′) we obtain (e) $(\imath_{\underline{u}} \underline{x}) \Phi(\underline{x}) =^{\cap} (\imath \underline{x}) \Phi(\underline{x})$, whence on the one hand we have (f) $\Phi[(\imath \underline{x}) \Phi(\underline{x})] \supset \Phi[(\iota_{\underline{u}} \underline{x}) \Phi(\underline{x})]$. By $\mid_{\underline{u}}$ and $\sim$(b) we have $\diamond [\mid_{\underline{u}} \sim (\exists_1^{\cap} \underline{x}) \Phi(\underline{x})]$, which by $(\overline{92})$ yields $\sim \diamond [\mid_{\underline{u}} (\exists_1^{\cap} \underline{x}) \Phi(\underline{x})]$. Thence by Theor. 50.1(IV) we deduce $(\iota_{\underline{u}} \underline{x}) \Phi(\underline{x}) =^{\cap} \underline{a}*$. Then by (a) we have, on the other hand, (g) $\sim\Phi[(\iota_{\underline{u}} \underline{x}) \Phi(\underline{x})]$.

Since $\mid_{\underline{u}}$, (a), and $\sim$(b) yield (f) and (g), we easily see that $\mid_{\underline{u}}$, (a), and $\sim$(b) yield the R.H.S. of $\vdash$ in (IV). Then, remembering that the analogue holds for $\mid_{\underline{u}}$, (a), and (b), we conclude that (IV) holds.

To prove (V), we first start with $\mid_{\underline{u}}$, (h) $\Phi(\underline{x})$ $(\exists_1 \underline{x}) \Phi(\underline{x})$, and (i) $(\exists_1^{\cap} \underline{x}) \Phi(\underline{x})$. By (III), from $\mid_{\underline{u}}$ and (i) we deduce (l) $\Phi[(\imath_{\underline{u}} \underline{x}) \Phi(\underline{x})]$. From (l) and (h), by Def. 11.3 we easily obtain (m) $\underline{x} = (\imath_{\underline{u}} \underline{x}) \Phi(\underline{x})$. So we conclude that $\mid_{\underline{u}}$, (h), (i) $\vdash$ (m).

Now we start with $\mid_{\underline{u}}$, (h), and $\sim$(i). By $(8')_1$ and $(\overline{92})$, $\mid_{\underline{u}}$ and $\sim$(i) yield $\sim \diamond [\mid_{\underline{u}} (\exists_1^{\cap} \underline{x}) \Phi(\underline{x})]$, whence by (II) we deduce $(\imath_{\underline{u}} \underline{x}) \Phi(\underline{x}) =^{\cap} (\imath \underline{x}) \Phi(\underline{x})$. So by (h) and AS38.1(I) we have (m) again. We conclude that $\mid_{\underline{u}}$, (h), $\sim$(i) $\vdash$ (m).

Since $\mid_{\underline{u}}$, (h), (i) $\vdash$ (m) also holds, we have $\mid_{\underline{u}}$, (h) $\vdash$ (m). Then by Theor. 33.2, (V) holds.

(V′) is an immediate consequence of (V) and AS38.1(I).

(VI) is an easy consequence of (I), (II), Theor. 50.1(V), and $(30)_2$.

(VII) is implied by (V′) and $(31)_1$.

In order to prove (VIII) and (IX) we write these parts in the forms $\vdash \mid_{\underline{u}} \underline{p}_8 \supset \underline{q}$ and $\vdash \mid_{\underline{u}} \underline{p}_9 \supset \underline{q}$ respectively, so that $\underline{q}$ is $\Phi[(\imath_{\underline{u}} \underline{x}) \Phi(\underline{x})]$.

By (III) we obviously have (p) $\vdash \mid_{\underline{u}} (\exists_1^{\cap} \underline{x}) \Phi(\underline{x}) \wedge (\underline{p}_8 \vee \underline{p}_9) \supset \underline{q}$. Furthermore $(31)_2$ and $(35)_2$ yield $\vdash (\underline{p}_8 \vee \underline{p}_9) \supset \Phi[(\imath \underline{x}) \Phi(\underline{x})]$. So by

(II′) we obviously have $\vdash \mathsf{I}_{\underline{u}} \sim(\exists^{\frown}_{1}\underline{x})\underline{p} \wedge (\underline{p}_8 \vee \underline{p}_9) \supset \underline{q}$. Thence by (p) we obtain $\vdash \mathsf{I}_{\underline{u}} (\underline{p}_8 \vee \underline{p}_9) \supset \underline{q}$. So (VIII) and (IX) hold.

To prove (X) we start with $\mathsf{I}_{\underline{u}}$ and (r) $\sim(\exists_1\underline{x})\underline{p}$. By $(27)_1$ (r) yields (r′) $\sim(\exists^{\frown}_{1}\underline{x})\underline{p}$. By As38.$\overline{1}$(II), (r) yields (s) $(\imath\underline{x})\underline{p} = \underline{a}^*$, and by (II′), $\mathsf{I}_{\underline{u}}$ and (r′) yield (s′) $(\imath_{\underline{u}}\underline{x})\underline{p} =^{\frown} (\imath\underline{x})\underline{p}$. $\mathsf{I}_{\underline{u}}$ and (r) yield (s) $\wedge$ (s′), hence $(\imath_{\underline{u}}\underline{x})\underline{p} = \underline{a}^*$. Thus by Theor. 33.2, (X) holds.

QED

N52. An axiom that introduces a constant representing the real elementary case. The calculus MC^{ν}_{ρ} and the modal operator $\mathfrak{R}$.

The introduction of the constant ρ allows us, among other things, to turn $\iota_{\underline{u}}$ and $\imath_{\underline{u}}$ into operators independent of any parameters. Furthermore the operators $\iota_{\underline{u}}$ and ι_{ρ} allow us to solve [N55] certain logical puzzles on the basis of the interpreted calculus MC^{ν}_{ρ}, to be defined shortly.

The sentence I_{ρ} is the analogue for MC^{ν}_{ρ} of the propositional constant introduced by Meredith and Prior in [27], p. 215, in a very different way [N53]. So $\mathsf{I}_{\underline{u}}$ appears to be a generalization of the latter constant.

The sentence I_{ρ} is useful in dealing with counterfactual conditionals and with some nonmodal sentences of ordinary language that are of interest in sciences such as geology, astronomy, and so on. We prefer to talk more extensively about this in the forthcoming sections, after some formulas have been presented.

We might identify the constant ρ representing the real case with the constant $\underline{c}_{((1))1}$ and assume $\rho \in \underline{\text{ElR}}$ [Def. 47.3] as an axiom. However it is equivalent and (on the whole) simpler to identify ρ with $\underline{c}_{(((1)))1}$ and to accept the following axiom:

A52.1. $\rho \in \underline{\mathrm{El}}$ [Def. 48.1]

We denote the calculus that is based on MC^{ν} and that includes A52.1 as an additional axiom by MC^{ν}_{ρ}. By A52.1 from $(83)_4$ and (89) we deduce

(93) $\vdash \Diamond |_{\rho}, \quad \vdash \Diamond (|_{\rho} \underline{p}) \equiv |_{\rho} \supset^{\cap} \underline{p},$ for every matrix $\underline{p}$.

It is useful to define the modal operator $\mathfrak{R}$ for the real case as follows:

DEF. 52.1. $\mathfrak{R}\, \underline{p} \equiv_D \Diamond (|_{\rho} \underline{p})$, whence $\mathfrak{R}\, \underline{p} \equiv^{\cap} (|_{\rho} \supset^{\cap} \underline{p})$—cf. $(93)_2$.

From A52.1, Def. 52.1, and Theor. 49.3—cf. $(\overline{92})$—(and Theor. 32.2) we easily deduce the following:

THEOR. 52.1. For all matrices $\underline{p}_1, \ldots, \underline{p}_n$ in ML^{ν} (possibly including $\underline{N}$), let $\phi(\underline{p}_1, \ldots, \underline{p}_n)$ be a matrix built up in a way independent of $\underline{p}_1, \ldots, \underline{p}_n$, starting out from $\underline{p}_1, \ldots, \underline{p}_n$, (,), $\sim$, $\wedge$, and the universal quantifiers $(\underline{x}_1), \ldots, (\underline{x}_r)$. Then

(94) $\vdash \mathfrak{R}\, \phi(\underline{p}_1, \ldots, \underline{p}_n) \equiv^{\cap} \phi(\mathfrak{R}\underline{p}_1, \ldots, \mathfrak{R}\, \underline{p}_n)$.

Furthermore from Def. 52.1, (93), and (47′) we easily deduce

$$\begin{array}{ll} & \vdash \Diamond\, \mathfrak{R}\, \underline{p} \equiv \underline{N}\, \mathfrak{R}\, \underline{p}, \quad \vdash \mathfrak{R}\, \underline{Np} \equiv \underline{Np}, \quad \vdash \underline{Np} \supset \mathfrak{R}\, \underline{p}, \\ (95) & \vdash (\underline{p} \equiv^{\cap} \underline{q}) \supset (\mathfrak{R}\, \underline{p} \equiv \mathfrak{R}\, \underline{q}), \quad \vdash \mathfrak{R}\, \mathfrak{R}\, \underline{p} \equiv \mathfrak{R}\, \underline{p}, \\ & \vdash \underline{x} =^{\cap} \Delta\, \mathfrak{R}\, \phi(\underline{x}) \supset \mathfrak{R}\, \phi(\Delta). \end{array}$$

By Def. 52.1 $\mathfrak{R}\, \underline{p}$ can be read as that $\underline{p}$ occurs in the real case; for explanations of the intuitive meaning to be given to $\mathfrak{R}\, \underline{p}$ see N53, in particular fn. 1.

By $(95)_1$ $\mathfrak{R}\, \underline{p}$ is noncontingent. Furthermore, since $\vdash \underline{N}(\underline{p} \vee \sim \underline{p})$, from $(95)_3$ and (94) we deduce $\vdash \mathfrak{R}\, \underline{p} \vee \mathfrak{R} \sim \underline{p}$. Now let $\underline{p}$ be a contingent matrix and let e.g. $\mathfrak{R}\, \underline{p}$ hold. Then both $\sim \underline{p} \supset^{\cap} \mathfrak{R} \sim \underline{p}$ and $\mathfrak{R}\, \underline{p} \supset^{\cap} \underline{p}$ are false. So $\underline{p}$ and $\mathfrak{R}\, \underline{p}$ appear to be very different matrices.

It is natural to replace $\underline{u}$ with ρ in the preceding description operators $\iota_{\underline{u}}$ and $\gamma_{\underline{u}}$ [Ns50, 51].

THEOR. 52.2. For every matrix $\Phi(\underline{x})$, or $\underline{p}$, and every variable $\underline{x}$, in MC^{ν}_{ρ} we have

(I) $\vdash \mathfrak{R}(\exists_1^{\frown}\underline{x})\Phi(\underline{x}) \supset \mathfrak{R}\Phi[(\iota_{\rho}\underline{x})\Phi(\underline{x})]$,

(II) $\vdash \sim\mathfrak{R}(\exists_1^{\frown}\underline{x})\Phi(\underline{x}) \supset (\iota_{\rho}\underline{x})\Phi(\underline{x}) =^{\frown} \underline{a}^*$,

(III) $\vdash \mathfrak{R}(\exists_1^{\frown}\underline{x})\underline{p} \supset (\gamma_{\rho}\underline{x})\underline{p} =^{\frown} (\iota_{\rho}\underline{x})\underline{p}$,

(IV) $\vdash \sim\mathfrak{R}(\exists_1^{\frown}\underline{x})\underline{p} \supset (\gamma_{\rho}\underline{x})\underline{p} =^{\frown} (\gamma\underline{x})\underline{p}$,

(V) $\vdash \mathfrak{R}[\Phi(\underline{x})(\exists_1^{\frown}\underline{x})\Phi(\underline{x})] \supset \mathfrak{R}[\underline{x} = (\gamma_{\rho}\underline{x})\Phi(\underline{x}) = (\gamma\underline{x})\Phi(\underline{x})]$.

Proof. The parts (I) to (IV) of Theor. 52.2 are, respectively, the analogues for ρ of Theors. 50.1(II), 50.1(IV), 51.1(I) and 51.1(II) for $\underline{u}$. Each of these parts appears shorter than the corresponding theorem—which is why we wrote (I) to (IV) explicitly—and can be immediately deduced from the same theorem, from A52.1, and from Def. 52.1.

(V) follows from Theor. 51.1(V), (V′) and Def. 52.1.

QED

N53. Semantics for MC^{ν}_{ρ}. Comparison of MC^{ν}_{ρ} with a theory by Prior and Meredith.

From the semantical point of view the admissible models of MC^{ν}_{ρ} obviously are those models of MC^{ν} that satisfy A52.1 [Def. 9.2]. Then it is a matter of routine to see that there is a Γ-case γ' such that for every matrix $\underline{p}$, value assignment $\mathscr{V}$, and admissible model $\mathscr{M}$, $\mathfrak{R}\,\underline{p}$ $(\sim\mathfrak{R}\underline{p})$ is Γ-true [Def. 9.6(c)] at $\mathscr{M}$ and $\mathscr{V}$ if and only if $\underline{p}$ $(\sim\underline{p})$ is true at $\mathscr{M}$ and $\mathscr{V}$ in γ.

Obviously, γ' coincides with the Γ-case $\gamma_{\mathfrak{R}}$ considered in N11 in observing that for some sciences, e.g. mechanics, physics, or chemistry, all possible cases are on a par, whereas for others, such as astronomy, the real case $\gamma_{\mathfrak{R}}$ is privileged (peculiar).

In connection with astronomy, let us observe that in various cases the same nonmodal sentence $\underline{p}$ of a standard formal modal language is used to translate sentences of ordinary languages that (may coincide but) have quite different meanings. For instance let us consider the following sentence of ordinary language:

(a) The artificial satellite $\underline{S}$ has the distance $\underline{d}$ from Venus at the instant $\underline{t}$.

If in accordance with common practice A52.1 and Def. 52.1 are not taken into account, then the natural translation of (a) into MC^{ν} is a nonmodal sentence $\underline{p}_{\underline{a}}$ of the following form:

$$\underline{p}_{\underline{a}} \equiv_D \underline{S} \in \underline{\text{Art}}\,\underline{\text{Sat}} \wedge \underline{\text{dist}}_V(\underline{S}, \underline{t}) = \underline{d}.$$

However on the one hand (a) may be asserted as a factual sentence (e.g. in case $\underline{t}$ is a past instant and $\underline{d}$ is known), so that (a) can be replaced by ". . . $\underline{S}$ has . . . $\underline{d}$ . . . at . . . $\underline{t}$ in the real case." On the other hand (a) may be considered as a counterfactual assumption that could be realized—e.g. in case $\underline{t}$ is a past instance and (a) is known to be false. For instance this could be done in order to prove, using the modal rule $\underline{G}$—cf. Def. 33.1 and fn. 14 in N33—that if (a) held, then a certain useful observation ω could be made (whereas in fact ω was not made).

In certain natural languages the counterfactual assertion of (a) is usually made by changing the mood of the verb. Let (a′) be the result of this change. More precisely (a) is replaced by (a′) in asserting a counterfactual conditional such as:

(b) If . . . $\underline{S}$ had the distance $\underline{d}$ . . . , then the observation ω would be made.

Usually the translation of (b) into a formal language has the following form:

$$\underline{p}_{\underline{a}} \supset^{\frown} \underline{r} \quad \text{where} \quad \underline{r} \equiv_{D} \underline{\text{Made}}\,(\omega).$$

Thus both of the ordinary sentences (a) and (a′) are translated into the formal sentence $\underline{p}_{\underline{a}}$.

Let us remark that in asserting (b) we use (a′) and not (a). However if we want to prove (b) in an ordinary language using the modal rule $\underline{G}$ and the deduction theorem, we usually begin by asserting (a), not (a′), as an assumption, understanding that (a) should hold in any possible case. We might also know that (a) is false in the real case—cf. fn. 14 in N33.

This ambiguity, present in standard modal language, disappears when MC^{ν}_{ρ} is used. Indeed, in order to say that (a) is true or false (in the real case) we assert in MC^{ν}_{ρ} $\mathfrak{R}\underline{p}_{\underline{a}}$ and $\sim\mathfrak{R}\underline{p}_{\underline{a}}$—cf. (94)—respectively (of course we understand that $\underline{\text{ArtSat}}$ is the absolute concept of an artificial satellite).

Furthermore we translate (b) into $\underline{p}_{\underline{a}} \supset^{\frown} \underline{r}$ as in the standard formal languages. Lastly, if we want to prove $\underline{p}_{\underline{a}} \supset^{\frown} \underline{r}$ using the modal rule $\underline{G}$, we assume $\underline{p}_{\underline{a}}$, an assumption which is compatible with $\sim\mathfrak{R}\underline{p}_{\underline{a}}$ (and $\mathfrak{R}\sim\underline{p}_{\underline{a}}$), provided, of course, that the two assumptions are each consistent (not $\underline{L}$-false).

In connection with the preceding considerations it is interesting to compare MC^{ν}_{ρ} with the theory by Prior and Meredith that involves an analogue, $\underline{n}$, of $|_{\rho}$, is presented in [27], § 7 and is based on S5 and the "special postulates" (i) to (iv) in [27], p. 215.

The analogue $\underline{n}$ of I_{ρ} is a propositional constant. So we may identify $\underline{n}$ with I_{ρ} itself. Then we write the special postulates (i) to (iv) in the following equivalent form, where $\underline{p}$ is an arbitrary matrix:

(I) $\vdash \mathsf{I}_{\rho}$—the world is the case.

(II) $\vdash \underline{N}\,\mathsf{I}_{\rho} \supset^{\cap} \underline{p}$—the world is unnecessary.

(III) $\vdash \underline{p} \supset^{\cap} [(\mathsf{I}_{\rho} \supset^{\cap} \underline{p}) \supset^{\cap} (\underline{p} \supset^{\cap} \mathsf{I}_{\rho})]$—nothing but impossibility is less likely.

(IV) $\vdash \underline{p} \to \mathsf{I}_{\rho} \supset^{\cap} \underline{p}$—the world is everything that is the case.

The special postulates (II) and (III) are easy consequences (in MC^{ν}_{ρ}) of (96) below:

THEOR. 53.1. The following holds in MC^{ν}_{ρ} for every matrix $\underline{p}$ (possibly containing $\underline{u}$):

(96) $\vdash \underline{u} \in \underline{El} \supset \sim \underline{N}\,\mathsf{I}_{\underline{u}}$, $\quad \vdash \diamond\, \underline{p} \supset^{\cap} [(\underline{p} \supset^{\cap} \mathsf{I}_{\underline{u}}) \supset^{\cap} (\mathsf{I}_{\underline{u}} \supset^{\cap} \underline{p})]$.

Proof. (I) We start with $\underline{u} \in \underline{El}$. Thence by Defs. 48.1 and 18.13, we easily deduce $(\iota \underline{x})\underline{u}(\underline{x}) \in \underline{ElR}$, which by Def. 47.2 yields $(\iota \underline{x})\underline{u}(\underline{x}) \in \underline{PR}$, whence by Def. 47.1 we have $(\iota \underline{x})\underline{u}(\underline{x}) =^{\cup} 0$. Thence we have $\diamond \sim (\iota \underline{x})\underline{u}(\underline{x}) = 1$, which by Def. 48.2 yields $\sim \underline{N}\,\mathsf{I}_{\underline{u}}$. By Theor. 33.2 we conclude that $(96)_1$ holds.

(II) We start with (a) $\diamond\, \underline{p}$ and (b) $\underline{p} \supset^{\cap} \mathsf{I}_{\underline{u}}$. Thence we deduce (c) $\diamond\, (\mathsf{I}_{\underline{u}}\, \underline{p})$. By $(\overline{89})$ (c) yields (d) $\mathsf{I}_{\underline{u}} \supset^{\cap} \underline{p}$. Thus (a), (b) $\vdash$ (d), and by Theor. 33.2 and the modal part of the generalization theorem [Theor. 32.2] we have $(96)_2$.

QED

The special postulates (I) and (IV)—unlike (II) and (III)—are not provable in MC^{ν}_{ρ} unless they are suitably modified. This occurs because factual sentences such as (a) have different translations into Meredith and Prior's formal language and into MC^{ν}_{ρ}: e.g. for (a) these translations are $\underline{p}_{\underline{a}}$ and $\mathfrak{R}\,\underline{p}_{\underline{a}}$ respectively (in case (a) cannot be replaced by (a′)).[1] So the analogues for MC^{ν}_{ρ} of the special postulates (i) and (iv) are

(I′) $\vdash \mathfrak{R} \mid_{\rho}$,

(IV′) $\vdash \mathfrak{R}\,\underline{p}$ yields $\vdash \mid_{\rho} \supset^{\cap} \underline{p}$.

By Def. 52.1 $\mathfrak{R} \mid_{\rho}$ is $\diamond(\mid_{\rho} \mid_{\rho})$. So (I′) follows from A52.1 and $(83)_4$. Since $\mathfrak{R}\,\underline{p}$ is $\diamond(\mid_{\rho} \underline{p})$ [Def. 52.1], from A52.1 and (89) we deduce $\vdash \mathfrak{R}\,\underline{p} \equiv^{\cap} \mid_{\rho} \supset^{\cap} \underline{p}$. Then (IV′) is obviously a metatheorem in MC^{ν}_{ρ}.

We conclude that all of the above special postulates (i) to (iv), by Meredith and Prior, which substantially are (I) to (IV), are in effect provable in MC^{ν}_{ρ}.

Let us remark that $\mid_{\rho}$ and Meredith and Prior's analogue of $\mid_{\rho}$, i.e. $\underline{n}$, are introduced in quite different ways: $\underline{n}$ is considered as a primitive constant that satisfies four special postulates, (i) to (iv), whereas $\mid_{\rho}$ is a special case of the matrix $\mid_{\underline{u}}$, which concerns the arbitrary elementary possible case $\underline{u}$ and which is defined in MC^{ν} on the basis of the strong axiom AS12.19. To define $\mid_{\rho}$ a single very simple additional axiom, A52.1, is required.

N54. Discussion of $\iota_{\underline{u}}$ and $\daleth_{\underline{u}}$.

Our intensional description operator $\iota_{\underline{u}}$ [Def. 50.1] depends on the parameter $\underline{u}$, and the conditions under which the usual replacement property of descriptions holds for $\iota_{\underline{u}}$ include the matrix $\mid_{\underline{u}}$, which has no analogue in ordinary language. We now present a twofold justification of these facts. First we show that the use of an intensional description, $(\iota \underline{x})\underline{p}$, independent of any parameter and satisfying the usual replacement property of descriptions as follows:

(a) $\vdash (\exists_1^{\frown} \underline{x}) \underline{p} \wedge \underline{p} \supset \underline{x} =^{\frown} (\iota \underline{x}) \underline{p}$ for every matrix $\underline{p}$,

is incompatible with the axioms of MC^{ν}, particularly by those of the lower predicate calculus of MC^{ν}, the axioms for identity ASs12.10–12, the general instantiation axiom AS12.8, and the modal axiom AS12.23. Then we prove the efficiency of the operators $\iota_{\underline{u}}$ and ι_{ρ} by showing, in N55, how certain philosophical puzzles can be solved with their help.

In order to prove the incompatibility of (a) above, by AS12.23 and rule $\underline{C}$ we deduce

(b) $\vdash \diamond \overline{\underline{p}} \diamond \sim \overline{\underline{p}}$ where $\overline{\underline{p}} \equiv_D \underline{v}_{11} = \underline{v}_{(1)1}$.

Then, setting $V =_D (\lambda \underline{x}) \underline{x} = \underline{x}$ and using $\underline{x}$ for $\underline{v}_{(1)2}$, we easily deduce

(c) $\vdash \underline{N} (\exists_1^{\frown} \underline{x}) \Phi (\underline{x})$, where $\Phi (\underline{x}) \equiv_D \underline{x} =^{\frown} \Lambda \wedge \overline{\underline{p}} \vee \underline{x} =^{\frown} V \wedge \sim \overline{\underline{p}}$ $(\vdash \underline{N} \Lambda \neq V)$.

Then from (a) and (c) we have $\underline{N} \Phi [(\iota \underline{x}) \Phi (\underline{x})]$ whence by $(c)_2$,

(d) $\vdash \overline{\underline{p}} \supset \xi =^{\frown} \Lambda, \sim \overline{\underline{p}} \supset \xi =^{\frown} V$, where $\xi =_D (\iota \underline{x}) \Phi (\underline{x})$.

Then by (b), $\vdash \xi =^{\frown} \Lambda \wedge \xi =^{\frown} V$. Thence by ASs12.8 and 12.10–12 we deduce $\Lambda =^{\frown} V$, which contradicts $(c)_3$.

We conclude that a description operator with the property (a) cannot be accepted when one wants a logical interpreted calculus, such as MC^{ν}, that solves the problem (mentioned in [12], pp. 195–96) of combining modalities and variables in such a way that the customary inferences of the logic of quantification remain valid.

This problem is not solved in most theories of modal logic, even when only logical types of very low levels are involved. This occurs e.g. in Montague's work [28], where the semantical rules are such that, in particular, the specification rule

$(\underline{x})\,\Phi(\underline{x}) \supset \Phi(\underline{a})$, where $\underline{a}$ is a constant or a description,

does not hold in connection with any logical type.

Let us remark that as far as descriptions are concerned, the nonvalidity of the above rule is due to the fact that in [28] an intensional description satisfying property (a) is used, and that by the preceding syntactical considerations about (a), it would be no use changing Montague's semantical rules to solve the above problem mentioned in [12] while keeping an intensional description operator independent of any parameter and satisfying (a). (Incidentally intensional descriptions are useless in connection with extensional properties, so an extensional description seems necessary to me to apply modal logic to physics, for example.)

N55. Solution of some philosophical puzzles by means of $\mathfrak{I}$, $\mathfrak{I}_\rho$, and ι_u. Comparison with some solutions by Thomason and Stalnaker.

Let us consider the following assertion in ordinary language:

(a) The (modally prefixed) number of living presidents (living at the instant $\underline{t}$) could be larger than it is (in reality).

The most natural translation of (a) into MC^ν_ρ is

$(\bar{a})\quad \Diamond\,(\mathfrak{I}\underline{n})\underline{q} > (\iota_\rho\underline{n})\underline{q}$, where $\underline{q} \equiv_D \underline{n} \in \underline{Nn} \wedge \underline{LivPres} \in \underline{n}$,

and where $\vdash \underline{Nn} \in \underline{Abs}$ [Theor. 45.1(II)] and $\underline{LivPres}$ denotes (in ML^ν) the extensional concept of living president.

Indeed, $(\exists\underline{n})\underline{q}$ is true. Thence $(\exists_1\underline{n})\underline{q}$, and (since $\vdash \underline{Nn} \in \underline{Abs}$—cf. Def. 18.8) $(\exists_1^{\frown}\underline{n})\underline{q}$ can easily be deduced—cf. Theor. 41.2(II). So in every elementary possible case $\underline{u}$, $(\iota_\rho\underline{n})\underline{q}$ means the number of presidents living in the real case ρ [Def. 50.1]; furthermore in the case $\underline{u}$, $(\text{ɿ}\,\underline{n})\underline{q}$ is (nonstrictly) equal to that natural number $\underline{n}$ for which $\underline{q}$ holds (in the case $\underline{u}$)—cf. rule (δ_9) in N11 or AS12.18(I)—i.e. the number of presidents living in case $\underline{u}$.

The following assertion of ordinary language is similar to and more complex than (a):

(b) Even if no presidents died (one might say that) the number of living presidents could be larger than in that case.

A translation of (b) into MC^{ν}_{ρ} is

$$(\bar{b}) \qquad (\underline{u})\,\{\Diamond\,(|_{\underline{u}} \wedge \underline{\text{NoPresDied}}) \supset \Diamond\,(\text{ɿ}\,\underline{n})\underline{q} > (\iota_{\underline{u}}\underline{n})\underline{q}\}.$$

Let us reword (a) as follows:

(a′) The number of living presidents is smaller than a possible number of living presidents.

Now it is natural to "generalize" (a′), hence (a), into the following sentence equivalent to (b):

(b′) Even if no president ever died, the number of living presidents would be smaller than a possible number of living presidents.

The natural translations of (a′) and (b′) into MC^{ν}_{ρ} are

$$(\bar{a}') \qquad (\exists\underline{n})\,[\Diamond\,\underline{q} \wedge (\text{ɿ}\underline{n})\underline{q} > (\text{ɿ}_{\rho}\underline{n})\underline{q}] \text{ where } \underline{q} \equiv_{D} \underline{n} \in \underline{\text{Nn}} \wedge \text{LivPres} \in \underline{n},$$

and

$$(\bar{b}') \qquad (\underline{u})\,\{\Diamond\,(|_{\underline{u}} \wedge \text{NoPresDied}) \supset (\exists\underline{n})\,[\Diamond\,\underline{q}(\text{ɿ}\underline{n})\underline{q} > (\iota_{\underline{u}}\underline{n})\underline{q}]\}$$

respectively. The translation of (b) into MC^{ν}_{ρ}—realized by means of $(\bar{b})$ or $(\bar{b}')$—seems to require the use of $|_{\underline{u}}$ and $\iota_{\underline{u}}$ (or $\text{ɿ}_{\underline{u}}$).

Many writers understand the ordinary sentence

(c) If the number of living presidents is 3, then it is necessarily greater than 2

in such a way that the antecedent can be true and the consequent false. Then the problem is to construct an interpreted logical calculus in which (c) can be expressed but not deduced. Such a problem is solved by Carnap in [12], p. 194, where a "paradox" similar to (c) is considered. An analogue of this solution for MC^{ν}, or MC^{ν}_{ρ}, and (c) consists in translating (c) into the sentence

$(\bar{c}_1)$ $(\imath \underline{n})\underline{q} = 3 \supset \underline{N}(\imath \underline{n})\underline{q} > 2$ where $\underline{q} \equiv_D \underline{n} \in \underline{Nn} \wedge \underline{LivPres} \in \underline{n}$.

Another solution of paradox (c) consists in interpreting "the number of living presidents" as the number of living presidents in the real case. Then (c) is true. The translation of this interpretation of (c) into MC^{ν}_{ρ} is

$(\bar{c}_2)$ $\mathfrak{R}(\iota_{\rho}\underline{n})\underline{q} = 3 \supset \underline{N}(\iota_{\rho}\underline{n})\underline{q} > 2$, or
$(\bar{\bar{c}}_2)$ $\mathfrak{R}(\imath\underline{n})\underline{q} = 3 \supset \underline{N}(\iota_{\rho}\underline{n})\underline{q} > 2$ [Def. 52.1].

THEOR. 55.1. Both $(\bar{c}_2)$ and $(\bar{\bar{c}}_2)$ are provable in MC^{ν}_{ρ}.

Proof. First we start with (α) $\mathfrak{R}(\iota_{\rho}\underline{n})\underline{q} = 3$, whence by Def. 52.1 we have (β) $(\iota_{\rho}\underline{n})\underline{q} =^{\cup} 3$.

Since $3 \neq \underline{a}^*$, by Theor. 52.2(II), (α) yields (γ) $\mathfrak{R}(\exists_1^{\frown}\underline{n})\Phi(\underline{n})$ where $\Phi(\underline{n})$ is $\underline{q}$—cf. $(\bar{c}_1)_2$. Thence by Theor. 52.2(I) we deduce $\mathfrak{R}\,\Phi[(\iota_{\rho}\underline{n})\underline{q}]$, which by Def. 52.1 yields $\diamond\,\Phi[(\iota_{\rho}\underline{n})\underline{q}]$. Thence we have, in particular, (δ) $\diamond\,(\iota_{\rho}\underline{n})\underline{q} \in \underline{Nn}$. Since $\vdash \underline{Nn} \in \underline{Abs}$ [Theor. 45.1(II)], by Defs. 18.8 and 18.7, (β) and (δ) yield (ϵ) $\underline{N}(\iota_{\rho}\underline{n})\underline{q} = 3$. So $(\alpha) \vdash (\epsilon)$. Now it is easy to conclude that $\vdash (\bar{c}_2)$.

In order to prove $(\bar{\bar{c}}_2)$ we now start with (α') $\mathfrak{R}(\imath\underline{n})\underline{q} = 3$, whence by AS38.1(II) (and Def. 52.1) we have $\mathfrak{R}(\exists_1\underline{n})\underline{q}$. Thence by reasoning as we did after $(\bar{a})$, we deduce $\mathfrak{R}(\exists_1^{\frown}\underline{n})\underline{q}$, which by

Theor. 52.2(V) easily yields $\mathfrak{R}(\imath \underline{n})\underline{q} = (\iota_\rho \underline{n})\underline{q}$, whence by ASs12.11, 12, (α'), and (94) we have $\mathfrak{R}(\iota_\rho \underline{n})\underline{q} = 3$. We conclude [Theor. 33.2] that $\vdash \mathfrak{R}(\imath \underline{n})\underline{q} = 3 \supset \mathfrak{R}(\iota_\rho \underline{n}) = 3$. This result and ($\overline{c}_2$) imply ($\overline{\overline{c}}_2$).
QED

In [34] Thomason solves paradox (a) in substantially the two ways used above, using two interpreted calculi of the first order, QI and Q3, where modalities and variables are combined in such a way that the customary inferences of the (extensional) logic of quantification do not remain valid. In particular specification and existential generalization do not hold. Furthermore only extensional properties can be considered and $\vdash \underline{x} =^{\cup} \underline{y} \equiv \underline{x} =^{\cap} \underline{y}$ holds.

We have shown that paradox (a) can be solved in two ways using the interpreted calculus MC_ρ^ν, which does not have the features of Q1 or Q3 mentioned above. This is of interest because these features appear restrictive especially in applying modal logic to physics. Furthermore MC_ρ^ν allows us to solve the generalization (b) of (a).

Here are some other philosophical puzzles considered by Thomason and Stalnaker in [35]:

(d) Socrates can run while not running.

(e) The largest state in the Union might be smaller than some state in the Union.

(f) It is necessarily the case that the president of the U.S. is a citizen of the U.S.

(g) The president of the U.S. has the property of being necessarily a citizen of the U.S. (false)

(h) The man sitting in the corner can ride a horse.

(k) If I had more money (than I have), I'd buy a car.

These sentences can be translated into MC_ρ^ν as follows. First we remark that (d) can be rephrased into

(d_1) Socrates can run at an—or at every—instant $\underline{t}$ at which he is not running in reality.

Then, denoting the absolute concept of instant by $\underline{Inst}$ we translate (d_1) into MC_ρ^ν by

(d_1') $(\exists \underline{t}) \{\underline{t} \in \underline{Inst} \wedge \mathfrak{R} \sim \underline{Run}(\underline{Socr}, \underline{t}) \wedge \diamond\ \underline{Run}(\underline{Socr}, \underline{t})\}$

or by

(d_1'') $(\underline{t}) \{\underline{t} \in \underline{Inst} \wedge \mathfrak{R} \sim \underline{Run}(\underline{Socr}, \underline{t}) \supset \diamond\ \underline{Run}(\underline{Socr}, \underline{t})\}$.

Now we rephrase (e) into

(e_1) The state that in reality belongs to (i.e. is a state of) the Union and is larger or equal to every state of the Union might be smaller than some state in the Union ($\underline{State}_{\underline{u}}$).

Understanding $\underline{State} \in \underline{Abs}$ and $\underline{State}_{\underline{u}} \in \underline{Ext}$, our translation of ($e_1$) into MC_ρ^ν is

(e_1') $\diamond (\exists \underline{z}) \{\underline{z} \in \underline{State}_{\underline{u}} \wedge \underline{z} > (\iota_\rho \underline{x}) [\underline{x} \in \underline{State} \cap \underline{State}_{\underline{u}} \wedge (\underline{y}) (\underline{y} \in \underline{State}_{\underline{u}} \supset \underline{x} \geq \underline{y})]\}$.

Understanding $\underline{Money}$, $\underline{Man} \in \underline{Abs}$, and $\underline{Pres}_{\underline{u}}$, $\underline{Citz}_{\underline{u}} \in \underline{Ext}$—where $\underline{Pres}_{\underline{u}}$ means president of the United States and $\underline{Citz}_{\underline{u}}$ means citizen of the United States—we translate (f) to (k) into

($\bar{f}$) $\underline{Npres}_{\underline{u}} \in \underline{Citz}_{\underline{u}}$ $\quad (\underline{pres}_{\underline{u}} =_D (\imath \underline{x}) \underline{x} \in \underline{Pres}_{\underline{u}})$.

($\bar{g}$) $\underline{N}(\iota_\rho \underline{x}) (\underline{x} \in \underline{Man} \cap \underline{Pres}_{\underline{u}}) \in \underline{Citz}_{\underline{u}}$.

($\bar{h}$) $\diamond\ \underline{Ride}_{\underline{H}}[(\iota_\rho \underline{x}) (\underline{x} \in \underline{Man} \cap \underline{\text{sitting in the corner}})]$.

($\bar{k}$) $(\imath \underline{x}) (\underline{x} \in \underline{Money} \wedge \underline{\text{I have}}\ \underline{x}) > (\iota_\rho \underline{x}) (\underline{x} \in \underline{Money} \wedge \underline{\text{I have}}\ \underline{x}) \supset ^\frown (\underline{\text{I buy a car}})$.

In [35] Thomason and Stalnaker sketch a theory of modal logic that cannot be embodied in any rigorous formal theory of theirs—as they in effect say on p. 11. In particular they characterize a certain abstraction operator semantically, and by means of it they translate (d) to (k). Thomason and Stalnaker's translations of (f) and (g)—i.e. (12) and (13) in [35]—can be given the following respective counterparts in MC_{ρ}^{ν}:

(f*) $\underline{N}\,\underline{Citz}_{\underline{u}}[(\imath \underline{x})\underline{x} \in \underline{Pres}_{\underline{u}}]$, or $\underline{Citz}_{\underline{u}}{}^{\cap}[(\imath \underline{x})\underline{x} \in \underline{Pres}_{\underline{u}}]$ [Def. 6.6],

and

(g*) $(\iota_{\rho}\underline{x})\,[\underline{x} \in \underline{Man} \wedge \underline{Pres}_{\underline{u}}] \in \underline{Citz}_{\underline{u}}{}^{\cap}$.

The other philosophical puzzles (d) to (k) can be translated in a similar way.

Denoting the president of the United States by $\underline{a}$, let us remark that in [35], Ns4, 9, Thomason and Stalnaker substantially express (f) by "applying $\underline{N}$ to $\underline{Citz}_{\underline{u}}(a)$" and express (g) by "applying $\underline{NCitz}$ to $\underline{a}$." They denote the result of this operation by $\hat{\underline{x}}\ \underline{NCitz}(\underline{x})\underline{a}$, where $\hat{\underline{x}}$ is Thomason's abstraction operator. Of course $\hat{\underline{x}}\ \underline{NCitz}(\underline{x})\underline{a} \equiv \underline{NCitz}(\underline{a})$, or the principle of abstraction $(\lambda\,\underline{x})\,\underline{F}(\underline{x})\,(\underline{x}) \equiv \underline{F}(\underline{x})$—cf. $(45)_1$—does not hold in Thomason and Stalnaker's theory.

From our point of view Thomason and Stalnaker's interpretation of (f) and (g) can be schematized by the application of one property, $\underline{Citz}_{\underline{u}}{}^{\cap}$, to two different individual concepts of president of the United States, the president of the United States and the man who is the president of the United States in the real case. This we did by means of (f*) and (g*). The fact that Thomason and Stalnaker consider only one concept of president, $\underline{a}$, and in particular their use of only one description operator, is largely responsible for the failure of their theory: First, they say, [35], N8, "This attempt to define abstracts fails, and in fact it can be proved that any such attempt will fail; abstraction, as we have characterized it semantically, is not definable in the modal logic Q3"; second, they end [35] asserting that "in [34], it was argued that it was a failure to dispense with the principle of instantiation"—i.e. AS12.8—"which hindered a successful generalization of classical logic; here we have shown that this may equally well be regarded as a failure to discard the principle . . . of abstraction."

We may say that, unlike the language used in [35] to solve the puzzles (d) to (k), the interpreted modal logical calculus MC_ρ^ν within which we gave our solutions of those puzzles has foundations that are stated explicitly and thoroughly; it generalizes classical—i.e. extensional—logic in that all principles of extensional logic, and in particular the principle of instantiation AS12.8 and the principle of abstraction—cf. theorem $(45)_1$—remain valid (possibly in slightly modified forms). Furthermore MC_ρ^ν allows us to solve in various ways all the philosophical puzzles considered in [35] that concern traditional modal logic, or alethic logic, according to Von Wright's paper [38]—and not for example deontic logic or belief theory. These methods substantially include Thomason and Stalnaker's solutions of the same puzzles. It can be shown that MC_ρ^ν allows us to solve the puzzles that generalize some of the puzzles (d) to (k) the way (b) generalizes (a).

11

Definition within MC^{ν} of Certain Analogues of the QIs for ML^{ν}; The Translation $\Delta \to \Delta^*$ of EL^{ν} and $EL^{\nu+1}$ into ML^{ν}

N56. A set of absolute concepts in MC^{ν} that can characterize the object system used in our semantical theory for $EL^{\nu+1}$.

Keeping in mind our translation $\Delta \to \Delta^*$ of $EL^{\nu+1}$ into ML^{ν} [N46], we want to characterize in ML^{ν} the semantical system constructed for $EL^{\nu+1}$ in N6 and based on the classes $\underline{O}_{\underline{t}}^{\nu+1}$ ($\underline{t} \in \tau^{\nu+1}$). We want to do this through the replacement of $\underline{O}_{\underline{t}}^{\nu+1}$ with a suitable absolute concept $\Omega_{\underline{t}}^{\nu+1}$ ($\underline{t} \in \tau^{\nu+1}$).

To get an intuitive idea about our forthcoming formal definition of $\Omega_{\underline{t}}^{\nu+1}$ in ML^{ν}, we consider AS25.1 which, taking Defs. 4.6 and 18.9 into account, can be put into the form

AS56.1. $(\exists \underline{F}) [\underline{F} \in \underline{Abs}\ \underline{a}^* \in {}^{\frown}\underline{F}\ \underline{N}(\underline{x})\underline{x} \in \underline{F}^{(\underline{e})}]$

where $\underline{x}$ and $\underline{F}$ are variables of the respective types $\underline{r}$ and $(\underline{r})$ ($\underline{r} = 1, \ldots, \nu$).

Consequently we can (intuitively) identify $\Omega_{\underline{r}}^{\nu+1}$ with an absolute concept of type $(\underline{r})$ for $\underline{r} = 1, \ldots, \nu$.

Now we consider the natural numbers of type ((1)) and decide that from now on we shall understand the absolute (concept or) class $\underline{El}$ only as defined by means of Defs. 47.1, 48.1, where 0 and 1 are just numbers of the type ((1)). Thus from now on the elements of El shall have the type (((1))).

It would be desirable to identify $\Omega_{\nu+1}^{\nu+1}$ with $\underline{El}$ if $\vdash \underline{a}^* \in \underline{El}$ held, but this is not the case. Moreover, we have $\vdash \sim \Diamond\ \underline{a}^* \in \underline{El}$, for by Def. 48.1 every element $\underline{u}$ of $\underline{El}$ is a set that necessarily has one element, whereas $a^*_{\underline{t}}$ is the empty set when $\underline{t}$ is the type of $\underline{u}$.

Then intuitively, we obtain an absolute property from the class $\underline{El}$ by substituting $\underline{a}^*$ for any element of $\underline{El}$. The resulting class $\underline{El}'$ has the same properties (82) as $\underline{El}$. Furthermore, $\underline{El}'$ and $\underline{El}$ are in a natural one-to-one correspondence. This correspondence transforms the matrix $|_{\underline{u}}$ [Def. 48.2] into a matrix, $|'_{\underline{u}}$, for which the analogues of the theorems in (83) on $|_{\underline{u}}$ hold. In addition, for $\underline{El}'$ and $|'_{\underline{u}}$ the analogues of theorems $(\overline{89})$ to (92) and $(\overline{92})$ hold. Technical details will be given shortly.

Lastly, for $\underline{t} \in \tau^{\nu}$ we want to identify $\Omega_{\underline{t}}^{\nu+1}$ with the absolute concept built up out of $\Omega_1^{\nu+1}, \ldots, \Omega_{\nu+1}^{\nu+1}$ as $\underline{O}_{\underline{t}}^{\nu+1}$ is built up out of $\underline{O}_1^{\nu+1}, \ldots, \underline{O}_{\nu+1}^{\nu+1}$ according to conditions (a) to (c) considered in N6. From the formal or technical point of view, for $\underline{t} \in \tau^{\nu}$ we translate "$\Omega_{\underline{t}}^{\nu+1}$" into a variable (in ML^{ν}) of a suitable type ($\underline{t}^*$), and for forthcoming applications [N57] we assume that this variable has an odd index.

DEF. 56.1. For $\underline{t} \in \tau^{\nu+1}$ we define $\underline{t}^*$ recursively by means of the following conditions—cf. (2) in N2:

(a) $\underline{t}^* = \underline{t}$ for $\underline{t} = 1$ to ν, and $\underline{t}^* = (((1)))$ for $\underline{t} = \nu+1$,

(b) $\underline{t}^* = (\underline{t}_1^*, \ldots, \underline{t}_{\underline{n}}^*)$ for $\underline{t} = (\underline{t}_1, \ldots, \underline{t}_{\underline{n}})$,

(c) $\underline{t}^* = (\underline{t}_1^*, \ldots, \underline{t}_{\underline{n}}^* : \underline{t}_0^*)$ for $\underline{t} = (\underline{t}_1, \ldots, \underline{t}_{\underline{n}} : \underline{t}_0)$.

Let us remark that <u>we have $t^* = t$ if and only if $t \in \tau^{\nu}$</u> [N2]; furthermore <u>$t^* \in \tau^{\nu}$ and $(t)^* = (t^*)$ for $t \in \tau^{\nu+1}$</u>.

We introduce the notations $\underline{v}_{\underline{t}}^*$ and $\underline{c}_{\underline{t}}^*$ as follows:

(97) $\underline{v}_{\underline{t}}^* =_D \underline{v}_{\underline{t},1}$ and $\underline{c}_{\underline{t}}^* =_D \underline{c}_{\underline{t},1}$ for $\underline{t} = \underline{t}^*$, i.e. $\underline{t} \in \tau^{\nu}$,

(98) $\underline{v}_{\underline{t}}^* =_D \underline{v}_{\underline{t}^*,3}$ and $\underline{c}_{\underline{t}}^* =_D \underline{c}_{\underline{t}^*,3}$ for $\underline{t} \neq \underline{t}^*$ and $\underline{t} \in \tau^{\nu+1}$.

By rule $\underline{C}$ with $\underline{v}_{(\underline{r})}^*$, we obtain from AS56.1

(99) $\vdash_{\underline{c}} \underline{v}_{(\underline{r})}^* \in \mathrm{Abs}\, \underline{a}_{\underline{r}}^* \in {}^{\frown} \underline{v}_{(\underline{r})}^* \; \underline{N}(\underline{v}_{\underline{r},1}) \underline{v}_{\underline{r},1} \in \underline{v}_{(\underline{r})}^{*(\underline{e})}$ $(\underline{r} = 1, \ldots, \nu)$ [Def. 18.9].

Furthermore, using rule $\underline{C}$ with $\underline{v}^*_{\nu+1^*,5}$ and $\underline{v}^*_{(\nu+1)}$ where $\nu+1^*$ means $\underline{t}^*$ for $\underline{t} = \nu+1$, we easily deduce from $(83)_1$ and $(46)_1$

$$(100)\quad \vdash_{\underline{C}} \underline{v}_{\nu+1^*5} \in \underline{El},$$
$$\vdash_{\underline{C}} (\underline{u}) [\underline{u} \in \underline{v}^*_{(\nu+1)} \equiv^\cap \underline{u} =^\cap \underline{a}^* \vee \underline{u} \in \underline{El} \wedge \sim \underline{u} =^\cap \underline{v}_{\nu+1^*,5}].$$

We shall shortly identify $\underline{v}^*_{(\nu+1)}$ with the class $\underline{El}'$ $(=_D \underline{el}'(\underline{v}_{\nu+1^*5}))$ mentioned above.

From $(46)_{1,2}$, using rule $\underline{C}$ with the variable $\underline{v}^*_{(\underline{t})}$ for $\underline{t} = (\underline{t}_1, \ldots, \underline{t}_n)$ and $\underline{t} = (\underline{t}_1, \ldots, \underline{t}_n : \underline{t}_0)$ where $\underline{t}_0, \underline{t}_1, \ldots, \underline{t}_n \in \tau^{\nu+1}$ we respectively deduce—cf. Def. 6.9—

$$(101)\quad \vdash_{\underline{C}} (\underline{F}) \{\underline{F} \in \underline{v}^*_{(\underline{t})} \equiv^\cap \underline{F} \in \underline{MConst}\, (\forall \underline{x}_1, \ldots, \underline{x}_n) [\underline{F}(\underline{x}_1, \ldots, \underline{x}_n) \supset \bigwedge_{\underline{i}=1}^{\underline{n}} \underline{x}_{\underline{i}} \in \underline{v}^*_{(\underline{t}_{\underline{i}})}]\} \text{ for } \underline{t} = (\underline{t}_1, \ldots, \underline{t}_n),$$

$$(102)\quad \vdash_{\underline{C}} (\underline{f}) \{\underline{f} \in \underline{v}^*_{(\underline{t})} \equiv^\cap (\forall \underline{x}_0, \ldots, \underline{x}_n) [\underline{x}_0 =^\cap \underline{f}(\underline{x}_1, \ldots, \underline{x}_n) \supset (\underline{x}_0 =^\cap \underline{a}^*_{\underline{t}_0} \vee \bigwedge_{\underline{i}=0}^{\underline{n}} \underline{x}_{\underline{i}} \in \underline{v}^*_{(\underline{t}_{\underline{i}})})]\} \text{ for } \underline{t} = (\underline{t}_1, \ldots, \underline{t}_n : \underline{t}_0).$$

By Defs. 18.7 and 18.8, $\vdash \underline{F} \in \underline{Abs} \wedge \underline{G} \in \underline{Abs} \supset \underline{F} \times \underline{G} \in \underline{Abs}$ where "$\times$" denotes Cartesian product. Hence from (101) we easily deduce

$$(103)\quad \vdash_{\underline{C}} \underline{F} \in \underline{v}^*_{(\underline{t})} \bigwedge_{\underline{i}=1}^{\underline{n}} \underline{v}^*_{(\underline{t}_{\underline{i}})} \in \underline{Abs} \supset^\cap \underline{F} \in \underline{Abs} \quad \text{for } \underline{t} = (\underline{t}_1, \ldots, \underline{t}_n) \in \tau^{\nu+1}.$$

Now we introduce $\underline{El}'$ and $|'_{\underline{u}}$ technically, by means of some abbreviating definitions that, unlike most, are based on (100) and some results of rule $\underline{C}$:

$$(100')\qquad \underline{El}' =_D \underline{v}^*_{(\nu+1)} \quad \text{whence} \quad \vdash_{\underline{C}} \underline{a}^* \in^\cap \underline{El}' \quad (\vdash \sim \underline{a}^* \in \underline{El}).$$

DEF. 56.2. $|'_{\underline{u}} \equiv_D \underline{u} \in \underline{El}' [\sim \underline{u} =^\cap \underline{a}^* |_{\underline{u}} \vee \underline{u} = {}^\cap \underline{a}^* |_{\underline{w}}]$
where $\underline{w}$ is $\underline{v}_{\nu+1^*5}$.

The definitions (100′) and 56.2, and (82), (83), (89) to (92), $(\overline{92})$, and (99) imply the following theorem:

THEOR. 56.1. Let the results (99) to (102) of rule $\underline{C}$ hold. Then

(82′) $\vdash_{\underline{c}} \underline{El}' \in \underline{MConst}$, $\vdash_{\underline{c}} \underline{El}' \in \underline{MSep}$, $\vdash_{\underline{c}} \underline{El}' \in \underline{Abs}$,

(83′) $\vdash_{\underline{c}} (\exists \underline{u}) |'_{\underline{u}}$, $\vdash_{\underline{c}} (\exists^{(1)} \underline{u}) |'_{\underline{u}}$, $\vdash_{\underline{c}} (\exists_{1} \underline{u}) |'_{\underline{u}}$, $\vdash_{\underline{c}} \underline{u} \in \underline{El}' \supset \Diamond |'_{\underline{u}}$,

$\vdash_{\underline{c}} (\exists^{(1)^{\frown}} \underline{u}) |'_{\underline{u}}$, $\vdash_{\underline{c}} (\exists^{\frown}_{1} \underline{u}) |'_{\underline{u}}$,

(89′) $\vdash \underline{u} \in \underline{El}' \supset [\Diamond (|'_{\underline{u}} \underline{p}) \equiv |'_{\underline{u}} \supset^{\frown} \underline{p}]$,

$\vdash_{\underline{c}} \Diamond (|'_{\underline{u}} \underline{p}) \equiv {}^{\frown}\underline{u} \in \underline{El}' (|'_{\underline{u}} \supset^{\frown} \underline{p})$.

If, in addition, the variable $\underline{u}$ does not occur free in the matrix $\underline{p}$, then

(90′) $\vdash_{\underline{c}} (\underline{u}) [\underline{u} \in \underline{El}' \supset \Diamond (|'_{\underline{u}} \underline{p})] \equiv \underline{Np} \equiv (\underline{u}) (|'_{\underline{u}} \supset^{\frown} \underline{p})$,

(91′) $\vdash_{\underline{c}} (\exists \underline{u}) [\underline{u} \in \underline{El}' (|'_{\underline{u}} \supset^{\frown} \underline{p})] \equiv \Diamond \underline{p} \equiv (\exists \underline{u}) \Diamond (|'_{\underline{u}} \underline{p})$.

Furthermore, let the matrix $\phi(\underline{p}_1, \ldots, \underline{p}_m)$ be built up starting out of $\underline{p}_1, \ldots, \underline{p}_m, \sim, \wedge, (\forall \underline{x}_1), \ldots, (\forall \underline{x}_n)$ independently of the (arbitrary) matrices $\underline{p}_1, \ldots, \underline{p}_m$; then

(92′) $\underline{u} \in \mathrm{El}' \vdash_{\underline{c}} |'_{\underline{u}} \supset^{\frown} \phi(\underline{p}_1, \ldots, \underline{p}_m) \equiv \phi(|'_{\underline{u}} \supset^{\frown} \underline{p}_1, \ldots, |'_{\underline{u}} \supset^{\frown} \underline{p}_m)$,

$\vdash_{\underline{c}} \Diamond [|'_{\underline{u}} \phi(\underline{p}_1, \ldots, \underline{p}_m)] \equiv \underline{u} \in \underline{El}' \phi[\Diamond (|'_{\underline{u}} \underline{p}_1), \ldots, \Diamond (|'_{\underline{u}} \underline{p}_m)]$.

THEOR. 56.2. By convention (97), the results (99) to (102) of rule $\underline{C}$ imply, for $\underline{t} \in \tau^{\nu+1}$,

(104) $\vdash_{\underline{c}} \underline{v}^*_{((\underline{t}))} \subseteq \underline{MConst}_{(\underline{t})^*}$, $\vdash_{\underline{c}} \underline{v}^*_{(\underline{t})} \in \underline{MConst}$,

(105) $\vdash_{\underline{c}} \underline{v}^*_{(\underline{t})} \in \underline{Abs}$, $\vdash_{\underline{c}} \underline{F} \in \underline{v}^*_{((\underline{t}_1, \ldots, \underline{t}_n))} \supset \underline{F} \in \underline{Abs}_{((\underline{t}_1, \ldots, \underline{t}_n))}$.

Proof. For every type $\underline{t} \in \tau^{\nu+1}$, (101)—with $\underline{t}$ replaced by $(\underline{t})$—yields $(104)_1$. For every $\underline{t} \in \tau^{\nu+1}$, $(105)_1$ yields $(104)_2$. So now we want to prove $(105)_1$ by induction.

For $\underline{t} = 1, \ldots, \nu$, $(105)_1$ holds by (99). It holds for $\underline{t} = \nu + 1$ by $(100')_1$ and $(82')_3$.

As the hypothesis of the induction we assume that $(105)_1$ holds for $\underline{t}=\underline{t}_1, \ldots, \underline{t}=\underline{t}_n$. Furthermore we take $\underline{t}$ to be $(\underline{t}_1, \ldots, \underline{t}_n)$.

Then by Theor. 41.1(II) the R.H.S. of the equivalence written in (101) is strictly equivalent to a modally closed matrix. So $(104)_2$ holds for $\underline{t}=(\underline{t}_1, \ldots, \underline{t}_n)$, which by $(104)_1$ and Theor. 41.1(II), (VI) yields $(105)_1$ for $\underline{t}=(\underline{t}_1, \ldots, \underline{t}_n)$.

Now let $(105)_1$ hold for $\underline{t}=\underline{t}_0, \ldots, \underline{t}=\underline{t}_n$ as the hypothesis of the induction, and take $\underline{t}$ to be $(\underline{t}_1, \ldots, \underline{t}_n : \underline{t}_0)$. Furthermore add the assumptions (a) $\underline{f}, \underline{g} \in \underline{v}_{(\underline{t})}$ and (b) $\underline{f} =^{\cup} \underline{g}$.

Then by AS40.2 we obtain $\diamond (\forall \underline{x}_1, \ldots, \underline{x}_n) [\underline{f}(\underline{x}_1, \ldots, \underline{x}_n) = g(\underline{x}_1, \ldots, \underline{x}_n)]$, whence $(\forall \underline{x}_1, \ldots, \underline{x}_n) \underline{f}(\underline{x}_1, \ldots, \underline{x}_n) =^{\cup} \underline{g}(\underline{x}_1, \ldots, \underline{x}_n)$. Thence we have (d) $\underline{f}(\underline{x}_1, \ldots, \underline{x}_n) =^{\cup} \underline{g}(\underline{x}_1, \ldots, \underline{x}_n)$, which by (a) and (102) yields either $\underline{f}(\underline{x}_1, \ldots, \underline{x}_n) =^{\cap} \underline{g}(\underline{x}_1, \ldots, \underline{x}_n) =^{\cap} \underline{a}^*_{\underline{t}_0 *}$ or $\underline{f}(\underline{x}_1, \ldots, \underline{x}_n), \underline{g}(\underline{x}_1, \ldots, \underline{x}_n) \in \underline{v}^*_{(\underline{t}_0)}$, so that by the inductive hypothesis ($\underline{v}^*_{(\underline{t}_0)} \in \underline{\mathrm{Abs}}$) and Defs. 18.7 and 18.8, from (d) we deduce (e) $\underline{f}(\underline{x}_1, \ldots, \underline{x}_1) =^{\cap} \underline{g}(\underline{x}_1, \ldots, \underline{x}_n)$ in either case.

As a consequence, by rule $\underline{G}$ [Def. 33.1] we have $(\forall \underline{x}_1, \ldots, \underline{x}_n)$ $\underline{f}(\underline{x}_1, \ldots, \underline{x}_n) =^{\cap} \underline{g}(\underline{x}_1, \ldots, \underline{x}_n)$, hence $\underline{f} =^{\cap} \underline{g}$ by AS40.2.

By Theor. 33.3 and rule $\underline{G}$ we easily conclude that $\vdash_{\underline{c}} (\forall \underline{f}, g)$ $[\underline{f}, \underline{g} \in \underline{v}^*_{(\underline{t})} \underline{f} =^{\cup} \underline{g} \supset \underline{f} =^{\cap} \underline{g}]$, i.e. $\vdash_{\underline{c}} \underline{v}^*_{(\underline{t})} \in \underline{\mathrm{MSep}}$ [Def. 18.7]. Furthermore the hypothesis of the induction and (102) yield $(104)_2$. Then by Def. 18.8 we have $(105)_1$ for $\underline{t}=(\underline{t}_1, \ldots, \underline{t}_n : \underline{t}_0)$.

By the principle of induction, $(105)_1$ holds for $\underline{t} \in \tau^{\nu+1}$.

From (103) and $(105)_1$ we have $(105)_2$.

QED

In the preceding as well as in most forthcoming theorems, assertions like $\vdash_{\underline{c}} \underline{q}$ or $\underline{p} \vdash_{\underline{c}} \underline{q}$ are made where $\vdash_{\underline{c}}$ is used (instead of $\vdash$) in that the results (99) to (102) of rule $\underline{C}$ are accepted. Denoting the conjunction of the matrices in ML^{ν} introduced in (99) to

(102) by $\rho_{\underline{c}}$, these assertions imply $\rho_{\underline{c}} \vdash \underline{q}$ and $\rho_{\underline{c}}, \underline{p} \vdash \underline{q}$ respectively. Furthermore by (82), Theor. 41.$\bar{1}$(II) and ($\bar{99}$) to (102), $\vdash \Diamond \rho_{\underline{c}} \equiv \underline{N}\rho_{\underline{c}}$. Hence the following theorem holds:

THEOR. 56.3. In connection with results (99) to (102) of rule $\underline{C}$, $\vdash_{\underline{c}} \underline{q}$ and $\underline{p} \vdash_{\underline{c}} \underline{q}$ imply $\vdash_{\underline{c}} \underline{Nq}$ and $\underline{Np} \vdash_{\underline{c}} \underline{Nq}$ respectively.

N57. The translation $\Delta \rightarrow \Delta^*$ of EL^{ν} and $EL^{\nu+1}$ into ML^{ν}.

The translation of $EL^{\nu+1}$ into ML^{ν} which we are going to consider will be called the star translation and obviously includes a translation of EL^{ν} into ML^{ν}.

For every designator Δ in $EL^{\nu+1}$ its star translation Δ^* into ML^{ν} is determined recursively by the following translation rules ($\underline{T}_1^*$) to ($\underline{T}_7^*$), where the results (99) to (102) of rule $\underline{C}$ and the presence of the variables $\underline{v}^*_{(\underline{t})}$ ($\underline{t} \in \tau^{\nu+1}$) in them are taken into account:

Rule	If Δ is	then Δ^* is
($\underline{T}_1^*$)	$\underline{v}_{\underline{tn}}$,	$\underline{v}_{\underline{t4n}}$ for $\underline{t}^* = \underline{t}$ and $\underline{v}_{\underline{t}^*4\underline{n}+2}$ for $\underline{t}^* \neq \underline{t}$.
($\underline{T}_2^*$)	$\underline{c}_{\underline{tn}}$,	$\underline{c}_{\underline{t4n}}$ for $\underline{t}^* = \underline{t}$ and $\underline{c}_{\underline{t}^*4\underline{n}+2}$ for $\underline{t}^* \neq \underline{t}$.
($\underline{T}_3^*$)	$\Delta_1 = \Delta_2$,	$\Delta_1^* = {}^{\cap}\Delta_2^*$.
($\underline{T}_4^*$)	the matrix or term $\Delta_0(\Delta_1, \ldots, \Delta_{\underline{n}})$,	the matrix or term $\Delta_0^*(\Delta_1^*, \ldots, \Delta_{\underline{n}}^*)$.
($\underline{T}_5^*$)	$\sim\Delta_1$ or $\Delta_1 \wedge \Delta_2$,	$\sim\Delta_1^*$ or $\Delta_1^* \wedge \Delta_2^*$ respectively.
($\underline{T}_6^*$)	$(\underline{v}_{\underline{tn}})\, \Delta_1$,	$(\underline{v}^*_{\underline{tn}})(\underline{v}^*_{\underline{tn}} \in \underline{v}^*_{(\underline{t})} \supset \Delta_1^*)$.
($\underline{T}_7^*$)	$(\iota\, \underline{v}_{\underline{tn}})\Delta_1$, where Δ_i is a matrix that does not necessarily contain the variable $\underline{v}_{\underline{tn}}$ free,	$(\iota\, \underline{v}^*_{\underline{tn}})(\underline{v}^*_{\underline{tn}} \in \underline{v}^*_{(\underline{t})}\, \Delta_1^*)$.

In rule ($\underline{T}_3^*$) we could write that Δ^* should simply be $\Delta_1^* = \Delta_2^*$ instead of $\Delta_1^* = {}^\cap\Delta_2^*$, as far as this section is concerned. However in N58, where the introduction of constants through formal definitions is considered, the latter version of rule ($\underline{T}_3^*$) is required.

Let us further remark that the above translation is a one-to-one correspondence between the designators in $EL^{\nu+1}$ and some designators in ML^ν; furthermore by (97), (98), and rule ($\underline{T}_1^*$), the latter designators do not include $\underline{v}_{\underline{t}}^*$ ($\underline{t} \in \tau^{\nu+1}$). In addition the type of $\underline{v}_{\underline{t}n}^*$ is $\underline{t}^*$—cf. (97), (98)—so that by Def. 56.1, $\underline{v}_{\underline{t}n}^* \in \underline{v}_{(\underline{t})}^*$ is well formed.

The translation rules of EL^ν into ML^ν are ($\underline{T}_1^*$) to ($\underline{T}_7^*$) for $\underline{t} \in \tau^\nu$. Then for every designator Δ in EL^ν, Δ^* is a designator of the same type $\underline{t} \in \tau^\nu$.

Incidentally let us remark that the use of the absolute concepts $\underline{v}_{(\underline{t})}^*$ ($\underline{t} \in \tau^\nu$) considered in (99) to (102) allows us to translate the extensional language EL^ν into the modal language ML^ν in a way different from the translation of an extensional language into a modal language proposed by Carnap in [13]. His proposal is based on extensional concepts; e.g. his translation of $(\forall \underline{F})\, \underline{F}(\underline{x})$ into a modal language is $(\forall \underline{F})\, [\underline{Ext}(\underline{F}) \supset \underline{F}(\underline{x})]$.

It may be added that our translation of $EL^{\nu+1}$ and EL^ν into ML^ν will also be considered from the syntactical point of view [N58].

THEOR. 57.1. The results (99) to (102) of rule $\underline{C}$ imply in MC^ν

(106) $$\vdash_{\underline{C}} \underline{a}_{\underline{t}^*}^* \in \underline{v}_{(\underline{t})}^* \quad \text{for} \quad \underline{t} \in \tau^{\nu+1}.$$

Proof. We prove (106) using induction on $\underline{t} \in \tau^{\nu+1}$.

Case A. $\underline{t} \in \{1, \ldots, \nu\}$ or $\underline{t} = \nu + 1$. Then $(106)_1$ holds by (99) or $(100)_2$ respectively.

Case B. $\underline{t} = (\underline{t}_1, \ldots, \underline{t}_n)$. Hence $\underline{t}^* = (\underline{t}_1^*, \ldots, \underline{t}_n^*)$ by Def. 56.1(b). Then AS12.21 holds for $\underline{a}^*_{\underline{t}^*}$. This implies $\vdash \underline{a}^*_{\underline{t}^*} \in \underline{\mathrm{MConst}}$ by Def. 13.2, and also $(106)_1$ by (101).

Case C. $\underline{t} = (\underline{t}_1, \ldots, \underline{t}_n : \underline{t}_0)$. We deduce $(106)_1$ from AS12.22 and (102). So by the principle of induction, (106) holds for every $\underline{t} \in \tau^{\nu+1}$.

QED

Incidentally, from (102) for $\underline{t} = (\underline{t}_1, \ldots, \underline{t}_n : \underline{t}_0)$ and from (106) we deduce

$$(107) \quad \vdash_{\underline{c}} (f)\,[\underline{f} \in \underline{v}^*_{(\underline{t})} \supset {}^{\cap}(\forall \underline{x}_1, \ldots, \underline{x}_n)\underline{f}(\underline{x}_1, \ldots, \underline{x}_n) \in \underline{v}^*_{(\underline{t}_0)}].$$

The following theorem reflects the fact that we are dealing with the translation into ML^{ν} of an extensional language:

THEOR. 57.2. Let the results (99) to (102) of rule $\underline{C}$ hold. Let Δ be a designator in $\mathrm{EL}^{\nu+1}$, let π be the conjunction of the matrices $\underline{v}^*_{\underline{t}n} \in \underline{v}^*_{(\underline{t})}$ corresponding to all variables $\underline{v}^*_{\underline{t}n}$ that occur free in Δ^*, and let π' be the conjunction of the matrices $\underline{c}^*_{\underline{t}n} \in \underline{v}^*_{(\underline{t})}$ for all constants $\underline{c}^*_{\underline{t}n}$ occurring in Δ^*, where π $[\pi']$ is understood to be $\underline{p} \vee \sim\underline{p}$ in case Δ^* has no free variables [no constants]. Then

$$(108) \qquad \pi\pi' \vdash_{\underline{c}} \Delta^* \in \underline{v}^*_{(\underline{t})} \quad \text{when } \Delta \text{ is a term of type } \underline{t}.$$

$$(109) \qquad \pi\pi' \vdash_{\underline{c}} \Diamond\, \Delta^* \equiv \underline{N}\Delta^* \quad \text{when } \Delta \text{ is a matrix.}$$

Furthermore for every matrix $\underline{p}$ in $\mathrm{EL}^{\nu+1}$ we have

$$(110) \quad \vdash [\exists \underline{v}_{\underline{t}n})\underline{p}]^* \equiv (\exists \underline{v}^*_{\underline{t}n})(\underline{v}^*_{\underline{t}n} \in \underline{v}^*_{(\underline{t})}\underline{p}^*),$$

$$\vdash [\exists_1 \underline{v}_{\underline{t}n})\underline{p}]^* \equiv (\exists_1 \underline{v}^*_{\underline{t}n})(\underline{v}^*_{\underline{t}n} \in \underline{v}^*_{(\underline{t})}\underline{p}^*).$$

Proof. (I) Let $\underline{m}$ be the length (i.e. the number of the symbols) of the designator Δ in $\mathrm{EL}^{\nu+1}$. The formation rules of $\mathrm{EL}^{\nu+1}$ are the rules $(\underline{f}_1)$ to $(\underline{f}_7)$ and $(\underline{f}_9)$ [N3]. Therefore for $\underline{m} = 1$, Δ is either a variable or a constant, whence by $(\underline{T}^*_1)$ and $(\underline{T}^*_2)$, (108) holds.

Now let $\underline{r}$ be any integer greater than 1. As our inductive hypothesis we assume that (108) and (109) hold for every $\underline{m} < \underline{r}$.

<u>Case 1</u>. Δ has the form $\Delta_1 = \Delta_2$. Then by rule $(\underline{T}^*_3)$, (109) obviously holds.

<u>Case 2</u>. Δ is $\Delta' (\Delta_1, \ldots, \Delta_{\underline{n}})$ where the type $\underline{t}'$ of Δ' is either $(\underline{t}_1, \ldots, \underline{t}_{\underline{n}})$ or $(\underline{t}_1, \ldots, \underline{t}_{\underline{n}} : \underline{t})$. By our hypothesis of induction

(111) $$\pi\pi' \vdash_{\underline{c}} \Delta'^* \in \underline{v}^*_{(\underline{t}')}.$$

For $\underline{t}' = (\underline{t}_1, \ldots, \underline{t}_{\underline{n}})$, by $(104)_1$ we obtain $\Delta'^* \in \underline{v}^*_{(\underline{t}')} \vdash_{\underline{c}} \Delta'^* \in$ <u>MConst</u>. Furthermore, by rule $(\underline{T}^*_4)$ Δ^* is $\Delta'^*(\Delta^*_1, \ldots, \Delta^*_{\underline{n}})$. So by (111) and Def. 13.2 we have (109).

For $\underline{t}' = (\underline{t}_1, \ldots, \underline{t}_{\underline{n}} : \underline{t})$, by (107) we have $\Delta'^* \in \underline{v}^*_{(\underline{t}')} \vdash_{\underline{c}} \Delta'^*(\Delta^*_1, \ldots, \Delta^*_{\underline{n}}) \in \underline{v}^*_{(\underline{t})}$. So by (111) we have (108).

<u>Case 3</u>. Δ is $\sim\Delta_1$ or $\Delta_1 \wedge \Delta_2$. Since (by our hypothesis of induction) (109) holds when Δ is Δ_1 or Δ_2, (109) obviously holds in case 3.

<u>Case 4</u>. Δ is $(\underline{v}_{\underline{tn}})\Delta_1$. Since (109) holds when Δ is equal to Δ_1, by $(104)_2$ and rule $(\underline{T}^*_6)$ it holds in case 4.

<u>Case 5</u>. Δ is $(\imath \underline{v}_{\underline{tn}})\Delta_1$. Since (109) holds when Δ is equal to Δ_1—hence, under the hypothesis $\pi\pi'$, the matrix $\underline{v}^*_{\underline{tn}} \in \underline{v}^*_{(\underline{t})} \supset \Delta_1$ is noncontingent by $(104)_2$—the matrix $\Phi(\underline{v}^*_{\underline{tn}})$, which is defined to be $\underline{v}^*_{\underline{tn}} \in \underline{v}^*_{(\underline{t})} \Delta^*_1$, is also noncontingent. Then

(112) $$\pi\pi', \underline{N}(\exists_1 \underline{v}^*_{\underline{tn}}) \Phi(\underline{v}^*_{\underline{tn}}) \vdash_{\underline{c}} (\exists \underline{v}^*_{\underline{tn}}) \underline{N} \Phi(\underline{v}^*_{\underline{tn}}),$$

$$\pi\pi', \vdash_{\underline{c}} \underline{N}(\exists_1 \underline{v}^*_{\underline{tn}}) \Phi(\underline{v}^*_{\underline{tn}}) \vee \underline{N} \sim (\exists_1 \underline{v}^*_{\underline{tn}}) \Phi(\underline{v}^*_{\underline{tn}}) \quad \text{where}$$

$$\Phi(\underline{v}^*_{\underline{tn}}) \equiv_D \underline{v}^*_{\underline{tn}} \in \underline{v}^*_{(\underline{t})} \Delta^*_1.$$

Since $\pi\pi' \vdash_{\underline{c}} \diamond \Phi(\underline{v}^*_{tn}) \equiv \underline{N}\Phi(\underline{v}^*_{tn})$, by $(105)_1$, $(112)_3$, and Theor. 41.1(V), $\pi\pi' \vdash_{\underline{c}} (\lambda \underline{v}^*_{tn}) \overline{\Phi}(\underline{v}^*_{tn}) \in \underline{\text{Abs}}$. Hence by Theor. 41.2(III) $\vdash_{\underline{c}} (\exists_1 \underline{v}^*_{1-tn}) \Phi(\underline{v}^*_{tn}) \supset \Phi(\Delta^*)$. Thence by $(112)_2$ and AS38.1(II) we have (a) $\pi\pi' \vdash_{\underline{c}} \Phi(\Delta^*) \vee \Delta^* = {}^{\frown}\underline{a}^*$. Furthermore $(112)_3$ yields (b) $\vdash \Phi(\Delta^*) \supset \Delta^* \in \underline{v}^*_{(t)}$.

From (a), (b), and (106) we deduce the validity of (108) in case 5.

Thus (108) and (109) have been completely proved.

(II) By rules $(\underline{T}^*_5)$ and $(\underline{T}^*_6)$ [N57] and by well-known theorems of modal logic (holding in MC^{ν}),

$$\vdash [(\exists \underline{v}_{tn})\underline{p}]^* \equiv [\sim(\underline{v}_{tn})\sim\underline{p}]^* \equiv \sim(\underline{v}^*_{tn})(\underline{v}^*_{tn} \in \underline{v}^*_{(t)} \supset \sim\underline{p}^*),$$

whence $(110)_1$.

(III) By Defs. 34.1 and 2, $(105)_1$, rules $(\underline{T}^*_3)$, $(\underline{T}^*_5)$, and $(\underline{T}^*_6)$, [N57], and by well-known theorems, for $\underline{m} \neq \underline{n}$ we have

$$\vdash [(\exists^{(1)} \underline{v}_{tn})\phi(\underline{v}_{tn})]^* \equiv (\forall \underline{v}^*_{tm}, \underline{v}^*_{tn})[\underline{v}^*_{tm}, \underline{v}^*_{tn} \in \underline{v}^*_{(t)} \wedge \phi(\underline{v}_{tm})^* \wedge \phi(\underline{v}_{tn})^* \supset \underline{v}^*_{tm} = {}^{\frown}\underline{v}^*_{tn}] \equiv (\exists^{(1)\frown} \underline{v}^*_{tn})[\underline{v}^*_{tn} \in \underline{v}^*_{(t)} \phi(\underline{v}_{tn})^*].$$

By $(14)_{1,2,3}$, this result and $(110)_1$ yield $(110)_2$.

QED

N58. Invariance of the entailment relation for the translation $\Delta \to \Delta^*$ of $EL^{\nu+1}$ into ML^{ν}.

We consider a theory $\mathscr{T}$ in $EC^{\nu+1}$ which fulfills the following two conditions:

(a) The closures of the axioms of $\mathscr{T}$ are axioms of $\mathscr{T}$.

(b) The constants $\underline{c}_1, \underline{c}_2, \ldots$ of the respective types $\underline{t}_1, \underline{t}_2, \ldots$ are introduced in $\mathscr{T}$ by definitions of the form $\underline{c}_i = \Delta_i$, where Δ_i is a closed designator which contains at most the constants $\underline{c}_1, \ldots \underline{c}_{i-1}$.

Of course any designator in $\mathscr{T}$ contains only defined constants.

By the translation $\mathscr{T}^*$ of $\mathscr{T}$ (into ML^ν) we mean that theory based on MC^ν whose definitions and nonlogical axioms are the translations into ML^ν of the definitions and nonlogical axioms of $\mathscr{T}$ respectively.

THEOR. 58.1. Let Δ be a designator in the theory $\mathscr{T}$ (in $EC^{\nu+1}$). Then, accepting results (99) to (102) of rule $\underline{C}$, we have

$$(113) \qquad \vdash_{\underline{c}} \underline{c}^*_{\underline{i}} \in \underline{v}^*_{(\underline{t}_{\underline{i}})} \qquad (\underline{i} = 1, 2, \ldots)$$

and, in case Δ is a term of type $\underline{t}$ or a matrix, (108) and (109) become, respectively,

$$(114) \qquad \pi \vdash_{\underline{c}} \Delta^* \in \underline{v}^*_{(\underline{t})}, \qquad \pi \vdash_{\underline{c}} \Diamond \Delta^* \equiv \underline{N}\,\Delta^*.$$

Proof. By rule $(\underline{T}^*_3)$ $\vdash [\underline{c}_{\underline{i}} = \Delta_{\underline{i}}]^*$ is (a) $\vdash \underline{c}^*_{\underline{i}} = {}^\frown\Delta^*_{\underline{i}}$ $(\underline{i} = 1, 2, \ldots)$. For $\underline{i} = 1$, $\Delta_{\underline{i}}$ contains no constants or variables, whence by (a) and (108), (113) holds for $\underline{i} = 1$.

Let (113) hold for $\underline{i} = 1, \ldots, \underline{n} - 1$. Then in connection with $\Delta_{\underline{n}}$ we have $\vdash_{\underline{c}} \pi'$ (and π is $\underline{p} \vee \sim\underline{p}$). Here as well as in the sequel we understand that π and π' are defined as in Theor. 57.2. Since $\vdash_{\underline{c}} \pi'\pi$, by (108) we have $\vdash_{\underline{c}} \Delta^* \in \underline{v}^*_{(\underline{t})}$, which by (a) for $\underline{i} = \underline{n}$ yields (113) for $\underline{i} = \underline{n}$.

By the principle of induction, (113) holds for $\underline{i} = 1, 2, \ldots$.

In connection with any designator Δ in $\mathscr{T}$, $\vdash_{\underline{c}} \pi'$ holds in $\mathscr{T}^*$ by (113). Then (108) and (109) become $(114)_1$ and $(114)_2$ respectively.

QED

THEOR. 58.2. Let $\underline{p}$ be a closed axiom in $EC^{\nu+1}$—i.e. a closed instance in $EL^{\nu+1}$ of either AS12.$\underline{r}$ for some $\underline{r}$ = 1, 2, 3, 4, 6, 8, 10–12, 14–18, 20–22, AS12.13', or else AS45.1 for natural numbers. Then its star translation $\underline{p}^*$ into ML^ν [N57] is a theorem in MC^ν.

Proof.

Case 1. For some $\underline{r} = 1, 2, 3$ $\underline{p}$ is an instance of AS12.$\underline{r}$ in $EL^{\nu+1}$, so that $(\underline{N})$ in $\underline{p}$ is (. . .). If this (. . .) is not empty, $\underline{p}^*$ can be put into the form—i.e. $\underline{p}^*$ is strictly equivalent to a matrix of the form—$(. . .)\pi \supset \underline{q}$ where $\underline{q}$ is an instance of AS12.$\underline{r}$ in ML^{ν} (and π is defined in connection with $\underline{q}$ as in Theor. 57.2). We conclude that $\vdash \underline{p}^*$ (in MC^{ν})— $\underline{p}^*$ is $\underline{q}$ for (. . .) empty.

Case 2. $\underline{p}$ is an instance of AS12.4. Then—cf. rule $(\underline{T}_6^*)$ and the well-known equivalence between $\pi_{\underline{x}} \supset (\underline{r} \supset \underline{s})$ and $\pi_{\underline{x}}\underline{r} \supset \underline{s}$— $\underline{p}^*$ can be put into the form

(115) $(\underline{N})\pi \supset \underline{q}$ for $\underline{q} \equiv_D (\underline{x})(\pi_{\underline{x}}\underline{r} \supset \underline{s}) \supset [(\underline{x})(\pi_{\underline{x}} \supset \underline{r}) \supset (\underline{x})(\pi_{\underline{x}} \supset \underline{s})]$

where $\pi_{\underline{x}}$ is $\underline{x} \in \underline{v}^*_{(\underline{t})}$.

Since, as is well known, $\vdash (\pi_{\underline{x}}\underline{r} \supset \underline{s}) \supset [(\pi_{\underline{x}} \supset \underline{r}) \supset (\pi_{\underline{x}} \supset \underline{s})]$, by Theor. 32.2 and AS12.4 we have $\vdash \underline{q}$. Moreover $\underline{q}$ yields matrix $(115)_1$. So we have $\vdash p^*$.

Case 3. $\underline{p}$ is an instance of AS12.6. Then—cf. rule $(\underline{T}_6^*)$—$\underline{p}^*$ can be put into the form

(116) $(\underline{N})\pi \supset [\underline{q} \supset (\underline{x})(\pi_{\underline{x}} \supset \underline{q})]$ where $\underline{x}$ has no free occurrences in $\underline{q}$.

Since $\vdash \underline{q} \supset (\underline{x})\underline{q}$ by AS12.6 and $\vdash (\underline{x})\underline{q} \supset (\underline{x})(\pi_{\underline{x}} \supset \underline{q})$, as is well known, we easily deduce $\vdash \underline{p}^*$.

Case 4. $\underline{p}$ is an instance of AS12.8. Then $\underline{p}^*$ can be put into the form

(117) $$(\underline{N})\pi \supset \{(\underline{x})[\underline{x} \in \underline{v}^*_{(\underline{t})} \supset \Phi(\underline{x})] \supset \Phi(\Delta)\}$$

where the term Δ is free for $\underline{x}$ in $\Phi(\underline{x})$—hence also in $\underline{x} \in \underline{v}^*_{(\underline{t})} \supset \Phi(\underline{x})$ by (97), (98), and rule $(\underline{T}_6^*)$ in N57; where $\Phi(\Delta)$ is obtained from $\Phi(\underline{x})$ by replacing the free occurrences of $\underline{x}$ with occurrences of Δ—cf. convention (a) in N30; and where π is the conjunction of the matrices $\underline{x}_i \in \underline{v}^*_{(\underline{t}_i)}$ in correspondence with all variables $\underline{x}_i$ that occur free in $\Phi(\underline{x})$ and $\Phi(\Delta)$.

These variables include those that occur free in Δ. Then by (114) (with Δ^* replaced by Δ) and Theor. 33.3 we have $\vdash_{\underline{c}} \pi \supset \Delta \in \underline{v}^*_{(\underline{t})}$, so that, using rule $\underline{C}$, matrix (117)—hence $\underline{p}^*$—can be proved to be strictly equivalent to

$$(118)\ (\underline{N})\pi \supset \underline{q} \quad \text{where} \quad \underline{q} \equiv_D (\underline{x})\,[\underline{x} \in \underline{v}^*_{(\underline{t})} \supset \Phi(\underline{x})] \supset [\Delta \in \underline{v}^*_{(\underline{t})} \supset \Phi(\Delta)].$$

So, since $\vdash \underline{q}$ by AS12.8 in MC^ν, we have $\vdash_{\underline{c}} \underline{p}^*$.

Case 5. $\underline{p}$ is an instance of either AS12.$\underline{r}$ for some $\underline{r} \in \{10, 11, 12\}$ or AS12.13′. Then $\underline{p}^*$ can be put into the form $(\underline{N})\pi \supset \underline{q}$ where $\underline{q}$ is an instance of either the analogue of AS12.$\underline{r}$ for strict identity or AS12.13 (in MC^ν). So we obviously have $\vdash_{\underline{c}} \underline{p}^*$ (in MC^ν).

Case 6. $\underline{p}$ is a closed instance of AS12.14. Then, since $\vdash \pi \supset [(\underline{q} \equiv \underline{r}) \equiv^\cap (\pi\underline{q} \equiv \pi\underline{r})]$, $\underline{p}^*$ can be put into the form

$$(119)\ (\underline{N})\,\underline{F}, \underline{G} \in \underline{v}^*_{(\underline{t})} \wedge \pi \supset \{(\forall \underline{x}_1, \ldots, \underline{x}_n)\,[\pi\,\underline{F}(\underline{x}_1, \ldots, \underline{x}_n) \equiv \pi\,\underline{G}(\underline{x}_1, \ldots, \underline{x}_n)] \equiv \underline{F} =^\cap \underline{G}\} \text{ where } \pi \equiv_D \bigwedge_{\underline{i}=1}^{\underline{n}} \underline{x}_i \in \underline{v}^*_{(\underline{t}_i)}$$

and where $\underline{t} = (\underline{t}_1, \ldots \underline{t}_n)$. By (101) we can delete π in $(119)_1$. Furthermore by $(104)_1$ $\underline{F}, \underline{G} \in \underline{v}^*_{(\underline{t})} \vdash_{\underline{c}} \underline{F}, \underline{G} \in \mathrm{MConst}$. So by Def. 13.2, and $(104)_2$, matrix $(119)_1$—hence $\underline{p}^*$—can be put into the form

$$(119')\ (\underline{N})\,\underline{F}, \underline{G} \in \underline{v}^*_{(\underline{t})} \supset \{(\forall \underline{x}_1, \ldots, \underline{x}_n)\,[\underline{F}(\underline{x}_1, \ldots, \underline{x}_n) \equiv^\cap \underline{G}(\underline{x}_1, \ldots, \underline{x}_n)] \equiv \underline{F} =^\cap \underline{G}\}.$$

Hence by ASs12.14 and 12.5 we have $\vdash_{\underline{c}} \underline{p}^*$.

Case 7. $\underline{p}$ is a closed instance of AS12.15. Then $\underline{p}^*$ can be put into the form

(120) $(\underline{N})\underline{f}, \underline{g} \in \underline{v}^*_{(\underline{t})} \supset \{(\forall \underline{x}_1, \ldots, \underline{x}_{\underline{n}}) [\pi \supset \underline{f}(\underline{x}_1, \ldots, \underline{x}_{\underline{n}}) =^{\cap} \underline{g}(\underline{x}_1, \ldots, \underline{x}_{\underline{n}})] \equiv \underline{f} =^{\cap} \underline{g}\}$

where π is as in $(119)_2$.

By (102) we have

$$\underline{f}, \underline{g} \in \underline{v}^*_{(\underline{t})} \vdash_{\underline{c}} [\pi \supset \underline{f}(\underline{x}_1, \ldots) =^{\cap} \underline{g}(\underline{x}_1, \ldots)] \equiv^{\cap} \underline{f}(\underline{x}_1, \ldots) =^{\cap} \underline{g}(\underline{x}_1, \ldots)$$

so that matrix (120)—hence $\underline{p}^*$—can be put into the form

$$(\underline{N})\underline{f}, \underline{g} \in \underline{v}^*_{(\underline{t})} \supset [(\forall \underline{x}_1, \ldots)\underline{f}(\underline{x}_1, \ldots) =^{\cap} \underline{g}(\underline{x}_1, \ldots) \equiv \underline{f} =^{\cap} \underline{g}].$$

Then by ASs12.15 and 12.5 we have $\vdash \underline{p}^*$.

Case 8. $\underline{p}$ is an instance of AS12.16. Then by $(110)_1$, $\underline{p}^*$ can be put into the form

(121) $(\exists \underline{F}) \{\underline{F} \in \underline{v}^*_{(\underline{t})} (\forall \underline{x}_1, \ldots) [\pi \underline{F}(\underline{x}_1, \ldots) \equiv \pi \underline{q}]\}$ where

$$\pi =_D \bigwedge_{\underline{i}=1}^{\underline{n}} \underline{x}_{\underline{i}} \in \underline{v}^*_{(\underline{t}_{\underline{i}})}.$$

From A12.19 (in MC^{ν}) and Def. 13.2, by rule $\underline{C}$ for the variable $\underline{F}$ we obtain (a) $\underline{F} \in \underline{MConst} (\forall \underline{x}_1, \ldots) [\underline{F}(\underline{x}_1, \ldots) \equiv \pi \underline{q}]$, which yields $\underline{F} \in \underline{MConst} \wedge (\forall \underline{x}_1, \ldots) [\underline{F}(\underline{x}_1, \ldots) \supset \pi]$. Thence by (101) we have (b) $\underline{F} \in \underline{v}^*_{(\underline{t})}$; furthermore by (a) we deduce $(\forall \underline{x}_1, \ldots) [\underline{F}(\underline{x}_1, \ldots) \equiv \pi \underline{F}(\underline{x}_1, \ldots)]$, which together with (b) and (a) yields $(121)_1$, and we precisely have $\vdash_{\underline{c}} \underline{p}^*$.

Case 9. $\underline{p}$ is an instance of AS12.17. Then by $(110)_1$ $\underline{p}^*$ can be put into the form

(122) $(\exists \underline{f}) \{\underline{f} \in \underline{v}^*_{(\underline{t})} (\forall \underline{x}_1, \ldots) [\pi \supset \underline{f}(\underline{x}_1, \ldots) =^{\cap} \Delta]\}$ where

$$\pi =_D \bigwedge_{\underline{i}=1}^{\underline{p}} \underline{x}_{\underline{i}} \in \underline{v}^*_{(\underline{t}_{\underline{i}})}$$

where $\underline{p} \geq \underline{n}$ and $\underline{x}_1, \ldots, \underline{x}_{\underline{p}}$ include all variables free in Δ.

From $(46)_2$, by rule $\underline{C}$ for the variable $\underline{f}$ we deduce (c) $(\forall \underline{x}, \ldots)$ $\underline{f}(\underline{x}_1, \ldots) =^{\cap} (\imath \underline{z})(\pi \underline{z} = \Delta)$ where the variable $\underline{z}$ is distinct from $\underline{x}_1, \ldots \underline{x}_p$, and does not occur free in Δ.

Since $\underline{x}_1, \ldots, \underline{x}_p$ include all variables occurring free in Δ and Δ has the form Δ_0^* with Δ_0 of type $\underline{t}_0$, by Theor. 58.1—cf. $(114)_1$—we have (d) $\vdash_{\underline{c}} \pi \supset \Delta \in \underline{v}^*_{(t_0)}$.

By $(104)_2$ and Def. 13.2 we have $\vdash_{\underline{c}} \pi \equiv \underline{N}\pi$, whence we easily deduce

$$(123) \quad \vdash_{\underline{c}} \pi \supset (\imath \underline{z})(\pi \underline{z} = \Delta) =^{\cap} \Delta, \quad \vdash_{\underline{c}} \sim\pi \supset (\imath \underline{z})(\pi \underline{z} = \Delta) =^{\cap} \underline{a}^*_{\underline{t}_0^*}.$$

Then by (c) (and $\vdash \pi \vee \sim\pi$) we have $(\forall \underline{x}_1, \ldots)[\pi \underline{f}(\underline{x}_1, \ldots) =^{\cap} \Delta \vee \sim\pi \underline{f}(\underline{x}_1, \ldots) =^{\cap} \underline{a}^*_{\underline{t}_0^*}]$, whence by (d) and (102), we easily deduce (e) $\underline{f} \in \underline{v}^*_{(t)}$.

From (c) and $(123)_1$ we obtain $\vdash (\forall \underline{x}_1, \ldots, \underline{x}_n)[\pi \supset \underline{f}(\underline{x}_1, \ldots, \underline{x}_n) =^{\cap} \Delta]$, which together with (e) easily yields $(\bar{1}22)_1$, hence $\underline{p}^*$. We conclude that $\vdash_{\underline{c}} \underline{p}^*$.

Case 10. $\underline{p}$ is a closed instance of either part of AS12.18. Then $\underline{p}^*$ can be put into the first or second of the forms

$$(124) \; (\underline{N})\pi_1 (\exists_1 \underline{x})(\pi_{\underline{x}} \underline{q}) \underline{q} \supset \underline{x} =^{\cap} (\imath \underline{x})(\pi_{\underline{x}} \underline{q}),$$
$$(\underline{N})\pi_1 \sim (\exists_1 \underline{x})(\pi_{\underline{x}} \underline{q}) \supset (\imath \underline{x})(\pi_{\underline{x}} \underline{q}) = \underline{a}^*$$

respectively, where $\pi_{\underline{x}}$ has the form $\underline{x} \in \underline{v}^*_{(t')}$ and $\pi_1 \pi_{\underline{x}}$ is the matrix π defined by $(122)_2$ in terms of all variables $\underline{x}_1, \ldots, \underline{x}_n$ which are free in $\underline{q}$.

By $(114)_2$, with Δ^* identified with $\underline{q}$ we have $\pi \vdash_{\underline{c}} \Diamond \underline{q} \equiv \underline{N}\underline{q}$, and by $(104)_2$, $\vdash_{\underline{c}} \Diamond \pi \equiv \underline{N}\pi$. Thence by $(105)_1$ and Theor. $\bar{4}1.1$(V) we obtain $\pi_1 \vdash_{\underline{c}} (\lambda \underline{x})(\pi_{\underline{x}} \underline{q}) \in \underline{\mathrm{Abs}}$. Furthermore $\vdash \underline{\mathrm{Abs}} \subset \underline{\mathrm{MSep}}$. Then by Theor. 41.$\bar{2}$(II) and $\bar{(27)}_{2,1}$ we easily deduce $\pi_1 \vdash_{\underline{c}} (\exists_1 \underline{x})(\pi_{\underline{x}} \underline{q}) \equiv \underline{N}(\exists_1 \underline{x})(\pi_{\underline{x}} \underline{q})$. Now we easily see that

(125) $\vdash_{\underline{c}} \pi_1 \sim (\exists_1 \underline{x})(\pi_{\underline{x}} \underline{q}) \equiv^{\frown} \underline{N}[\pi_1 \sim (\exists_1 \underline{x})(\pi_{\underline{x}} \underline{q})]$,

$\vdash_{\underline{c}} (\underline{N}) \pi_1 (\exists_1 \underline{x})(\pi_{\underline{x}} \underline{q}) \pi_{\underline{x}} \underline{q} \supset \underline{N}[\pi_1 (\exists_1 \underline{x})(\pi_{\underline{x}} \underline{q}) \pi_{\underline{x}} \underline{q}]$.

In case $\underline{p}^*$ has the form $(124)_2$ we have $\vdash_{\underline{c}} \underline{p}^*$ by AS12.18(II), $(125)_1$, and AS12.5.

Now remark that by ASs12.18(I) and 12.5, $\vdash \underline{N}[\pi_1 (\exists_1 \underline{x})(\pi_{\underline{x}} \underline{q}) \pi_{\underline{x}} \underline{q}] \supset \underline{x} =^{\frown} (\imath \underline{x})(\pi_{\underline{x}} \underline{q})$. Then by $(125)_2$, matrix $(124)_1$ can be proved (using rule $\underline{C}$). So $\vdash_{\underline{c}} \underline{p}^*$ also when $\underline{p}^*$ has the form $(124)_1$.

Case 11. $\underline{p}$ is an instance of AS12.20 (Zermelo's axiom) in $EC^{\nu+1}$. Then the proof of $\vdash \underline{p}^*$ is a matter of routine on the basis of AS12.20 in ML^{ν}, (104), (105), and $(110)_1$.

Case 12. $\underline{p}$ is an instance of ASs12.21 or 22. Since by Def. 11.2 $\underline{a}^*_{\underline{t}}$ is $(\imath \underline{v}_{\underline{t}1}) \underline{v}_{\underline{t}1} \neq \underline{v}_{\underline{t}1}$, by rule $(\underline{T}^*_7)$ the star translation of $\underline{a}^*_{\underline{t}}$ is $(\imath \underline{v}^*_{\underline{t}1}) [\underline{v}^*_{\underline{t}1} \in \underline{v}^*_{(\underline{t})} \wedge \sim \underline{v}^*_{\underline{t}1} =^{\frown} \underline{v}^*_{\underline{t}1}]$. Hence (f) $(\underline{a}^*_{\underline{t}})^* =^{\frown} \underline{a}^*_{\underline{t}^*}$ in MC^{ν}.

Let $\underline{p}$ be an instance of AS12.21 in $EC^{\nu+1}$, so that $\underline{t}$ has the form $(\underline{t}_1, \ldots, \underline{t}_{\underline{n}})$ and $\underline{t}^* = (\underline{t}^*_1, \ldots, \underline{t}^*_{\underline{n}})$ [Def. 56.1(b)]. Furthermore by AS12.21 in MC^{ν}, $\vdash \underline{q}$ where $\underline{q}$ is $\sim \underline{a}^*_{\underline{t}^*}(\underline{x}_1, \ldots, \underline{x}_{\underline{n}})$, and $\underline{p}^*$ has the form $(\underline{N}) \pi \supset \underline{q}$ by (f) and $(\underline{T}^*_6)$. We conclude that ($\vdash \underline{p}^*$ hence) $\vdash_{\underline{c}} \underline{p}^*$ in MC^{ν}.

Now let $\underline{p}$ be an instance of AS12.22 in $EC^{\nu+1}$, so that $\underline{t}$ has the form $(\underline{t}_1, \ldots, \underline{t}_{\underline{n}} : \underline{t}_0)$. Then by rules $(\underline{T}^*_1)$, $(\underline{T}^*_3)$, $(\underline{T}^*_4)$, and $(\underline{T}^*_6)$ and by (f), $\underline{p}^*$ can be put into the form $(\underline{N}) \pi \supset \underline{q}$ where $\underline{q}$ is an instance of AS12.22 in MC^{ν}. Hence $\vdash_{\underline{c}} \underline{p}^*$ in MC^{ν}.

Case 13. $\underline{p}$ is an instance of AS45.1 in EC^{ν}. To give a hint for proving $\vdash_{\underline{c}} \underline{p}^*$, we remark that $\underline{p}$ contains the constant $\underline{Nn}$, which is defined by means of other constants, e.g. $\underline{Clos}$ (closure). The constant $\underline{Nn}^*$ in ML^{ν} obviously differs from $\underline{Nn}$ in meaning. However on the basis of the consequence

(126) $$\vdash_{\underline{c}} (\underline{v}_{\underline{r}1})\, \underline{v}_{\underline{r}1} \in \underline{v}^{*(\underline{e})}_{(\underline{r})}$$

of (99) we can prove

(127) $\vdash_{\underline{c}} (\underline{m}^*) (\exists \underline{m}, \underline{F}) [\underline{m}^* \in \underline{Nn}^* \supset \underline{m} \in \underline{Nn} \wedge \underline{F} \in \underline{m} \cap \underline{m}^*]$.

So for $\underline{m}^* \in \underline{Nn}^*$, $\underline{m}^*$ is not empty, and this allows us to prove, in MC^{ν}, the analogue $\underline{q}$ of AS45.1 for $\underline{Nn}^*$. So since $\underline{p}$ can be put into the form $(\underline{N})\, \pi \supset \underline{q}$ we have $\vdash_{\underline{c}} \underline{p}^*$.

QED

THEOR. 58.3. Let the theory $\mathscr{T}$, based on $EC^{\nu+1}$, fulfill conditions (a) and (b) considered at the beginning of this section, and let results (99) to (102) of rule $\underline{C}$ hold. Then the following theses hold:

(a) For every closed theorem $\underline{p}$ of $\mathscr{T}$, $\vdash_{\underline{c}} \underline{p}^*$ holds in the translation $\mathscr{T}^*$ of $\mathscr{T}$ (defined at the outset of N58).

(b) If $\underline{p}_1, \ldots, \underline{p}_{\underline{m}} \vdash \underline{p}_0$ in $\mathscr{T}$, then $\pi, \underline{p}^*_1, \ldots, \underline{p}^*_{\underline{m}} \vdash_{\underline{c}} \underline{p}^*_0$ in $\mathscr{T}^*$ where π is the conjunction of the matrices $\underline{v}^*_{\underline{tn}} \in \underline{v}^*_{(\underline{t})}$ corresponding to all variables $\underline{v}_{\underline{tn}}$ that occur free in $\underline{p}_0 \wedge \ldots \wedge \underline{p}_{\underline{m}}$.

<u>Proof</u>. Since the theorem $\underline{p}$ of $\mathscr{T}$ is closed, by the assumed condition (a) there is a proof $\underline{p}_1, \ldots, \underline{p}_{\underline{n}}$ of $\underline{p}$ in $\mathscr{T}$ where $\underline{p}_1, \ldots, \underline{p}_{\underline{n}}$ are closed. So by Def. 29.2 $\underline{p}_{\underline{n}}$ is $\underline{p}$, hence $\underline{p}^*_{\underline{n}}$ is $\underline{p}^*$, and in addition, for $\underline{i} = 1, \ldots, \underline{n}$ one of the following cases holds.

<u>Case 1</u>. $\underline{p}_{\underline{i}}$ is a nonlogical axiom or a definition in $\mathscr{T}$, whence—since the definitions of $\mathscr{T}$ have the form $\underline{c}_{\underline{r}} = \Delta_{\underline{r}}$—$\underline{p}^*_{\underline{i}}$ is the analogue for $\mathscr{T}^*$. So $\vdash \underline{p}_{\underline{i}}$.

<u>Case 2</u>. $\underline{p}_{\underline{i}}$ is a logical axiom of $EC^{\nu+1}$, whence $\vdash_{\underline{c}} \underline{p}_{\underline{i}}$ by Theor. 58.2.

<u>Case 3</u>. $\underline{p}_{\underline{i}}$ is an earlier $\underline{p}_{\underline{j}}$ $(\underline{j} < \underline{i})$, whence $\underline{p}^*_{\underline{i}}$ is $\underline{p}^*_{\underline{j}}$.

Case 4. For some $\kappa < \underline{i}$ and $\underline{j} < \underline{i}$, $\underline{p}_\kappa$ is $\underline{p}_{\underline{j}} \supset \underline{p}_{\underline{i}}$. So by rule ($\underline{T}_5^*$) $\underline{p}_\kappa^*$ is $\underline{p}_{\underline{j}}^* \supset \underline{p}_{\underline{i}}^*$.

We conclude that from $\underline{p}_1^*, \ldots, \underline{p}_{\underline{n}}^*$ we obtain a proof of $\underline{p}^*$ in MC^ν by inserting before $\underline{p}_{\underline{i}}^*$ a proof of $\vdash_{\underline{c}} \underline{p}_{\underline{i}}^*$ when case 2 holds for $\underline{p}_{\underline{i}}$ ($\underline{i} = 1, \ldots, \underline{n}$). Then $\vdash_{\underline{c}} \underline{p}^*$ holds in $\mathscr{T}^*$ and thesis (a) is proved.

If $\underline{p}_1, \ldots, \underline{p}_{\underline{n}} \vdash_{\underline{c}} \underline{p}_0$ holds in $\mathscr{T}$, by Theors. 33.2, 32.2 a closed matrix $\underline{p}$ of the form $(\ldots)\underline{p}_1 \wedge \ldots \wedge \underline{p}_{\underline{m}} \supset \underline{p}_0$ is provable in $\mathscr{T}$. Then by thesis (a) $\vdash_{\underline{c}} \underline{p}^*$ holds in $\mathscr{T}^*$, i.e. $\vdash_{\underline{c}} (\underline{N})\pi \wedge \underline{p}_1^* \wedge \ldots \wedge \underline{p}_{\underline{n}}^* \supset \underline{p}^*$, whence thesis (b) follows.

QED

12

A Semantical System for ML^ν Defined in ML^ν Itself; Invariance of the Entailment Relation under the Extensional Translation $\Delta \rightarrow \Delta^\eta$ in Both Senses

N59. An assignment of analogues of QIs made within MC^ν.

Let $\underline{y}$ be a variable of type $\underline{t} \epsilon \tau^\nu$ and let θ be $(\underline{t}^\eta)^*$, i.e. $[\underline{t}^\eta]^*$—cf. N6. Intuitively we want to define within ML^ν $\bar{\underline{y}}$ as the quasi intension, or briefly the intension of $\underline{y}$, based on the absolute concepts $\underline{v}_{(1)1}, \ldots, \underline{v}_{(\nu)1}$—cf. (99)—whose extensionalizations are universal classes of the respective types (1), . . . , (ν), and on the element $\underline{v}_{\nu+1*5}$ of $\underline{El}'$ —cf. $(100)_2$.

For the sake of simplicity we accept, throughout the remainder of the book, results (99) to (102) of rule $\underline{C}$, and we formalize the above underlined assertion by

(128) $\underline{I}_{\underline{t}}(\bar{\underline{y}}, \underline{y}, \underline{v}_{(1)1}, \ldots, \underline{v}_{(\nu)1}\, \underline{v}_{\nu+1*5})$ or briefly $\underline{I}_{\underline{t}}(\bar{\underline{y}}, \underline{y})$

where $\bar{\underline{y}}$ is a variable of the type $\theta = \underline{t}^{\eta *}$ $(\underline{t} \epsilon \tau^{\nu})$—hence different from $\underline{y}$—and we define $\underline{I}_{\underline{t}}(\bar{\underline{y}}, \underline{y})$ for $\underline{t} \epsilon \tau^{\nu}$ recursively, by means of the formulas (129) to (131) below involving the matrix $|'_{\underline{u}}$, which is simply related to $|_{\underline{u}}$ and which has the analogues of the properties (83) and (89) to (92) of $|_{\underline{u}}$ —cf. (83′) and (89′) to (92′) in Theor. 56.1:

(129) $\underline{I}_{\underline{t}}(\bar{\underline{y}}, \underline{y}) \equiv_D \bar{\underline{y}} \epsilon \underline{v}^*_{(\underline{t}\eta)}\, (\underline{u}) [|'_{\underline{u}} \supset {}^\cap \bar{\underline{y}}(\underline{u}) = \underline{y}]$—cf. Def. 56.2, $(98)_1$—

for $\underline{t} = 1, \ldots, \nu$,

(130) $\underline{I}_{\underline{t}}(\bar{\underline{y}}, \underline{y}) \equiv_D (\forall \bar{\underline{x}}_1, \ldots, \bar{\underline{x}}_{\underline{n}}, \underline{u}) \{ \bar{\underline{y}}(\bar{\underline{x}}_1, \ldots, \bar{\underline{x}}_{\underline{n}}, \underline{u})$

$\equiv {}^\cap (\exists \underline{x}_1, \ldots, \underline{x}_{\underline{n}}) \diamond [|'_{\underline{u}} \underline{y}(\underline{x}_1, \ldots, \underline{x}_{\underline{n}}) \bigwedge_{\underline{i}=1}^{\underline{n}} \underline{I}_{\underline{t}_{\underline{i}}}(\bar{\underline{x}}_{\underline{i}}, \underline{x}_{\underline{i}})]\}$

for $\underline{t} = (\underline{t}_1, \ldots, \underline{t}_{\underline{n}})$,

(131) $\underline{I}_{\underline{t}}(\bar{\underline{y}}, \underline{y}) \equiv_D (\forall \bar{\underline{x}}_1, \ldots, \bar{\underline{x}}_{\underline{n}}) \bar{\underline{y}}(\bar{\underline{x}}_1, \ldots \bar{\underline{x}}_{\underline{n}}) = {}^\cap (\imath \bar{\underline{x}}_0) (\exists \underline{x}_0, \ldots,$

$\underline{x}_{\underline{n}}) [\underline{x}_0 = {}^\cap \underline{y}(\underline{x}_1, \ldots, \underline{x}_{\underline{n}}) \bigwedge_{\underline{i}=0}^{\underline{n}} \underline{I}_{\underline{t}_{\underline{i}}}(\bar{\underline{x}}_{\underline{i}}, \underline{x}_{\underline{i}})]$ for $\underline{t} = (\underline{t}_1, \ldots,$

$\underline{t}_{\underline{n}} : \underline{t}_0)$.

It must be added that from now on $\underline{u}$ is assumed to be κ^*, i.e. $\underline{v}_{\nu+1*6}$—cf. N15 and rule $(\underline{T}^*_1)$ in N57— and in (130) [(131)] the $\underline{i}$-th of the variables $\underline{x}_1, \bar{\underline{x}}_1, \ldots, \underline{x}_{\underline{n}}, \bar{\underline{x}}_{\underline{n}}$ $[\underline{x}_0, \bar{\underline{x}}_0, \ldots, \underline{x}_{\underline{n}}, \bar{\underline{x}}_{\underline{n}}]$ is assumed to be the first among the variables that make (130) [(131)] well formed and that are different from $\underline{u}$ and the preceding variables.

THEOR. 59.1. Accepting results (99) to (102) of rule $\underline{C}$, we have

(132) $\vdash_{\underline{C}} \diamond \underline{I}_{\underline{t}}(\bar{\underline{y}}, \underline{y}) \equiv \underline{I}_{\underline{t}}(\bar{\underline{y}}, \underline{y}) \equiv \underline{N}\underline{I}_{\underline{t}}(\bar{\underline{y}}, \underline{y})$, $\vdash_{\underline{C}} \underline{I}_{\underline{t}}(\bar{\underline{y}}, \underline{y}) \supset \bar{\underline{y}} \epsilon \underline{v}^*_{(\underline{t}\eta)}$

$(\underline{t} \epsilon \tau^{\nu})$.

Proof. From (129) to (131) and $(104)_2$ we deduce $(132)_{1,2}$.

To prove $(132)_3$ we first remark that for $\underline{t} \in \{1, \ldots, \nu\}$ it is included in (129). Now we assume the equality $\underline{t} = (\underline{t}_1, \ldots \underline{t}_n)$ and, as inductive hypothesis, the validity of $(132)_3$ for $\underline{t} = \underline{t}_1, \ldots, \underline{t} = \underline{t}_n$.

We start with (a) $\underline{I}_t(\bar{\underline{y}}, \underline{y})$. By (130) and Def. 13.2, this yields (b) $\bar{\underline{y}} \in \underline{\text{MConst}}$. Furthermore, since $(132)_{1,2,3}$ hold for $\underline{t} = \underline{t}_1, \ldots, \underline{t} = \underline{t}_n$, (a) and (130) yield (c) $(\forall \bar{\underline{x}}_1, \ldots, \bar{\underline{x}}_n, \underline{u}) [\bar{\underline{y}}(\bar{\underline{x}}_1, \ldots, \bar{\underline{x}}_n, \underline{u}) \supset \bar{\underline{x}}_i \in \underline{v}^*_{(\underline{t}_i \eta)} \diamond |'_{\underline{u}}] (\underline{i} = 1, \ldots, \underline{n})$.

From Def. 56.2, $(82')_1$, and $(100')_1$ we deduce (d) $\vdash_{\underline{c}} \diamond |'_{\underline{u}} \supset \underline{u} \in \underline{v}^*_{(\nu+1)}$.

By (101), matrices (b), (c), and (d) yield $\bar{\underline{y}} \in \underline{v}^*_{(\underline{t}\eta)}$. So we conclude [Theor. 33.3] that $(132)_3$ holds for $\underline{t} = (\underline{t}_0, \ldots, \underline{t}_n)$.

We now assume the equality $\underline{t} = (\underline{t}_1, \ldots, \underline{t}_n : \underline{t}_0)$ and, as inductive hypothesis, the validity of $(132)_3$ for $\underline{t} = \underline{t}_0, \ldots, \underline{t} = \underline{t}_n$. From this validity and $(132)_{1,2}$ we easily deduce (e) $\sim[\bar{\underline{x}}_0 \in \underline{v}^*_{(\underline{t}_0\eta)} \wedge \ldots \wedge \bar{\underline{x}}_n \in \underline{v}^*_{(\underline{t}_n\eta)}] \supset (\imath \bar{\underline{x}}_0)\underline{q} = {}^\frown\underline{a}^*$ where $(\imath \bar{\underline{x}}_0)\underline{q}$ is the description explicitly written in (131). So (a) $\vdash_{\underline{c}}$ (e). Hence by (131) and (102) we can easily deduce that $(132)_3$ holds for $\underline{t} = (\underline{t}_1, \ldots, \underline{t}_n : \underline{t}_0)$.

By the principle of induction $(132)_3$ holds for $\underline{t} \in \tau^\nu$.

QED

THEOR. 59.2. Denoting the matrix $\bar{\underline{y}} \in \underline{v}^*_{(\underline{t}\eta)}$ by $\bar{\pi}$ and accepting results (99) to (102) of rule $\underline{C}$,

(133) $\vdash_{\underline{c}} (\underline{y})(\exists^{(1)\frown} \bar{\underline{y}})\underline{I}_t(\bar{\underline{y}}, \underline{y}), \quad \vdash_{\underline{c}} (\bar{\underline{y}})(\exists^{(1)\frown} \underline{y})\underline{I}_t(\bar{\underline{y}}, \underline{y})$,

(134) $\vdash_{\underline{c}} (\underline{y})(\exists \bar{\underline{y}})\underline{I}_t(\bar{\underline{y}}, \underline{y}), \quad \vdash_{\underline{c}} (\bar{\underline{y}})(\exists \underline{y})[\bar{\pi} \supset \underline{I}_t(\bar{\underline{y}}, \underline{y})]$,

(135) $\vdash_{\underline{c}} (\underline{y})(\exists^\frown_1 \bar{\underline{y}})\underline{I}_t(\bar{\underline{y}}, \underline{y}), \quad \vdash_{\underline{c}} (\bar{\underline{y}})[\bar{\pi} \supset (\exists^\frown_1 \underline{y})\underline{I}_t(\bar{\underline{y}}, \underline{y})]$,

and in addition, the definientia of $\underline{I}_t(\bar{\underline{y}}, \underline{y})$ in (130) and (131) can be, so to speak, inverted according to the following formulas respectively:

(136) $\vdash_{c} I_t(\bar{y}, y) \equiv \bar{y} \in v^*_{(t\eta)} (\forall x_1, \ldots, x_n) \{y(x_1, \ldots, x_n) \equiv^\cap (\exists \bar{x}_1, \ldots, \bar{x}_n, u) [\vdash'_u \bar{y}(\bar{x}_1, \ldots \bar{x}_n, u) \bigwedge_{i=1}^{n} I_{t_i}(\bar{x}_i, x_i)]\}$ for $t = (t_1, \ldots t_n)$,

(137) $\vdash_{c} I_t(\bar{y}, y) \equiv \bar{y} \in v^*_{(t\eta)} (\forall x_1, \ldots, x_n) y(x_1, \ldots, x_n) =^\cap (\imath x_0)(\exists \bar{x}_0, \ldots, \bar{x}_n) [\bar{x}_0 =^\cap \bar{y}(\bar{x}_1, \ldots, \bar{x}_n) \bigwedge_{i=0}^{n} I_{t_i}(\bar{x}_i, x_i)]$ for $t = (t_1, \ldots, t_n : t_0)$.

Lastly we have, for $t = (t_1, \ldots, t_n : t_0)$,

(131′) $\vdash_{c} I_t(\bar{y}, y) \equiv (\forall \bar{x}_0, \ldots, \bar{x}_n) \{\bar{x}_0 =^\cap \bar{y}(\bar{x}_1, \ldots, \bar{x}_n) \equiv^\cap \sim \pi_{\bar{x}} \bar{x}_0 =^\cap a^* \vee \pi_{\bar{x}} (\exists x_0, \ldots, x_n) [x_0 =^\cap y(x_1, \ldots x_n) \bigwedge_{i=0}^{n} I_{t_i}(\bar{x}_i, x_i)]\}$ where $\pi_{\bar{x}} \equiv_D \bigwedge_{i=1}^{n} \bar{x}_i \in v^*_{(t\eta)}$,

(137′) $\vdash_{c} I_t(\bar{y}, y) \equiv \bar{y} \in v^*_{(t\eta)} (\forall x_0, \ldots, x_n) \{x_0 =^\cap y(x_1, \ldots x_n) \equiv^\cap (\exists \bar{x}_0, \ldots, \bar{x}_n) [\bar{x}_0 =^\cap \bar{y}(\bar{x}_1, \ldots, \bar{x}_n) \bigwedge_{i=0}^{n} I_{t_i}(\bar{x}_i, x_i)]\}$.

Proof. We prove Theor. 59.2 completely in most cases. In the other cases we simply give some directions for the proof.

To prove (133) to (137) induction on $t \in \tau^\nu$ is to be used.

Case A. $t \in \{1, \ldots, \nu\}$, so that (136) and (137) need not be taken into account.

(I) To prove $(133)_1$ we start with (a) $I_t(\bar{y}, y) \wedge I_t(\bar{y}', y)$. Then by (129) we have (b) $y, \bar{y}' \in v^*_{(t\eta)}$ and (c) $(u)[\vdash'_u \supset^\cap \bar{y}(u) = \bar{y}'(u)]$.

From (b), (107) with t replaced by t^η, and $(\overline{1}05)_1$, we have $\vdash_c \bar{y}(u) =^\cup \bar{y}'(u) \supset^\cap \bar{y}(u) =^\cap \bar{y}'(u)$, which by (c) and $(83')_4$ [in Theor. 56.$\overline{1}$] yields (c′) $(u)[u \in El' \supset^\cap \bar{y}(u) =^\cap \bar{y}'(u)]$.

By $(8)_2$ in N6, in case A we have $\underline{t}^{\eta} = (\nu+1:\underline{t})$. So by (b), $(100')_1$, and (102) we have $(\underline{u})[\sim \underline{u} \in \underline{El}' \supset^{\cap} \overline{y}(\underline{u}) =^{\cap} \underline{a}^* =^{\cap} \overline{y}'(\underline{u})]$, which together with (c′) yields $(\underline{u})\overline{y}(\underline{u}) =^{\cap} \overline{y}'(\underline{u})$. Thence by AS12.15 we have (d) $\overline{y} =^{\cap} \overline{y}'$.

We can obviously assert $(\forall \overline{y}, \overline{y}')[(a) \supset (d)]$, so that by Def. 34.2 we have $(133)_1$ in case A.

(II) To prove $(133)_2$ in case A, we assume (a) $\underline{I}_t(\overline{y}, \underline{y}) \wedge \underline{I}_t(\overline{y}, \underline{y}')$. Then by (129) we obtain $(\underline{u})(|'_{\underline{u}} \supset^{\cap} \underline{y} = \underline{y}')$, which together with the consequence $\vdash_{\underline{c}} \underline{N}(\exists \underline{u})\, |'_{\underline{u}}$ of $(8\overline{3}')_1$ yields $\underline{y} =^{\cap} \underline{y}'$. So by Theor. 33.3, rule $\underline{G}$, and Def. 34.2 we deduce $(133)_2$ in case A.

(III) To prove $(134)_1$ in case A, by rule $\underline{C}$ with the variable $\overline{y}$ of type $\theta = \underline{t}^{\eta *}$, we deduce from $(46)_2$

(138) $\vdash_{\underline{c}} (\underline{u})[\overline{y}(\underline{u}) =^{\cap} (\imath \underline{x})\Phi(\underline{x}, \underline{y}, \underline{u})]$ where

$$\Phi(\underline{x}, \underline{y}, \underline{u}) \equiv_D \underline{x} \in \underline{v}^*_{(t^{\eta})} \underline{u} \in \underline{El}' \diamond (|'_{\underline{u}} \underline{x} = \underline{y}).$$

Let us assume (a) $\underline{u} \in \underline{El}'$. Since $\underline{t} \in \{1, \ldots, \nu\}$, by (99) we have $\vdash_{\underline{c}} \underline{N}\underline{y} \in \underline{v}^{*(e)}_{(t^{\eta})}$, which by Def. 18.9 yields $\underline{N}(\exists \underline{x})[\underline{x} = \underline{y}\,\underline{x} \in \underline{v}^*_{(t^{\eta})}]$. Thence by $(83')_4$ and (a) we deduce $\diamond[|'_{\underline{u}} (\exists \underline{x})(\underline{x} = \underline{y}\,\underline{x} \in \underline{v}^*_{(t^{\eta})})]$ which—since $\underline{x}$ does not occur in $|'_{\underline{u}}$—by $(\overline{1}04)_2$ easily yields $(\exists \underline{x})[\underline{x} \in \underline{v}^*_{(t^{\eta})} \diamond (|'_{\underline{u}} \underline{x} = \underline{y})]$. Now it is easy to see that (b) $\underline{u} \in \underline{El}' \vdash_{\underline{c}} (\exists \underline{x})\Phi(\underline{x}, \underline{y}, \underline{u})$—cf. $(138)_2$.

It is also easy to see that by $(138)_2$, (a), and $(89')_1$, $\Phi(\underline{x}, \underline{y}, \underline{u})$ and $\Phi(\underline{x}', \underline{y}, \underline{u})$ yield (c) $\underline{x}, \underline{x}' \in \underline{v}^*_{(t^{\eta})}$, $|'_{\underline{u}} \supset^{\cap} \underline{x} = \underline{y}$, and $|'_{\underline{u}} \supset^{\cap} \underline{x}' = \underline{y}$, whence $|'_{\underline{u}} \supset^{\cap} \underline{x} = \underline{x}'$. So by (a) and $(8\overline{3}')_4$ we have $\underline{x} =^{\cup} \underline{x}'$, which by (c), $(1\overline{0}5)_1$, and Defs. 18.8 and 18.7 yields $\underline{x} =^{\cap} \underline{x}'$.

By Def. 34.2 we conclude that $\underline{u} \in \underline{El}' \vdash_{\underline{c}} (\exists^{(1)\cap} \underline{x})\Phi(\underline{x}, \underline{y}, \underline{u})$. So by (b), $(14)_{1,2,3}$, and Theor. 33.3 we have $\vdash_{\underline{c}} \underline{u} \in \underline{El}' \supset (\exists_1^{\cap} \underline{x})\Phi(\underline{x}, \underline{y}, \underline{u})$; and from $(138)_2$, $(104)_2$, and $(82')_1$ we deduce (d) $\vdash_{\underline{c}} \underline{N}\Phi(\underline{x}, \underline{y}, \underline{u}) \equiv \diamond\Phi(\underline{x}, \underline{y}, \underline{u})$.

Then by $(35)_6$ we easily deduce $\vdash_{\underline{c}} \underline{u} \in \underline{El}' \supset \Phi[(\imath \underline{x})\Phi(\underline{x}, \underline{y}, \underline{u}), \underline{y}, \underline{u}]$, which by $(138)_1$ and rule $\underline{G}$ yields (e) $(\underline{u})\{\underline{u} \in \underline{El}' \supset \Phi[\overline{\underline{y}}(\underline{u}), \underline{y}, \underline{u}]\}$.

From (138) we obviously have $(\underline{u})[\sim \underline{u} \in \underline{El}' \supset \overline{\underline{y}}(\underline{u}) =^{\frown} \underline{a}^*]$, which by (e) yields $(\underline{u})\{\Phi[\overline{\underline{y}}(\underline{u}), \underline{y}, \underline{u}] \vee \overline{\underline{y}}(\underline{u}) =^{\frown} \underline{a}^*\}$. Thence by $(138)_2$ and (106) we deduce $(\underline{u})[\overline{\underline{y}}(\underline{u}) \in \underline{v}^*_{(t)}]$, hence (f) $\overline{\underline{y}} \in \underline{v}^*_{(t\eta)}$ by (102).

By $(138)_2$ and $(89')$ $\Phi[\overline{\underline{y}}(\underline{u}), \underline{y}, \underline{u}] \supset^{\frown} [\,|'_{\underline{u}} \supset^{\frown} \overline{\underline{y}}(\underline{u}) = \underline{y}]$. So by the consequence $\vdash_{\underline{c}} (\underline{u})(\,|'_{\underline{u}} \supset^{\frown} \underline{u} \in \underline{El}')$ of Def. 56.2 and by (e) and (d) we have $(\underline{u})[\,|'_{\underline{u}} \supset^{\frown} \overline{\underline{y}}(\underline{u}) = \underline{y}]$. From this, (f), and (129), we obtain $\underline{I}_t(\overline{\underline{y}}, \underline{y})$, whence $(\exists \overline{\underline{y}})\underline{I}_t(\overline{\underline{y}}, \underline{y})$. In the last result the variable $\overline{\underline{y}}$ introduced by rule $\underline{C}$ at step (138) does not occur free. Then we easily have $(134)_1$.

(IV) To prove $(134)_2$ in Case A, from $\vdash \underline{x} =^{\frown} \underline{x}$ we easily deduce $\vdash (\underline{x})(\exists \underline{y})\underline{y} =^{\frown} \underline{x}$, which by AS38.2 and rule $\underline{C}$ wity $\underline{y}$ yields

$$(139) \qquad \underline{y} =^{\frown} (\imath \underline{x})(\underline{u})\Phi(\underline{u}, \underline{x}) \quad \text{where} \quad \Phi(\underline{u}, \underline{x}) \equiv_D \,|'_{\underline{u}} \supset \overline{\underline{y}}(\underline{u}) = \underline{x}.$$

From $(83')_6$ we easily obtain (a) $\vdash (\exists_1 \underline{x})(\underline{u})\Phi(\underline{u}, \underline{x})$, and since $\vdash (\underline{\text{ext}}\ \underline{x})(\underline{u})\Phi(\underline{u}, \underline{x})$ [Def. 35.1], by (a), $(31)_2$, and $(139)_1$ we have $(\underline{u})\Phi(\underline{u}, \underline{y})$. By rule $\underline{G}$ this implies $(\underline{u})\underline{N}\Phi(\underline{u}, \underline{y})$, which by $(139)_2$ and AS12.5 yields $(\underline{u})[\,|'_{\underline{u}} \supset^{\frown} \overline{\underline{y}}(\underline{u}) = \underline{y}]$. Thence by (129) we obtain $\overline{\pi} \supset (\exists \underline{y})\underline{I}_t(\overline{\underline{y}}, \underline{y})$ where $\overline{\pi}$ is $\overline{\underline{y}} \in \underline{v}^*_{(t\eta)}$. So by Theors. 33.1 and 32.2 we have $(\overline{1}34)_2$ in case A.

(VI) For $\underline{r} = 1, 2$ theorem $(135)_{\underline{r}}$ follows from $(133)_{\underline{r}}$ and $(134)_{\underline{r}}$ by $(14)_{1,2,3}$.

Case B. $\underline{t}$ is $(\underline{t}_1, \ldots, \underline{t}_n)$—so that (137) need not be proved and (133) to (137) hold for $\underline{t} = \underline{t}_1, \ldots, \underline{t} = \underline{t}_n$ as hypotheses of the induction.

(I) For proving $(133)_1$ in case B we assume (a) $\underline{I}_t(\overline{\underline{y}}, \underline{y}) \wedge \underline{I}_t(\overline{\underline{y}}', \underline{y})$. Then by (130) we deduce $(\forall \overline{\underline{x}}_1, \ldots, \overline{\underline{x}}_n, \underline{u})[\overline{\underline{y}}(\overline{\underline{x}}_1, \ldots, \overline{\underline{x}}_n, \underline{u})$

$\equiv^{\cap} \bar{y}'(\bar{x}_1, \ldots, \bar{x}_n, u)]$, whence by AS40.1, $\bar{y} =^{\cap} \bar{y}'$. Now we can easily conclude that $\vdash_{c} (\exists^{(1)\cap} \bar{y}) I_t(\bar{y}, y)$ [Def. 34.2], whence $(133)_1$ holds in case B.

(II) To prove $(134)_1$ we simply remark that by (130), $(134)_1$ in case B is an instance of AS40.3.

(III) We now want to prove (136), which is useful in proving $(133)_2$ and $(134)_2$ in case B. Therefore we assume (a) $I_t(\bar{y}, y)$, whence by (130) we deduce just the R.H.S. ρ_{130} of definition (130).

First we add to (a) the assumption (b) $y(x_1, \ldots, x_n)$. By $(83')_1$ and rule C with u we obtain $|'_u$. Since $(134)_1$ holds for $t = t_i$, by rule C with $\bar{x}_i$ we have (c) $I_{t_i}(\bar{x}_i, x_i)$ $(i = 1, \ldots, n)$. So by (b) and $|'_u$ we have $\diamond [|'_u\, y(x_1, \ldots, x_n) I_{t_1}(\bar{x}_1, x_1) \wedge \ldots \wedge I_{t_n}(\bar{x}_n, x_n)]$, which by ρ_{130} easily yields $\bar{y}(\bar{x}_1, \ldots, \bar{x}_n, u)$. Thence by (c) and $|'_u$, we deduce

$$(140) \quad (\exists \bar{x}_1, \ldots, \bar{x}_u, u)[\,|'_u\, \bar{y}(\bar{x}_1, \ldots, \bar{x}_n, u) \bigwedge_{i=1}^{n} I_{t_i}(\bar{x}_i, x_i)].$$

Then by Theor. 33.3, rule G, Theor. 56.3, $(132)_{1,2}$, and AS30.7 we conclude that

$$(141) \quad I_t(\bar{y}, y) \vdash_{c} (\forall x_1, \ldots, x_n) \{y(x_1, \ldots, x_n) \supset^{\cap} (\exists \bar{x}_1, \ldots, \bar{x}_n, u)[\,|'_u\, \bar{y}(\bar{x}_1, \ldots, \bar{x}_n, u) \bigwedge_{i=1}^{n} I_{t_i}(\bar{x}_i, x_i)]\}.$$

Now we add to (a) $I_t(\bar{y}, y)$ the assumption (140) instead of (b). Then by rule C we have (d_1) $|'_u$, (d_2) $\bar{y}(\bar{x}_1, \ldots, \bar{x}_n, u)$, and (d_3) $I_{t_i}(\bar{x}_i, x_i)$ $(i = 1, \ldots, n)$.

By rule C with $\xi_1, \ldots, \xi_n$, from (d_2) and ρ_{130} we obtain (e_1) $\diamond [|'_u\, y(\xi_1, \ldots, \xi_n)]$ and $\diamond I_{t_i}(\bar{x}_i, \xi_i)$, which by $(132)_1$ yields (e_2) $I_{t_i}(\bar{x}_i, \xi_i)$ $(i = 1, \ldots, n)$.

On the one hand (d_1), i.e. $|'_{\underline{u}}$, yields $\underline{u} \in \underline{El}'$ [Def. 56.2]. Thence by $(89')_2$ and (e_1), we deduce $|'_{\underline{u}} \supset^{\cap} \underline{y}(\xi_1, \ldots, \xi_{\underline{n}})$, which by (d_1) yields (f) $\underline{y}(\xi_1, \ldots, \xi_{\underline{n}})$.

On the other hand, $(\bar{e}_2)$ and $(132)_3$ yield $\bar{\underline{x}}_{\underline{i}} \in \underline{v}^*_{(\underline{t_i}\eta)}$, which is $\bar{\pi}$ for $\underline{t} = \underline{t}_{\underline{i}}$. Then by $(133)_2$ for $\underline{t} = \underline{t}_{\underline{i}}$ and Def. $3\bar{4}.2$, from (d_3) and (e_2) we deduce $\underline{x}_{\underline{i}} =^{\cap} \xi_{\underline{i}}$ $(\underline{i} = 1, \ldots, \underline{n})$.

The last result and (f) yield $\underline{y}(\underline{x}_1, \ldots, \underline{x}_{\underline{n}})$. So we can conclude that $\underline{I}_{\underline{t}}(\bar{\underline{y}}, \underline{y})$, (140) $\vdash_{\underline{c}} \underline{y}(\underline{x}_1, \ldots, \underline{x}_{\underline{n}})$. Then by Theor. 33.3, rule $\underline{G}$, $(13\bar{2})_{1,2}$, and AS$3\bar{0}$.7 we can reverse the strict implication written in (141). Hence we can replace $\supset^{\cap}$ in (141) with $\equiv^{\cap}$. After this replacement, by Theor. 33.3 and $(132)_3$ we easily deduce

(142) $\vdash_{\underline{c}} \underline{I}_{\underline{t}}(\bar{\underline{y}}, \underline{y}) \supset \rho_{136}$ where $\rho_{136} \equiv_D$ the R.H.S. of the equivalence (136).

Now we assume, instead of $\underline{I}_{\underline{t}}(\bar{\underline{y}}, \underline{y})$, just ρ_{136}, and we want to deduce $\underline{I}_{\underline{t}}(\bar{\underline{y}}, \underline{y})$ in case B—cf. $(1\bar{3}0)$. To this end we first add to ρ_{136} the assumption (g) $\bar{\underline{y}}(\bar{\underline{x}}_1, \ldots, \bar{\underline{x}}_{\underline{n}}, \underline{u})$. From ρ_{136} we deduce (h) $\bar{\underline{y}} \in \underline{v}^*_{(\underline{t}\eta)}$. By (101) and $(100')_1$, from (g) and (h) we deduce (i_1) $\bar{\underline{y}} \in \underline{\text{MConst}}$, (i_2) $\underline{u} \in \underline{El}'$, and (i_3) $\bar{\underline{x}}_{\underline{i}} \in \underline{v}^*_{(\underline{t_i}\eta)}$ $(\underline{i} = 1, \ldots, \underline{n})$.

By $(134)_2$ for $\underline{t} = \underline{t}_{\underline{i}}$ and $(132)_{1,2}$ from (i_3) we deduce $(\exists \underline{x}_{\underline{i}}) \underline{NI}_{\underline{t_i}}$ $(\bar{\underline{x}}_{\underline{i}}, \underline{x}_{\underline{i}})$, which by rule $\underline{C}$ with $\underline{x}_{\underline{i}}$ yields (l) $\underline{NI}_{\underline{t_i}}(\bar{\underline{x}}_{\underline{i}}, \underline{x}_{\underline{i}})$ $(\underline{i} = 1, \ldots, \underline{n})$.

By $(83')_4$, (i_2) yields $\diamond |'_{\underline{u}}$, which by (l), (i_1), and (g) yields $\diamond \rho'_{136}$, where

(143) $\rho'_{136} \equiv_D (\exists \xi_1, \ldots, \xi_{\underline{n}}, \underline{u}) [|'_{\underline{u}} \bar{\underline{y}}(\xi_1, \ldots, \xi_{\underline{n}}, \underline{u}) \bigwedge_{\underline{i}=1}^{\underline{n}} \underline{I}_{\underline{t}_{\underline{i}}}(\xi_{\underline{i}}, \underline{x}_{\underline{i}})]$.

By ρ_{136} the matrix $\diamond \rho'_{136}$ yields (m) $\diamond [\rho'_{136}\, \underline{y}(\underline{x}_1, \ldots, \underline{x}_{\underline{n}})]$. Then by rule $\underline{C}$ with $\xi_1, \ldots, \xi_{\underline{n}}$ we easily have

(144) $$\diamond [|'_{\underline{u}} \underline{y}(\underline{x}_1, \ldots, \underline{x}_{\underline{n}}) \bigwedge_{\underline{i}=1}^{\underline{n}} \underline{I}_{\underline{t}_{\underline{i}}}(\xi_{\underline{i}}, \underline{x}_{\underline{i}})].$$

From (144) and (1), by $(133)_1$ for $\underline{t} = \underline{t}_i$, we deduce $\xi_i =^{\cap} \underline{\overline{x}}_i$ ($\underline{i} = 1, \ldots, \underline{n}$), so that again by (144) we have ρ'_{130} where

(145) $\rho'_{130} \equiv_D (\exists \underline{x}_1, \ldots, \underline{x}_n) \diamond [\,|'_{\underline{u}}\, \underline{y}(\underline{x}_1, \ldots, \underline{x}_n) \bigwedge_{\underline{i}=1}^{\underline{n}} \underline{I}_{\underline{t}_i}(\underline{\overline{x}}_i, \underline{x}_i)]$.

Since by $(104)_2$ ρ_{136} is noncontingent, by Theor. 56.3 we conclude that

(146) $$\rho_{136} \vdash_{\underline{c}} \underline{\overline{y}}(\underline{\overline{x}}_1, \ldots, \underline{\overline{x}}_n, \underline{u}) \supset^{\cap} \rho'_{130}.$$

Now we add to ρ_{136} the assumption ρ'_{136}—cf. (143)—instead of (g). Thence by rule $\underline{C}$ we deduce (n_1) $|'_{\underline{u}}$, (n_2) $\underline{y}(\underline{x}_1, \ldots, \underline{x}_n)$, and (n_3) $\underline{I}_{\underline{t}_i}(\underline{\overline{x}}_i, \underline{x}_i)$ ($\underline{i} = 1, \ldots, \underline{n}$). From $(\overline{n}_2)$ and ρ_{136} we obtain—by rule $\underline{C}$ with $\xi_1, \ldots, \xi_{\underline{n}}$ and $\underline{v}$—$\underline{\overline{y}}(\xi_1, \ldots, \xi_{\underline{n}}, \underline{v})$, $\underline{I}_{\underline{t}_i}(\underline{\overline{x}}_i, \underline{x}_i)$ and $|'_{\underline{v}}$. Hence by $(83')_5$ and $(\overline{n}_1)$ we have $\underline{u} =^{\cap} \underline{v}$, and by $(\overline{n}_3)$ and $(133)_1$ for $\underline{t} = \underline{t}_i$ we deduce $\xi_i =^{\cap} \underline{\overline{x}}_i$ ($\underline{i} = 1, \ldots, \underline{n}$). Then by (n_2) we have the matrix (p) $\underline{\overline{y}}(\underline{\overline{x}}_1, \ldots, \underline{\overline{x}}_n, \underline{u})$. By assumption ρ_{136} (p) is a noncontingent matrix, as well as ρ'_{130}—cf. (145)—and ρ_{136}. So we conclude that

(146′) $$\rho_{136} \vdash_{\underline{c}} \rho'_{130} \supset^{\cap} \underline{\overline{y}}(\underline{\overline{x}}_1, \ldots, \underline{\overline{x}}_n, \underline{u}).$$

By (145) and (130), $\vdash_{\underline{c}} \underline{I}_{\underline{t}}(\underline{\overline{y}}, \underline{y}) \equiv^{\cap} (\forall \underline{x}_1, \ldots, \underline{x}_n) [\underline{\overline{y}}(\underline{\overline{x}}_1, \ldots, \underline{\overline{x}}_n, \underline{u}) \equiv^{\cap} \rho'_{130}]$. In addition $\underline{x}_1, \ldots, \underline{x}_n$ do not occur free in $\underline{I}_{\underline{t}}(\underline{\overline{y}}, \underline{y})$. So by rule $\underline{G}$, from (146) and (146′) we deduce $\rho_{136} \vdash_{\underline{c}} \underline{I}_{\underline{t}}(\underline{\overline{y}}, \underline{y})$. By Theor. 33.3 the last result and (142) yield (136) in case $\overline{B}$.

(IV) To prove $(133)_2$ in case B we assume (a) $\underline{I}_{\underline{t}}(\underline{\overline{y}}, \underline{y}) \wedge \underline{I}_{\underline{t}}(\underline{\overline{y}}, \underline{y}')$, whence by (136) and AS40.1 we deduce $\underline{y} =^{\cap} \underline{y}'$. By Theors. $\overline{3}3.3$, rule $\underline{G}$, and Def. 34.2 we have $\vdash_{\underline{c}} (\exists^{(1)\cap} \underline{y}) \underline{I}_{\underline{t}}(\underline{\overline{y}}, \underline{y})$, whence $(133)_2$.

(V) To prove $(134)_2$ in case B we assume $\overline{\pi}$, i.e. $\underline{\overline{y}} \in \underline{v}^*_{(\underline{t}\eta)}$, and from $(46)_1$ we deduce, by rule $\underline{C}$ with $\underline{y}$,

(147) $\vdash_{\underline{c}} (\forall \underline{x}_1, \ldots, \underline{x}_n) \{\underline{y}(\underline{x}_1, \ldots, \underline{x}_n) \equiv^{\cap} (\exists \underline{\overline{x}}_1, \ldots, \underline{\overline{x}}_n, \underline{u})$
$[\,|'_{\underline{u}}\, \underline{\overline{y}}(\underline{\overline{x}}_1, \ldots, \underline{\overline{x}}_n, \underline{u}) \bigwedge_{\underline{i}=1}^{\underline{n}} \underline{I}_{\underline{t}_i}(\underline{\overline{x}}_i, \underline{x}_i)]\}$.

From $\bar{\pi}$, (147), and (136) we deduce $\underline{I}_t(\bar{\underline{y}}, \underline{y})$. We conclude that $\vdash_{\underline{c}} \bar{\pi} \supset (\exists \underline{y})\, \underline{I}_t(\bar{\underline{y}}, \underline{y})$. So by rule $\underline{G}$, $(\overline{134})_2$ holds.

(VI) For $\underline{r} = 1, 2$, $(135)_{\underline{r}}$ follows from $(133)_{\underline{r}}$ and $(134)_{\underline{r}}$ in case B by $(14)_{1,2,3}$.

Case C. $\underline{t} = (\underline{t}_1, \ldots, \underline{t}_n : \underline{t}_0)$, so that (136) need not be considered. We only give some directions for the proof in case C.

(I) To prove (131′), (137′), and (137), we consider the definition $(131')_2$ of $\pi_{\bar{\underline{x}}}$. Then by $(132)_3$, by $(135)_2$ for $\underline{t} = \underline{t}_1, \ldots, \underline{t} = \underline{t}_n$, and by $(135)_1$ for $\underline{t} = \underline{t}_0$, $\pi_{\bar{\underline{x}}}$ is a necessary and sufficient condition for the description $(\imath \bar{\underline{x}}_0)\, \overline{(\ldots)}$ in (131) to fulfill its existence and uniqueness condition. Furthermore in (131) the operator $(\imath \bar{\underline{x}}_0)$ is applied to a noncontingent matrix—cf. $(132)_{1,2}$—so that by $(35)_3$ this description has the well-known corresponding replacement property if and only if $\pi_{\bar{\underline{x}}}$ holds. Hence (131) implies (131′).

In a similar way the R.H.S.'s ρ_{137} and $\rho_{137'}$ of the equivalences written in (137) and (137′) can be proved to be equivalent: $\vdash_{\underline{c}} \rho_{137} \equiv \rho_{137'}$.

The equivalence of the R.H.S.'s of the equivalences written in (131′) and (137′) can be proved on the basis of $(132)_3$, (134), and (135) through procedures used in case B (III). This equivalence and (131′) yield (137′). Since $\vdash_{\underline{c}} \rho_{137} \equiv \rho_{137'}$, (137′) yields (137).

(II) Theorems $(133)_1$ to $(\overline{135})_2$ in MC^{ν} can be proved in case C as in case B; of course (131) and (137) have to be used instead of (130) and (136) respectively.

QED

N60. Basic properties of $\Delta^{\eta *}$ for Δ in certain classes of designators in ML^{ν}.

We shall consider some properties for certain simple matrices and terms and for $\underline{a}^*_{\underline{t}}$, which expresses the nonexisting object in ML

THEOR. 60.1. Let $\underline{x}_1, \overline{\underline{x}}_1, \ldots, \underline{x}_{n+1}, \overline{\underline{x}}_{n+1}$ be distinct variables of the respective types $\underline{t}_1, \underline{t}_1^{\eta*}, \ldots, \underline{t}_{n+1}, \underline{t}_{n+1}^{\eta*}$. Furthermore let Δ and $\overline{\Delta}$ be the expressions $\underline{x}_{n+1}(\underline{x}_1, \ldots, \underline{x}_n)$ and $\overline{\underline{x}}_{n+1}(\overline{\underline{x}}_1, \ldots, \overline{\underline{x}}_n)$ respectively, which are assumed to be well formed and of the respective types $\underline{t}\ (\epsilon\overline{\tau}^{\nu})$ and $\underline{t}^{\eta*}$—cf. (1) in N2. Then, accepting results (99) to (102) of rule $\underline{C}$, we have

$$|'_{\underline{u}}, \bigwedge_{\underline{h}=1}^{\underline{n}+1} \underline{I}_{\underline{t}_{\underline{h}}}(\overline{\underline{x}}_{\underline{h}}, \underline{x}_{\underline{h}}) \vdash_{\underline{C}} \begin{cases} \overline{\Delta} \equiv \Delta \text{ for } \underline{t}_{n+1} = (\underline{t}_1, \ldots, \underline{t}_n), \\ \underline{I}_{\underline{t}}(\overline{\Delta}, \Delta) \text{ for } \underline{t}_{n+1} = (\underline{t}_1, \ldots, \underline{t}_n : \underline{t}_0), \\ \qquad\qquad\qquad\qquad \underline{t}_0 = \underline{t}. \end{cases}$$

Proof. We start with $|'_{\underline{u}}$ and (a) $\underline{I}_{\underline{t}_{\underline{h}}}(\overline{\underline{x}}_{\underline{h}}, \underline{x}_{\underline{h}})$ $(\underline{h} = 1, \ldots, \underline{n}+1)$.

Let Δ be a matrix, whence $\underline{t}_{n+1} = (\underline{t}_1, \ldots, \underline{t}_n)$. Then from the premise $|'_{\underline{u}}$, from (136) for $\underline{t} = \underline{t}_{n+1}$, and from (a) for $\underline{h} = \underline{n}+1$, we deduce

$$\text{(b) } (\forall \underline{x}_1, \ldots, \underline{x}_n)\{\underline{x}_{n+1}(\underline{x}_1, \ldots, \underline{x}_n) \equiv (\exists \overline{\underline{x}}_1, \ldots, \overline{\underline{x}}_n, \underline{u})$$
$$[|'_{\underline{u}}\overline{\underline{x}}_{n+1}(\overline{\underline{x}}_1, \ldots, \overline{\underline{x}}_n, \underline{u}) \bigwedge_{\underline{i}=1}^{\underline{n}} \underline{I}_{\underline{t}_{\underline{i}}}(\overline{\underline{x}}_{\underline{i}}, \underline{x}_{\underline{i}})]\}.$$

By $(133)_1$ for $\underline{t} = \underline{t}_1, \ldots, \underline{t} = \underline{t}_n$, $(83')_5$, and (29), from $|'_{\underline{u}}$, (b), and (a) for $\underline{h} = 1, \ldots, \underline{n}$, we easily deduce $\underline{x}_{n+1}(\underline{x}_1, \ldots, \underline{x}_n) \equiv \overline{\underline{x}}_{n+1}(\overline{\underline{x}}_1, \ldots, \overline{\underline{x}}_n, \underline{u})$, i.e. $\Delta \equiv \overline{\Delta}$. We conclude that the thesis (of the theorem) holds when Δ is a matrix.

Now let Δ be a term of type $\underline{t}(\epsilon\tau^{\nu})$, whence $\underline{t}_{n+1} = (\underline{t}_1, \ldots, \underline{t}_n : \underline{t}_0)$ where $\underline{t}_0$ is $\underline{t}$. Then by (137′), from (a) for $\underline{h} = \underline{n}+1$ we deduce both (c) $\overline{\underline{x}}_{n+1} \epsilon \underline{v}^*_{(\underline{t}_{n+1}\eta)}$ and

$$\text{(d) } (\forall \underline{x}_0, \ldots, \underline{x}_n)\{\underline{x}_0 =^\cap \underline{x}_{n+1}(\underline{x}_1, \ldots, \underline{x}_n) \equiv^\cap (\exists \overline{\underline{x}}_0, \ldots, \overline{\underline{x}}_n)$$
$$[\overline{\underline{x}}_0 =^\cap \overline{\underline{x}}_{n+1}(\overline{\underline{x}}_1, \ldots, \overline{\underline{x}}_n) \bigwedge_{\underline{i}=0}^{\underline{n}} \underline{I}_{\underline{t}_{\underline{i}}}(\overline{\underline{x}}_{\underline{i}}, \underline{x}_{\underline{i}})]\}.$$

By $(135)_1$ and (29), from (d) and (a) for $\underline{h} = 1, \ldots, \underline{n}$ we easily deduce, using rule $\underline{C}$ with $\underline{\bar{x}}_0$, the matrix

(e) $\Delta =^{\frown} \underline{x}_{n+1}(\underline{x}_1, \ldots, \underline{x}_n) \supset \underline{\bar{x}}_0 =^{\frown} \underline{\bar{x}}_{n+1}(\underline{\bar{x}}_1, \ldots, \underline{\bar{x}}_n)\, \underline{I}_{\underline{t}_0}(\underline{\bar{x}}_0, \Delta)$.

The antecedent in (e) is true by AS12.10 ($\vdash \underline{x} = \underline{x}$), and by definition we have $\bar{\Delta} = \underline{\bar{x}}_{n+1}(\underline{\bar{x}}_1, \ldots, \underline{\bar{x}}_n)$. Then (e) yields $\underline{I}_{\underline{t}_0}(\bar{\Delta}, \Delta)$ by (47′).

We conclude that the thesis of the theorem also holds when Δ is a term.

QED

THEOR. 60.2. For $\underline{t} \epsilon \tau^{\nu}$ we have: (I) $\vdash (\underline{a}^*_{\underline{t}})^{\eta} = \underline{a}^*_{t\eta}$ in $EC^{\nu+1}$, (II) $\vdash (\underline{a}^*_{\underline{t}})^{\eta *} =^{\frown} \underline{a}^*_{\underline{t}\eta *}$ in MC^{ν}, (III) $(\underline{N})\underline{I}_{\underline{t}}(\underline{a}^*_{\underline{t}\eta *}, \underline{a}^*_{\underline{t}})$ in MC^{ν}.

<u>Proof.</u> (I) Cf. (3) in N31.

(II) By substituting $\underline{t}^{\eta}$ for $\underline{t}$ in the result (f) obtained in dealing with case 12 in the proof of Theor. 58.2, we obtain $\vdash (\underline{a}^*_{t\eta})^* =^{\frown} \underline{a}^*_{t\eta *}$ in MC^{ν}. Furthermore (I) and rule $(\underline{T}^*_3)$ [N57] yield $\vdash (\underline{a}^*_{\underline{t}})^{\eta *} =^{\frown} (\underline{a}^*_{t\eta})^*$. Then (II) holds.

(III) <u>Case 1</u>. $\underline{t} \epsilon \{1, \ldots, \nu\}$. Then by $(8)_2$ in N6 $\underline{t}^{\eta} = (\nu + 1 : \underline{t})$ and by Def. 56.1(b), (a) $\underline{t}^{\eta *} = (\nu + 1^* : \underline{t}^*) = (\nu + 1 : \underline{t})$, so that by AS12.22, $\vdash (\underline{x})\, \underline{a}^*_{\underline{t}\eta *}(\underline{x}) =^{\frown} \underline{a}^*_{\underline{t}}$. This, the instance $\underline{a}^*_{\underline{t}\eta} \epsilon \underline{v}^*_{(\underline{t}\eta)}$ of (106) and (129) yield (III) in case 1.

<u>Case 2</u>. $\underline{t} = (\underline{t}_1, \ldots, \underline{t}_n)$. Then $\underline{t}^{\eta *} = (\underline{t}_1^{\eta *}, \ldots, \underline{t}_n^{\eta *}, \nu + 1^*)$ by $(8)_3$ in N6 and Def. 56.1(b), so that (III) follows by AS12.21 and (130).

<u>Case 3</u>. $\underline{t} = (\underline{t}_1, \ldots, \underline{t}_n : \underline{t}_0)$. Then $\underline{t}^{\eta *} = (\underline{t}_1^{\eta *}, \ldots, \underline{t}_n^{\eta *} : \underline{t}_0^{\eta *})$ by $(8)_4$ in N6 and Def. 56.1(c).

We assume $\underline{I}_{\underline{t}_0}(\underline{a}^*_{\underline{t}_0\eta *}, \underline{a}^*_{\underline{t}_0})$ as inductive hypothesis. Then from (106), $(134)_1$, AS12.22, and (131′) we deduce (III).

By the principle of induction (III) holds for $\underline{t} \epsilon \tau^{\nu}$.

QED

N61. The analogue $\stackrel{*}{=}\frac{t}{\underline{u}}$ of $=\frac{t}{\gamma}$ within ML^{ν}.

In N10 the equivalence $=\frac{t}{\gamma}$ between QIs for ML^{ν} of the type $\underline{t}$ was defined ($\underline{t}\epsilon\tau^{\nu}$). The analogous relation $=\frac{t}{\kappa}$ was defined within $EL^{\nu+1}$ for $\underline{t}\epsilon\tau^{\nu}$ by means of Def. 15.1, in order to translate ML^{ν} into $EL^{\nu+1}$ ($\Delta \rightarrow \Delta^{\eta}$). Now we denote by $\stackrel{*}{=}\frac{t}{\underline{u}}$, where $\underline{u}$ is κ^*, i.e. $\underline{v}_{\nu+1^*,6}$, what will turn out to be the star translation of $=\frac{t}{\kappa}$—obtained by rules ($\underline{T}_1^*$) to ($\underline{T}_7^*$) in N57—for $\underline{t}\epsilon\tau^{\nu}$:

DEF. 61.1. Let $\underline{t}\epsilon\tau^{\nu}$ and $\overline{\underline{y}}_1$ and $\overline{\underline{y}}_2$ be distinct variables in ML^{ν} of the type $\underline{t}^{\eta}*$. Then

(a) $\overline{\underline{y}}_1 \stackrel{*}{=}\frac{t}{\underline{u}} \overline{\underline{y}}_2 \equiv_D \overline{\underline{y}}_1(\underline{u}) =^{\cap} \overline{\underline{y}}_2(\underline{u})$ for $\underline{t}\epsilon\{1, \ldots, \nu\}$.

(b) $\overline{\underline{y}}_1 \stackrel{*}{=}\frac{t}{\underline{u}} \overline{\underline{y}}_2 \equiv_D \overline{\underline{y}}_1, \overline{\underline{y}}_2 \epsilon \underline{v}^*_{(\underline{t})} (\forall \overline{\underline{x}}_1, \ldots, \overline{\underline{x}}_{\underline{n}}) [\overline{\underline{y}}_1(\overline{\underline{x}}_1, \ldots, \overline{\underline{x}}_{\underline{n}}, \underline{u})$
$\equiv^{\cap} \overline{\underline{y}}_2(\overline{\underline{x}}_1, \ldots, \overline{\underline{x}}_{\underline{n}}, \underline{u})]$ for $\underline{t} = (\underline{t}_1, \ldots, \underline{t}_{\underline{n}})$.

(c) $\overline{\underline{y}}_1 \stackrel{*}{=}\frac{t}{\underline{u}} \overline{\underline{y}}_2 \equiv_D \overline{\underline{y}}_1, \overline{\underline{y}}_2 \epsilon \underline{v}^*_{(\underline{t})} (\forall \overline{\underline{x}}_1, \ldots, \overline{\underline{x}}_{\underline{n}}) [\overline{\underline{y}}_1(\overline{\underline{x}}_1, \ldots, \overline{\underline{x}}_{\underline{n}})$
$\stackrel{*}{=}\frac{t_0}{\underline{u}} \overline{\underline{y}}_2(\overline{\underline{x}}_1, \ldots, \overline{\underline{x}}_{\underline{n}})]$ for $\underline{t} = (\underline{t}_1, \ldots, \underline{t}_{\underline{n}} : \underline{t}_0)$.

Let us remark that for example by (102), (c) is equivalent to

(c′) $\overline{\underline{y}}_1 \stackrel{*}{=}\frac{t}{\underline{u}} \overline{\underline{y}}_2 \equiv_D \overline{\underline{y}}_1, \overline{\underline{y}}_2 \epsilon \underline{v}^*_{(\underline{t})} (\forall \overline{\underline{x}}_1, \ldots, \overline{\underline{x}}_{\underline{n}}) \{ \bigwedge_{\underline{i}=1}^{\underline{n}} \overline{\underline{x}}_{\underline{i}} \epsilon \underline{v}^*_{(\underline{t}_{\underline{i}})}$
$\supset [\overline{\underline{y}}_1(\overline{\underline{x}}_1, \ldots, \overline{\underline{x}}_{\underline{n}}) \stackrel{*}{=}\frac{t_0}{\underline{u}} \overline{\underline{y}}_2(\overline{\underline{x}}_1, \ldots, \overline{\underline{x}}_{\underline{n}})]\}$.

THEOR. 61.1. Accepting results (99) to (102) of rule $\underline{C}$, we have

(148) $\quad \Vdash'_{\underline{u}}, \underline{I}_{\underline{t}}(\overline{\underline{y}}_1, \underline{y}_1) \wedge \underline{I}_{\underline{t}}(\overline{\underline{y}}_2, \underline{y}_2) \vdash_{\underline{C}} \underline{y}_1 = \underline{y}_2 \equiv \overline{\underline{y}}_1 \stackrel{*}{=}\frac{t}{\underline{u}} \overline{\underline{y}}_2$

where $\underline{y}_1$ and $\underline{y}_2$ are distinct variables of type $\underline{t}\epsilon\tau^{\nu}$ and $\overline{\underline{y}}_1, \overline{\underline{y}}_2$ are distinct variables of the type $\underline{t}^{\eta}*$.

Proof. We use induction on $\underline{t}\epsilon\tau^{\nu}$ to prove (148).

Case A. $\underline{t} \in \{1, \ldots, \nu\}$. By Def. 61.1(a) the matrix $\overline{\underline{x}}_1 \overset{t}{\underset{u}{\stackrel{*}{=}}} \overline{\underline{x}}_2$ is equivalent to (a) $\overline{\underline{y}}_1(\underline{u}) =^{\cap} \overline{\underline{y}}_2(\underline{u})$.

By (129) our premises $|'_{\underline{u}}$ and $\underline{I}_{\underline{t}}(\overline{\underline{y}}_{\underline{i}}, \underline{y}_{\underline{i}})$—cf. (148)—yield $\underline{y}_{\underline{i}} \in \underline{v}^*_{(t\eta)}$ and $\overline{\underline{y}}_{\underline{i}}(\underline{u}) = \underline{y}_{\underline{i}}(\underline{i} = 1, \overline{2})$. So they yield the equivalence of (a) and $\underline{y}_1 = \underline{y}_2$. By Def. 61.1(a) we conclude that (148) holds.

Case B. $\underline{t} = (\underline{t}_1, \ldots, \underline{t}_n)$. We assume the premises in (148). Then by $(132)_3$, from $\underline{I}_{\underline{t}}(\overline{\underline{y}}_{\underline{i}}, \underline{y}_{\underline{i}})$ $(\underline{i} = 1, 2)$ we deduce $\overline{\underline{y}}_1, \overline{\underline{y}}_2 \in \underline{v}^*_{(t\eta)}$, which by $(104)_1$ yields $\overline{\underline{y}}_1, \overline{\underline{y}}_2 \in \underline{\text{MConst}}$ and by Defs. 13.2 and 61.1(b) makes the matrix (a) $\overline{\underline{y}}_1 \overset{t}{\underset{u}{\stackrel{*}{=}}} \overline{\underline{y}}_2$ equivalent to

$$(149) \quad (\forall \overline{\underline{x}}_1, \ldots, \overline{\underline{x}}_{\underline{n}}) [\overline{\underline{y}}_1(\overline{\underline{x}}_1, \ldots, \overline{\underline{x}}_{\underline{n}}, \underline{u}) \equiv \overline{\underline{y}}_2(\overline{\underline{x}}_1, \ldots, \overline{\underline{x}}_{\underline{n}}, \underline{u})].$$

By our premises in (148), on the one hand from (149) we easily deduce

$$(150) \qquad (\forall \underline{x}_1, \ldots, \underline{x}_{\underline{n}}) [\underline{y}_1(\underline{x}_1, \ldots, \underline{x}_{\underline{n}}) \equiv \underline{y}_2(\underline{x}_1, \ldots, \underline{x}_{\underline{n}})]$$

on the basis of (136), and on the other hand from (150) we easily deduce (149) on the basis of (130). So (a) is equivalent to (150), hence to the matrix $\underline{y}_1 = \underline{y}_2$ by AS40.1. Then (148) holds.

Case C. $\underline{t} = (\underline{t}_1, \ldots, \underline{t}_n : \underline{t}_0)$. By $(132)_3$, from our premises in (148), whose conjunction will be denoted by $\underline{p}_{148}$, we deduce (a) $\overline{\underline{y}}_1, \overline{\underline{y}}_2 \in \underline{v}^*_{t\eta}$. By the inductive hypothesis we have

$$(151) \quad |'_{\underline{u}} \bigwedge_{\underline{i}=1}^{2} \underline{I}_{\underline{t}_0} [\overline{\underline{y}}_{\underline{i}}(\overline{\underline{x}}_1, \ldots, \overline{\underline{x}}_{\underline{n}}), \underline{y}_{\underline{i}}(\underline{x}_1, \ldots, \underline{x}_{\underline{n}})] \vdash_{\underline{c}} \underline{y}_1(\underline{x}_1, \ldots, \underline{x}_{\underline{n}}) = \underline{y}_2(\underline{x}_1, \ldots, \underline{x}_{\underline{n}}) \equiv \overline{\underline{y}}_1(\overline{\underline{x}}_1, \ldots, \overline{\underline{x}}_{\underline{n}}) \overset{t_0}{\underset{u}{\stackrel{*}{=}}} \overline{\underline{y}}_2(\overline{\underline{x}}_1, \ldots, \overline{\underline{x}}_{\underline{n}}).$$

Now we add to $\underline{p}_{148}$ the premise (b) $\underline{y}_1 = \underline{y}_2$, which by AS40.2 is equivalent to (b′) $(\forall \underline{x}_1, \ldots, \underline{x}_{\underline{n}}) \underline{y}_1(\underline{x}_1, \ldots, \underline{x}_{\underline{n}}) = \underline{y}_2(\underline{x}_1, \ldots, \underline{x}_{\underline{n}})$, and (c) $\overline{\underline{x}}_{\underline{i}} \in \underline{v}^*_{(t_i\eta)}$ $(\underline{i} = 1, \ldots, \underline{n})$.

By $(134)_2$ and rule $\underline{C}$ with $\underline{x}_{\underline{i}}$, from (c) we obtain (d) $\underline{I}_{\underline{t}}(\overline{\underline{x}}_{\underline{i}}, \underline{x}_{\underline{i}})$ $(\underline{i} = 1, \ldots, \underline{n})$.

By Theor. 60.1, from $\underline{p}_{148}$ and (d) we deduce the premises in (151). Then by (151) the R.H.S. $\underline{r}_{151}$ of (151) holds. From $\underline{r}_{151}$ and (b′) we obtain (e) $\overline{y}_1(\overline{x}_1, \ldots, \overline{x}_n) \overset{*}{=}{}^{t_0}_{\underline{u}} \overline{y}_2(\overline{x}_1, \ldots, \overline{x}_n)$.

By Theor. 33.3 and rule $\underline{G}$ we conclude that $\underline{p}_{148}$, (b) $\vdash_{\underline{c}} (\forall \overline{x}_1, \ldots, \overline{x}_n) [(c) \supset (e)]$, which by Def. 61.1(c′) is equivalent with $\underline{p}_{148}$, (b) $\vdash_{\underline{c}} \overline{y}_1 \overset{*}{=}{}^{t}_{\underline{u}} \overline{y}_2$.

In a similar way, using $(134)_1$ instead of $(134)_2$, we can prove in addition that $\underline{p}_{148}, \overline{y}_1 \overset{*}{=}{}^{t}_{\underline{u}} \overline{y}_2 \vdash_{\underline{c}}$ (b). So by Theor. 33.3 we obtain (148) in case C.

By the principle of induction (148) holds for $\underline{t} \epsilon \tau^{\nu}$.

QED

By comparing the definitions of $=^{t}_{\kappa}$ (in $EL^{\nu+1}$) and $\overset{*}{=}{}^{t}_{\underline{u}}$ (in ML^{ν}) [Defs. 15.1, 61.1], and taking the translation rules $(\underline{T}^*_1)$ to $(\underline{T}^*_7)$ in N57 into account it is clear that the following theorem holds:

THEOR. 61.2. $\vdash (\Delta_1 =^{t}_{\kappa} \Delta_2)^* \equiv {}^\frown \Delta^*_1 \overset{*}{=}{}^{t}_{\underline{u}} \Delta^*_2$ for all terms Δ_1 and Δ_2 of type $\underline{t}(\epsilon\tau^{\nu+1})$ in $EC^{\nu+1}$.

From Def. 61.1 and $(104)_2$ we immediately infer

THEOR. 61.3. $\vdash_{\underline{c}} \Delta_1, \Delta_2 \epsilon \underline{v}^*_{(\underline{t})} \supset (\diamond \Delta_1 \overset{*}{=}{}^{t}_{\underline{u}} \Delta_2 \equiv \underline{N} \Delta_1 \overset{*}{=}{}^{t}_{\underline{u}} \Delta_2)$ in MC^{ν}.

N62. A basic conditioned equivalence between every designator Δ in MC^{ν} and $\Delta^{\eta *}$.

The following equivalence theorem is basic for proving the invariance of the entailment relation asserted in N61.

THEOR. 62.1. We accept results (99) to (102) of rule $\underline{C}$. Furthermore we assume that Δ is a designator in ML^{ν} of type $\underline{t}(\underline{t}\epsilon\overline{\tau}^{\nu})$ and that the distinct variables $\underline{z}_1, \ldots, \underline{z}_r$ of the respective types $\underline{t}_1, \ldots, \underline{t}_r$ are the free variables in Δ and that $\underline{c}_1, \ldots \underline{c}_s$

are the constants in Δ. Let $\tau_{\underline{j}}$ be the type of $\underline{c}_{\underline{j}}$ ($\underline{j} = 1, \ldots, \underline{s}$).

We also denote $\underline{z}_{\underline{i}}^{\eta *}$ by $\overline{\underline{z}}_{\underline{i}}$ ($\underline{i} = 1, \ldots, \underline{r}$) and $\underline{c}_{\underline{j}}^{\eta *}$ by $\overline{\underline{c}}_{\underline{j}}$ ($\underline{j} = 1, \ldots, \underline{s}$), and we set

$$(152) \qquad \underline{w} \equiv_D \bigwedge_{\underline{i}=1}^{\underline{r}} \underline{I}_{\underline{t}_{\underline{i}}} (\overline{\underline{z}}_{\underline{i}}, \underline{z}_{\underline{i}}), \quad \underline{w}' \equiv_D \bigwedge_{\underline{j}=1}^{\underline{s}} \underline{I}_{\tau_{\underline{j}}} (\overline{\underline{c}}_{\underline{j}}, \underline{c}_{\underline{j}}),$$

understanding e.g. that for $\underline{r} = 0$, "$\underline{w}, \underline{p} \vdash_{\underline{c}} \underline{q}$" stands for "$\underline{p} \vdash_{\underline{c}} \underline{q}$." Then

$$(153) \qquad \mid'_{\underline{u}}, \underline{w}, \underline{w}' \vdash_{\underline{c}} \begin{cases} \Delta^{\eta *} \equiv \Delta & \text{for } \underline{t} = 0 \\ \underline{I}_{\underline{t}}(\Delta^{\eta *}, \Delta) & \text{for } \underline{t} \in \tau^{\nu}. \end{cases}$$

In addition, let Δ be either a modally closed matrix or any term. Then

$$(154) \quad \underline{w}, \underline{w}' \vdash_{\underline{c}} \begin{cases} \Delta^{\eta *} \equiv {}^{\cap}\Delta & \text{for } \underline{t} = 0 \\ \underline{I}_{\underline{t}}(\Delta^{\eta *}, \Delta) & \text{for } \underline{t} \in \tau^{\nu}. \end{cases}$$

Proof. First we assume that (153) holds and Δ is either a term or a modally closed matrix. Thence we want to deduce (154). To this end we remark that by Theor. 15.1 on Δ^{η} and rules ($\underline{T}_1^*$) to ($\underline{T}_7^*$) [N57], for $\underline{t} \in \tau^{\nu}$, $\Delta^{\eta *}$ is certainly closed with respect to $\underline{u}$—i.e. κ^*—and for $\underline{t} = 0$, $\Delta^{\eta *}$ is closed with respect to $\underline{u}$ if and only if Δ is modally closed.

So, by our additional assumption on Δ, $\Delta^{\eta *}$ and the R.H.S. $\underline{q}$ of $\vdash_{\underline{c}}$ in (153) for $\underline{t} \in \overline{\tau}^{\nu}$—cf. (1) in N2—are closed with respect to $\underline{u}$. Furthermore, by $(132)_{1,2}$ we have (a) $\underline{w} \equiv \underline{Nw}$ and (b) $\underline{w}' \equiv \underline{Nw}'$. So by Theor. 33.3, rule $\underline{G}$, Theor. 56.3, and AS30.7, (153) implies $\underline{w}, \underline{w}' \vdash_{\underline{c}} (\underline{u}) (\mid'_{\underline{u}} \supset^{\cap} \underline{q})$. As a consequence $\underline{w}, \underline{w}' \vdash_{\underline{c}} (\exists \underline{u}) \mid'_{\underline{u}} \supset^{\cap} \underline{q}$.

Then by $(83')_1$ [in Theor. 56.1] and by (a) and (b) we have $\underline{w}$, $\underline{w}'$ $\vdash_{\underline{c}}$ $\underline{Nq}$. We conclude that if (153) holds, then (154) also holds, provided Δ is either a modally closed matrix or a term.

So it suffices to prove (153) for $\underline{t} \in \overline{\tau}^{\nu}$. For some parts of this proof only outlines will be given for the sake of brevity.

Our proof of (153) is based on rules $(\underline{T}_1)$ to $(\underline{T}_9)$ [N15] of extensional translation $(\Delta \rightarrow \Delta^{\eta})$, on rules $(\underline{T}_1^*)$ to $(\underline{T}_7^*)$ [N57] of star translation $(\Delta^{\eta} \rightarrow \Delta^{\eta *})$, and on the inductive procedure based on the length $\underline{l}$ of Δ. For $\underline{l} = 1$ only the following case can occur:

Case 1. Δ is $\underline{v}_{\underline{tn}}$ or $\underline{c}_{\underline{tn}}$. Then (153) obviously follows from the definitions of $\overline{\underline{x}}_{\underline{i}}$ and $\overline{\underline{c}}_{\underline{j}}$.

Now we consider a positive integer $\underline{l}_1$ and, as inductive hypothesis, we assume that (153) holds for every $\underline{l} < \underline{l}_1$. We want to prove (153) for $\underline{l} = \underline{l}_1$. Then one of the following cases must occur:

Case 2. Δ is $\Delta_1 = \Delta_2$. Then, denoting by θ the common type of the terms Δ_1 and Δ_2, $\Delta^{\eta *}$ is $\Delta_1^{\eta *} \stackrel{\theta}{\underset{\underline{u}}{\equiv}} \Delta_2^{\eta *}$.

We assume $\Vdash_{\underline{u}}'$, $\underline{w}$, and $\underline{w}'$. So by the inductive hypothesis, we deduce $\underline{I}_{\theta}(\Delta_{\underline{i}}^{\eta *}, \overline{\Delta}_{\underline{i}})$ $(\underline{i} = 1, 2)$, whence by (148) and the assumption $\Vdash_{\underline{u}}'$ we easily obtain $\Delta^{\eta *} \equiv \Delta$. So (153) holds in case 2.

Case 3. Δ is the atomic matrix or term $\Delta_{\underline{n}+1}(\Delta_1, \ldots, \Delta_{\underline{n}})$, where $\Delta_1, \ldots, \Delta_{\underline{n}+1}$ have the respective types $\underline{t}_1, \ldots, \underline{t}_{\underline{n}+1}$. Then by rule $(\underline{T}_4)$ in N15 and rule $(\underline{T}_4^*)$ in N57 we have (b) $\Delta^{\eta *}$ $= {}^{\cap}\Delta_{\underline{n}+1}^{\eta *}(\Delta_1^{\eta *}, \ldots, \Delta_{\underline{n}}^{\eta *})$.

The hypothesis of the induction is (c) $\underline{I}_{\underline{t}_{\underline{l}}}(\Delta_{\underline{l}}^{\eta *}, \Delta_{\underline{l}})$ $(\underline{l} = 1, \ldots, \underline{n} + 1)$.

On the basis of Theors. 60.1, 33.3, and rule $\underline{G}$, from (b) and (c) we easily deduce (153) in case 3.

Case 4. Δ is $\sim\Delta_1$ or $\Delta_1 \wedge \Delta_2$, or $(\forall \underline{v}_{tn})\Delta_3$. Then by rules $(\underline{T}_5)$ and $(\underline{T}_6)$ [N15] and $(\underline{T}_5^*)$ [N57], from the assumptions $|'_{\underline{u}}$, $\underline{w}$, and $\underline{w}'$ and our inductive hypothesis we easily deduce

$$(155) \qquad \Delta_{\underline{i}}^{\eta *} \equiv \Delta_{\underline{i}} \; (\underline{i} = 1, 2), \; \underline{I}_t(\underline{v}_{tn}^{\eta *}, \underline{v}_{tn}) \supset (\Delta_3^{\eta *} \equiv^{\cap} \Delta_3).$$

Thence by rules $(\underline{T}_5^*)$ and $(\underline{T}_6^*)$ [N57] and by (134), we easily deduce $\Delta^{\eta *} \equiv \Delta$. So (153) holds in case 4.

Case 5. Δ is $(\imath \underline{x})\Delta_1$. We denote Δ_1 by $\Phi(\underline{x})$, $\underline{x}^{\eta *}$ by $\overline{\underline{x}}$, $\underline{y}^{\eta *}$ by $\overline{\underline{y}}$, and $\underline{y}_1^{\eta *}$ by $\overline{\underline{y}}_1$ where $\underline{y}_1$ is a variable distinct from $\underline{y}$ but having the same type $\underline{t}$ as $\underline{y}$. Furthermore we assume, besides conventions (a) and (b) in N30, condition (c) in N30 on $\Phi(\underline{x})$ and $\Phi(\underline{y})$. We also accept results (99) to (102) of rule $\underline{C}$. Let us now set

$$(156) \qquad \underline{p}_{\underline{u}} \equiv_D [(\exists ! \underline{y}^{\eta}) \Phi(\underline{y})^{\eta}]^* \text{—cf. Def. 15.2,}$$

$$(157) \qquad \underline{q}_{\underline{u}} \equiv_D \underline{p}_{\underline{u}} (\overline{\underline{y}}) [\overline{\underline{y}} \in \underline{v}^*_{(t\eta)} \Phi(\underline{y})^{\eta *} \supset \overline{\underline{x}} \overset{*}{=}\tfrac{t}{\underline{u}} \overline{\underline{y}}] \vee \sim \underline{p}_{\underline{u}} \overline{\underline{x}} \overset{*}{=}\tfrac{t}{\underline{u}} (\underline{a}_t^*)^{\eta *}.$$

Then by rule $(\underline{T}_9)$ [N15] Δ^{η} is $(\imath \underline{x}^{\eta}) \lambda(\underline{x}^{\eta})$ where $\lambda(\underline{x}^{\eta})$ is a certain matrix in $EL^{\nu+1}$ closed with respect to κ. In addition, by rule $(\underline{T}_9)$, by rules $(\underline{T}_6^*)$ and $(\underline{T}_7^*)$ [N57], and by (157) and $(100')_1$ we have

$$(158) \; \vdash_{\underline{c}} \lambda(\underline{x}^{\eta})^* \equiv^{\cap} (\underline{u})(\underline{u} \in \underline{El}' \supset \underline{q}_{\underline{u}}), \quad \vdash_{\underline{c}} \Delta^{\eta *} =^{\cap} (\imath \overline{\underline{x}}) [\overline{\underline{x}} \in \underline{v}^*_{(t\eta)} \lambda(\underline{x}^{\eta})^*].$$

From (148) and Theor. 60.2(II), (III), we obtain

$$(159) \qquad |'_{\underline{u}}, \underline{I}_t(\overline{\underline{x}}, \underline{x}) \vdash_{\underline{c}} \overline{\underline{x}} \overset{*}{=}\tfrac{t}{\underline{u}} (\underline{a}_t^*)^{\eta *} \equiv \underline{x} = \underline{a}_t^*.$$

By the inductive hypothesis (153) on the matrix $\Phi(\underline{y})$—which is a shorter designator than Δ—and by Theor. 61.2, (134), and (148) we can prove

$$(160) \; |'_{\underline{u}}, \underline{I}_t(\overline{\underline{x}}, \underline{x}), \underline{w}, \underline{w}' \vdash_{\underline{c}} (\overline{\underline{y}}) [\overline{\underline{y}} \in \underline{v}^*_{(t\eta)} \Phi(\underline{y})^{\eta *} \supset \overline{\underline{x}} \overset{*}{=}\tfrac{t}{\underline{u}} \overline{\underline{y}}] \equiv (\underline{y}) [\Phi(\underline{y}) \supset \underline{x} = \underline{y}].$$

By (156), Def. 15.2, $(110)_1$, rules $(\underline{T}_5^*)$ and $(\underline{T}_6^*)$ [N57], and Theor. 61.2 we have

$$(161)\quad \vdash \underline{p}_{\underline{u}} \equiv^{\cap} (\exists \underline{\bar{y}}) \{\underline{\bar{y}} \in \underline{v}^*_{(\underline{t}\eta)} \Phi(\underline{y})^{\eta *} (\underline{\bar{y}}_1) [\underline{\bar{y}}_1 \in \underline{v}^*_{(\underline{t}\eta)} \Phi(\underline{y}_1)^{\eta *} \supset \underline{\bar{y}} \overset{*\underline{t}}{\underset{\underline{u}}{=}} \underline{\bar{y}}_1]\}.$$

Thence by the inductive hypothesis on $\Phi(\underline{y})$—i.e. $\underline{w}, \underline{w}'\ \underline{I}_{\underline{t}}(\underline{\bar{y}}, \underline{y}) \vdash_{\underline{c}} \Phi(\underline{y})^{\eta *} \equiv^{\cap} \Phi(\underline{y})$—and on the basis of $(132)_3$, (134), and $(14\bar{8})$ we can prove

$$(162)\quad |'_{\underline{u}}, \underline{w}, \underline{w}' \vdash_{\underline{c}} \underline{p}_{\underline{u}} \equiv (\exists \underline{y}) \{\Phi(\underline{y}) (\underline{y}_1) [\Phi(\underline{y}_1) \supset \underline{y} = \underline{y}_1]\} \equiv (\exists_1 \underline{y}) \Phi(\underline{y})$$

[Theor. 36.1].

From (157), (162), (160), and (159) we easily deduce

$$(163)\quad |'_{\underline{u}}\ \underline{w}, \underline{w}', \underline{I}_{\underline{t}}(\underline{\bar{x}}, \underline{x}) \vdash_{\underline{c}} \underline{q}_{\underline{u}} \equiv (\exists_1 \underline{y}) \Phi(\underline{y}) (\underline{y}) [\Phi(\underline{y}) \supset \underline{x} = \underline{y}] \vee \sim(\exists_1 \underline{y}) \Phi(\underline{y}) \underline{x} = \underline{a}^*_{\underline{t}}.$$

Thence by AS38.1 we have (a) $|'_{\underline{u}}\ \underline{w}\ \underline{w}'\ \underline{I}_{\underline{t}}(\underline{\bar{x}}, \underline{x}) \vdash_{\underline{c}} \underline{q}_{\underline{u}} \equiv \underline{x} = (\imath \underline{x}) \Phi(\underline{x})$, whence—since $\vdash (\imath \underline{x}) \Phi(\underline{x}) = (\imath \underline{x}) \Phi(\underline{\bar{x}})$—by Theor. $3\bar{3}.\bar{3}$, rule $\underline{G}$, and AS38.2 we easily infer

$$(164)\quad |'_{\underline{u}}, \underline{w}, \underline{w}' \vdash_{\underline{c}} \underline{I}_{\underline{t}}(\underline{\bar{x}}, \Delta) \supset \underline{q}_{\underline{u}} \qquad \text{where } \Delta =_{D} (\imath \underline{x}) \Phi(\underline{x}).$$

By (152) and $(132)_{1,2}$ we have (b) $\vdash \Diamond (\underline{w}\ \underline{w}') \equiv \underline{N}(\underline{w}\ \underline{w}')$. Furthermore, (154) holds in case Δ is $\Phi(\underline{y})$ as inductive hypothesis, and in case Δ is an equality matrix [case 2]. Then by $(104)_2$, (161), (157), $(132)_3$, (109), and Theor. 61.3 we easily obtain (c) $\underline{u} \in \underline{El}'\ \underline{w}\ \underline{w}'\ \underline{I}_{\underline{t}}(\underline{\bar{x}}, \underline{x}) \vdash_{\underline{c}} (\Diamond \underline{p}_{\underline{u}} \equiv \underline{Np}_{\underline{u}}) (\Diamond \underline{q}_{\underline{u}} \equiv \underline{Nq}_{\underline{u}})$.

In addition, by Def. 56.2 and $(83')_4$ in Theor. 56.1 we respectively have [Theor. 56.3]

$$(165)\qquad \vdash_{\underline{c}} (\underline{u}) [|'_{\underline{u}} \supset^{\cap} \underline{u} \in \underline{El}'],$$
$$\vdash_{\underline{c}} (\underline{u}) [\underline{u} \in \underline{El}' \supset^{\cap} \Diamond |'_{\underline{u}}].$$

By (165) and $(158)_1$, from $(132)_{1,2}$, $(82')_1$, (b), and (c) we easily deduce [Theor. 56.3]

$$(166) \qquad \underline{I}_{\underline{t}}(\overline{\underline{x}}, \underline{x}), \underline{w}, \underline{w}' \vdash_{\underline{c}} \lambda(\underline{x}^{\eta})^* \equiv^{\cap} (\underline{u})(|'_{\underline{u}} \supset^{\cap} \underline{q}_{\underline{u}}).$$

Since $\underline{x}$ does not occur in $|'_{\underline{u}}$ or $\underline{q}_{\underline{u}}$ or $\lambda(\underline{x}^{\eta})^*$, by $(164)_2$ the relation obtained from (166) by substituting Δ for $\underline{x}$ holds. Furthermore by (b), (c), and $(132)_{1,2}$, from $(164)_1$ we easily infer $\underline{w}, \underline{w}'$ $\vdash (\underline{u})[|'_{\underline{u}} \underline{I}_{\underline{t}}(\overline{\underline{x}}, \Delta) \supset^{\cap} \underline{q}_{\underline{u}}]$. Now by $(132)_3$ we easily see that

$$(167) \qquad \underline{w}, \underline{w}' \vdash_{\underline{c}} (\overline{\underline{x}}) \{\underline{I}_{\underline{t}}(\overline{\underline{x}}, \Delta) \supset \overline{\underline{x}} \in \underline{v}^*_{(t\eta)} \lambda(\underline{x}^{\eta})^*\}.$$

In our semantical system for ML^{ν} there is exactly one QI fulfilling requirements (b) and (c) in N11 for descriptions—cf. Theors. 11.1–3. Furthermore our extensional translation $\Delta \to \Delta^{\eta}$ of ML^{ν} into $EL^{\nu+1}$ is substantially equivalent with our semantical system for ML^{ν}. So—since Δ^{η} is $(\imath \underline{x}^{\eta})\lambda(\underline{x}^{\eta})$ where $\lambda(\underline{x}^{\eta})$ is determined by means of rule $(\underline{T}_9)$ in N15—we have $\vdash (\exists_1 \underline{x}^{\eta})\lambda(\underline{x}^{\eta})$ in $EC^{\nu+1}$.

As an easy consequence based on Theor. 58.3, (b) $[\vdash \diamond(\underline{w}\ \underline{w}') \equiv \underline{N}(\underline{w}\ \underline{w}')]$, and $(110)_2$, we have, in MC^{ν},

$$(168) \qquad \underline{w}\ \underline{w}' \vdash_{\underline{c}} \underline{N}(\exists_1 \overline{\underline{x}})\Psi(\overline{\underline{x}}) \quad \text{where} \quad \Psi(\overline{\underline{x}}) \equiv_D \overline{\underline{x}} \in \underline{v}^*_{(t\eta)} \lambda(\underline{x}^{\eta})^*.$$

We now start with (d) $\underline{w}\ \underline{w}'$. Thence by (152), $(168)_2$, $(104)_2$, and (109) we easily deduce (e) $(\overline{\underline{x}}) \diamond \Psi(\overline{\underline{x}}) \equiv \underline{N}\Psi(\overline{\underline{x}})]$.

By $(11)_2$ and $(14)_{1,2,3}$, from $(168)_1$, (d), and (e) we obtain (f) $\underline{N}(\exists^{\cap}_1 \overline{\underline{x}})\Psi(\overline{\underline{x}})$; and by $(\underline{T}^*_7)$ and $(168)_2$ $\Delta^{\eta *}$ is $(\imath \overline{\underline{x}})\Psi(\overline{\underline{x}})$. So (e), (f), and $(35)_6$ yield (g) $\Psi(\Delta^{\eta *})$.

From $(134)_1$ we have $\vdash_{\underline{c}} (\exists \overline{\underline{x}})\underline{I}_{\underline{t}}(\overline{\underline{x}}, \Delta)$, whence by rule $\underline{C}$ we deduce (h) $\underline{I}_{\underline{t}}(\overline{\underline{x}}, \Delta)$, which together with (167), $(168)_2$, and (d) yields $\Psi(\overline{\underline{x}})$. By $(14)_{1,2,3}$ and Def. 34.2 the matrices $\Psi(\overline{\underline{x}})$, (g), and (f) yield $\overline{\underline{x}} =^{\cap} \Delta^{\eta *}$. Thence by (h) and AS34.4 we obtain $\underline{I}_{\underline{t}}(\Delta^{\eta *}, \Delta)$. We conclude that (153) holds in case 5.

So (153) holds for $\underline{1} = \underline{1}_1$, and by the principle of induction (153) holds for every designator Δ in MC^ν.

QED

N63. Invariance of the entailment relation for the extensional translation of MC^ν into $EC^{\nu+1}$, in both senses.

Now we can reach the main goal of the last two chapters, and we present it in the forthcoming theorem. As a preliminary let us consider a theory $\mathscr{T}$ based on MC^ν, so that the designators in $\mathscr{T}$ belong to ML^ν and the logical axioms of $\mathscr{T}$ are the axioms of MC^ν. Of course, in saying e.g. that $\vdash \underline{p}$ in $\mathscr{T}$, $\underline{p}$ is understood to contain only constants that are defined in $\mathscr{T}$.

By $\overline{\mathscr{T}}^\eta$ we denote that theory based on $EC^{\nu+1}$ whose nonlogical axioms are the extensional translations of the nonlogical axioms of $\mathscr{T}$ and whose definitions are the definitions in $EL^{\nu+1}$ that are equivalent to the extensional translations of the definitions in $\mathscr{T}$ (and have the corresponding forms)—cf. fn. 15 in Memoir 2.

So $\overline{\mathscr{T}}^\eta$ differs from the theory $\mathscr{T}^\eta$—described at the outset of the proof of Theor. 38.1—in that the logical axioms of $\overline{\mathscr{T}}^\eta$ are the axioms of $EC^{\nu+1}$, while the logical axioms of $\mathscr{T}^\eta$ are the extensional translations of the axioms of MC^ν. However these translations are theorems in $EC^{\nu+1}$ [Theor. 30.1]. So $\vdash \underline{p}^\eta$ in $\mathscr{T}^\eta$ implies $\vdash \underline{p}^\eta$ in $\overline{\mathscr{T}}^\eta$.

Obviously if $\mathscr{T}$ is MC^ν, then $\overline{\mathscr{T}}^\eta$ is $EC^{\nu+1}$.

THEOR. 63.1. We assume that the theory $\mathscr{T}$ is based on MC^ν, contains the total closures of its (nonlogical) axioms, and has only definitions of the form $\underline{c}_i = {}^\frown\underline{\Delta}_i$ (where the constant $\underline{c}_i$ has an odd index).

Then $\underline{p}_1, \ldots, \underline{p}_{\underline{n}} \vdash \underline{p}_0$ in $\mathscr{T}$ (or MC^{ν}) if and only if $\underline{p}_1^{\eta}, \ldots, \underline{p}_{\underline{n}}^{\eta} \vdash \underline{p}_0^{\eta}$ in the theory $\bar{\mathscr{T}}^{\eta}$ mentioned above (or in $EC^{\nu+1}$ respectively).

Proof. We assume that $\underline{p}_1^{\eta}, \ldots, \underline{p}_{\underline{n}}^{\eta} \vdash \underline{p}_0^{\eta}$ in $\bar{\mathscr{T}}^{\eta}$ and we want to deduce that $\underline{p}_1, \ldots, \underline{p}_{\underline{n}} \vdash \underline{p}_0$ in $\mathscr{T}$. This suffices to prove the theorem, since the converse is an obvious consequence of Theor. 31.2.

Every definition in $\mathscr{T}$ has the form (a) $\underline{c}_{\underline{i}} = {}^{\cap}\Delta_{\underline{i}}$, and by the definition of $\bar{\mathscr{T}}^{\eta}$ the matrix $(\underline{c}_{\underline{i}} = {}^{\cap}\Delta_{\underline{i}})^{\eta}$ [N15] is equivalent in $EC^{\nu+1}$—cf. fn. 15 in Memoir 2—to its corresponding definition $\underline{c}_{\underline{i}}^{\eta} = \Delta_{\underline{i}}^{\eta}$ in $\bar{\mathscr{T}}^{\eta}$.

In addition, by a hypothesis on $\mathscr{T}$ and rules $(\underline{T}_1)$ to $(\underline{T}_9)$ [N15], the closures of the axioms of $\bar{\mathscr{T}}^{\eta}$ are axioms of $\bar{\mathscr{T}}^{\eta}$—so that the analogues of conditions (a) and (b) at the beginning of N58 hold for $\bar{\mathscr{T}}^{\eta}$.

Now let $\underline{p}$ be the total closure of $\underline{p}_1 \wedge \ldots \wedge \underline{p}_{\underline{n}} \supset \underline{p}_0$ for $\underline{n} > 0$ and the total closure of $\underline{p}_0$ for $\underline{n} = 0$. So by $(\underline{T}_1)$ to $(\underline{T}_9)$ [N15], $\underline{p}^{\eta}$ is the closure of $\underline{p}_1^{\eta} \wedge \ldots \wedge \underline{p}_{\underline{n}}^{\eta} \supset \underline{p}_0^{\eta}$ for $\underline{n} > 0$ and the closure of $\underline{p}_0^{\eta}$ for $\underline{n} = 0$. Then by our first assumption and the last two assertions on $\bar{\mathscr{T}}^{\eta}$, $\vdash \underline{p}^{\eta}$ in $\bar{\mathscr{T}}^{\eta}$.

Let $\underline{q}$ be the sentence without constants, obtained from $\underline{p}$ by replacing $\underline{c}_{\underline{i}}$ with $\Delta_{\underline{i}}$ ($\underline{i} = \ldots, 2, 1$) in an obvious way. Since $\vdash \underline{c}_{\underline{i}} = {}^{\cap}\Delta_{\underline{i}}$ in $\mathscr{T}$ ($\underline{i} = 1, 2, \ldots$), we have $\vdash \underline{p} \equiv {}^{\cap}\underline{q}$ in $\mathscr{T}$.

By rules $(\underline{T}_1)$ to $(\underline{T}_9)$ [N15] $\underline{q}^{\eta}$ is the matrix without constants obtained from $\underline{p}^{\eta}$ by replacing $\underline{c}_{\underline{i}}^{\eta}$ with $\Delta_{\underline{i}}^{\eta}$ ($\underline{i} = \ldots, 2, 1$). Since $\underline{c}_{\underline{i}}^{\eta} = \Delta_{\underline{i}}^{\eta}$ is a definition in $\bar{\mathscr{T}}^{\eta}$ ($\underline{i} = 1, 2, \ldots$), we also have $\vdash \underline{p}^{\eta} \equiv \underline{q}^{\eta}$, besides $\vdash \underline{p}^{\eta}$, in $\bar{\mathscr{T}}^{\eta}$. So we have $\vdash \underline{q}^{\eta}$ in $\bar{\mathscr{T}}^{\eta}$.

Since the analogues for $\bar{\mathscr{T}}^{\eta}$ of conditions (a) and (b) in N58 hold and $\underline{q}^{\eta}$ is a closed theorem in $\bar{\mathscr{T}}^{\eta}$, by Theor. 58.3 $\vdash_{\underline{c}} \underline{q}^{\eta *}$ in the star translation $\mathscr{T}^{\eta *}$ of $\mathscr{T}^{\eta}$, where $\vdash_{\underline{c}}$ (defined before Theor. 58.1) is used instead of $\vdash$ in that results (99) to (102) of rule $\underline{C}$ are accepted. Of course in a proof of $\underline{q}^{\eta *}$ only a finite number of these results of rule $\underline{C}$ are involved. Furthermore, among them only $\nu+1$ are essential, precisely results (99) and (100), in accordance with $(128)_1$. The other results—i.e. (101) and (102) for every nonindividual term type—can be replaced by metalinguistic definitions throughout the whole of our theory.

Since $\underline{q}$ is a modally closed sentence without constants, by Theor. 62.1—cf. (152), (154)—we have $\vdash_{\underline{c}} \underline{q}^{\eta *} \equiv {}^{\frown}\underline{q}$ in MC^{ν}, where $\vdash_{\underline{c}}$ is related with results (99) to (102) of rule $\underline{C}$ as well as in the assertion that $\vdash_{\underline{c}} \underline{q}^{\eta *}$. So we conclude that $\vdash_{\underline{c}} \underline{q}$ in $\mathscr{T}^{\eta *}$, and more precisely a proof of $\underline{q}$ can be found in $\mathscr{T}^{\eta *}$ where rule $\underline{C}$ is applied with some variables having odd indices to obtain a finite number of results (99) to (102). Furthermore the logical and nonlogical axioms of $\mathscr{T}$ imply those of $\mathscr{T}^{\eta *}$ by Theors. 58.2 and 62.1 respectively (and by Theor. 31.2).

Let us add that $\underline{q}$ is totally closed and without constants—for $\underline{p}$ is so. Then by Theor. 33.1 we have $\vdash \underline{q}$ in $\mathscr{T}$. Thence, remembering that $\vdash \underline{p} \equiv \underline{q}$ in $\mathscr{T}$, we deduce that $\vdash \underline{p}$ in $\mathscr{T}$.

As a consequence, by the definition of $\underline{p}$ we conclude that $\underline{p}_1, \ldots, \underline{p}_n \vdash \underline{p}_0$ in $\mathscr{T}$.

QED

N64. Relative completeness for MC^{ν}.

We want to show intuitively the meaning of Theor. 63.1 in connection with the question of completeness for the modal calculus MC^{ν} with respect to our semantical systems for ML^{ν} determined in Ns8, 11.

To this end let us remember that—as is especially evident from Theor. 16.1—our semantical system for ML^ν constructed in chapter 2 is virtually equivalent to our extensional translation ($\Delta \rightarrow \Delta^\eta$) of ML^ν into $EL^{\nu+1}$ [N15]. More specifically we mean the following: Let $\underline{s}$ be an arbitrary modally closed sentence in ML^ν. Then the condition that $\vdash \underline{s}^\eta$ in $EC^{\nu+1}$ is in effect equivalent with proving in our semantical metalanguage that $\underline{s}$ is $\underline{L}$-true ($\Vdash \underline{s}$) [Def. 9.5] for every choice of the infinite domains $\underline{D}_1, \ldots, \underline{D}_\nu$ and Γ [N6] on which our semantical system depends. Of course in asserting this we understand that in our metalanguage only the customary principles of logic can be used and that these principles are just the analogues (for this metalanguage) of the axioms of $EC^{\nu+1}$.

In addition, by Theor. 63.1, $\vdash \underline{s}$ in MC^ν if and only if $\vdash \underline{s}^\eta$ in $EC^{\nu+1}$. As a consequence $\vdash \underline{s}$ in MC^ν where $\underline{s}$ is a modally closed sentence in ML^ν if and only if we can prove in our semantical metalanguage the $\underline{L}$-truth of $\underline{s}$ ($\Vdash \underline{s}$) for an arbitrary choice of the infinite domains $\underline{D}_1, \ldots, \underline{D}_\nu$ and Γ [N6].

This assertion enables us to conclude intuitively that <u>the modal calculus MC^ν is (relatively) complete with respect to our extensional semantical system for ML^ν.</u>

We consider this completeness to be relative on the basis of the following fact: Since MC^ν embodies a theory of natural numbers [Ns27, 45], by a well-known theorem by Gödel there are infinitely many ways of choosing $\underline{s}$ for which $\underline{s}$ is undecidable in MC^ν, hence $\underline{s}^\eta$ is undecidable in $EC^{\nu+1}$ (which is equivalent to saying that either $\Vdash \underline{s}$ or $\Vdash \sim\underline{s}$ cannot be proved in our extensional semantical metalanguage). So (as an inference in our customary two-valued logic) $\underline{s}$ can be so chosen that $\underline{s}$ is $\underline{L}$-true in our semantical system for ML^ν; but on the one hand this cannot be proved in our extensional semantical metalanguage and on the other hand s is not provable in MC^{ν}.

Observe that axiom AS25.1 (which says that the number of individuals [at each type] is the same in all possible cases) and especially the strong axiom AS12.19 (which in effect asserts the existence of a modally constant attribute F which happens to equal an arbitrarily chosen attribute) have an interesting role in connection with the relative completeness of MC^{ν}. Indeed AS25.1 is independent of the other axioms of MC^{ν}, and it seems obvious that the same can be said of the strong axiom AS12.19, so that it is worthwhile presupposing this independence throughout the sequel. Furthermore ASs12.19 and 25.1 are the only axioms of MC^{ν} substantially different from the axioms of the common extensional and modal calculi. So it is natural to ask whether we can find another new axiom in ML^{ν} whose L-truth can be proved in our semantical metalanguage (so that it is not related to Gödel's undecidability theorem). By Theor. 63.1—that of relative completeness for MC^{ν}—the answer is no.

In N25 we introduced the new quasi intensions (QI′s) and showed that in connection with them AS25.1 does not hold while all other axioms of MC^{ν} do. Of course extensional translation can be redefined in such a way that it characterizes the new semantical system for ML^{ν} (based on the QI′s) just as extensional translation, according to its definition in N15, characterizes our semantical system for ML^{ν} based on the QIs.

In accordance with some earlier assertions I think that the theory developed in the last two chapters can be adapted to the new quasi intensions. Then we can prove an analogue of Theor. 63.1, which asserts that the modal calculus MC'^{ν} obtained from MC^{ν} by omitting (the only) AS25.1 is relatively complete with respect to the new semantical system for ML^{ν} based on the QI's and considered in N25.

As a consequence the strong modal axiom AS12.19 is the only axiom of MC'^{ν} that is substantially different from the axioms of the common extensional and modal calculi and is independent of the other axioms of MC'^{ν} whose L-truth in the new semantical system for ML^{ν} (based on the QI's) can be proved in our extensional semantical language for every admissible choice of the QI's.

13
The Nonrival Character of Various Modal Systems According to a Suggestion by Lemmon

N65. Introduction

These remarks are related to a suggestion made by Lemmon in 1959 [22] about how to show that various systems of modal logic are not rivals. A method for doing this in accordance with Lemmon's suggestion would consist in correlating the use of L (necessary) in each of these systems with the use of N (necessary) "in some established sphere of discourse"—cf. Hughes and Creswell, [18], pp. 79–80.

In order to carry out this suggestion from a general point of view, we take this sphere of discourse to be MC^{ν}; furthermore we consider Kripke's semantical theory [21] for Fey's modal system T. This theory, in which both validity and completeness hold for T, is based on the concept of an accessibility relation

R: a world (or Γ-case) u is accessible to a world w if and only if wRu.

This relation is reflexive. If it is also transitive or symmetric or an equivalence, then in the resulting semantical theory we can prove validity and completeness for S4, B (i.e. the Brouwer system), and S5 respectively. As far as S5 is concerned, R can be identified with the universal binary relation—cf. [18], p. 67.

We characterize the arbitrary accessibility relation R by means of a conceivability property, α, and define in ML^{ν} various classes, $Conc_T$ to $Conc_{S5}$, of these properties in correspondence with the kinds of accessibility relations used by Kripke in connection with the modal systems T, S4, B, and S5 [N66]. Furthermore for every conceivability property α, we define in ML^{ν} the corresponding necessity operator L_{α} and prove that if α belongs to any one of the classes $Conc_T$ to $Conc_{S5}$, then L_{α} fulfills the axioms on the necessity operator L in the corresponding modal system (which is T, S4, B, or S5) [Ns67, 68]. These results allow us to carry out Lemmon's suggestion from a general point of view [N70].

We also show that the Barcan formula holds for L_{α} if $\Vdash \alpha \in Conc_T$, and we explain why we do not accept the known objections to this formula, at least in connection with the modal calculus MC^{ν} [N69]. We also make some related remarks on the converse of this formula in N69.

I find it interesting that when $\alpha \in Conc_T$ holds, (1) the analogues for L_{α} of all axioms of MC^{ν} except one (substitution) hold, and (2) this exception pushes us naturally to consider certain problems (which are open as far as I know) concerning the modal systems that are weaker than S5 in connection with a general theory of identity, descriptions, and attributes (and classes).

N66. Definition in ML^{ν} of certain classes of conceivability properties that are possible analogues for various kinds of Kripke's acceptability relations.

Let wRu hold where R is any accessibility relation [N65]. Then u is said to be conceivable by someone living in w—cf. [18], p. 77. Briefly we shall say that u is conceivable in w. In addition, in case w is the actual world we shall simply say that u is conceivable.

Thus instead of considering T-, S4-, and S5-accessibility relations, on the one hand we shall consider T-, S4- and S5-conceivability properties. On the other hand we remember that the analogues for the semantical theory for ML^{ν} of Kripke's worlds considered above are Γ-cases, and that these cases were schematized within MC^{ν} into the (absolute) elementary possible cases (El s)—cf. Def. 48.1.

In accordance with the preceding considerations we define in MC^{ν} α is a T-conceivability property ($\alpha \in \underline{Conc}_{\underline{T}}$) as follows:

DEF. 66.1. $\alpha \in \underline{Conc}_{\underline{T}} \equiv_D \underline{N}(\alpha \subseteq \underline{El})\ (\underline{u})\ (|_{\underline{u}} \supset {}^{\frown}\underline{u} \in \alpha)$

where $|_{\underline{u}}$ [Def. 48.2] means that the elementary case u takes place.

Incidentally the condition $(\underline{u})\ (|_{\underline{u}} \supset {}^{\frown}\underline{u} \in \alpha)$ is the analogue for α of the reflexivity condition for R.

We define in MC^{ν} α is an S4-conceivability property by means of

DEF. 66.2. $\alpha \in \underline{Conc}_{S4} \equiv_D \alpha \in \underline{Conc}_{\underline{T}}\ (\forall \underline{u}, \underline{v})\ [\underline{u} \in \alpha \diamond (|_{\underline{u}}\underline{v} \in \alpha) \supset {}^{\frown} \underline{v} \in \alpha]$.

Incidentally the condition $(\forall \underline{u}, \underline{v})$ [. . .] in Def. 66.2 is the analogue for α of the transitivity condition for R.

It is useful to take the liberty of extending the above terminology and to define α is a B-conceivability property as follows:

DEF. 66.3. $\alpha \in \underline{\text{Conc}}_{\underline{B}} =_D \alpha \in \underline{\text{Conc}}_{\underline{T}} (\forall \underline{u}, \underline{v}) [\mathsf{I}_{\underline{u}} \underline{v} \in \alpha \supset^{\cap} \diamond (\mathsf{I}_{\underline{v}} \underline{u} \in \alpha)]$.

Incidentally the condition $(\forall \underline{u}, \underline{v}) [\ldots]$ in Def. 66.3 is the analogue for α of the symmetry condition for $\underline{R}$.

Now we can define the S5-conceivability property ($\underline{\text{Conc}}_{S5}$) simply as follows:

DEF. 66.4. $\underline{\text{Conc}}_{S5} =_D \underline{\text{Conc}}_{S4} \cap \underline{\text{Conc}}_{\underline{B}}$.

Incidentally from Defs. 66.1–4, 13.2, and $(82)_1$ in N48 we easily deduce (in MC^{ν})

(169) $\vdash \underline{\text{Conc} \in \text{MConst}}$ where $\underline{\text{Conc}}$ stands for any of $\underline{\text{Conc}}_{\underline{T}}$ to $\underline{\text{Conc}}_{S5}$.

Instead $\underline{\text{Conc}} \in \underline{\text{MSep}}$ [Def. 18.7] is false. Hence so is $\underline{\text{Conc}} \in \underline{\text{Abs}}$ [Def. 18.8].

From Defs. 66.2 and 3 we easily deduce

(170) $\vdash \underline{\text{Conc}}_{S4} \subset \underline{\text{Conc}}_{\underline{T}}, \quad \vdash \underline{\text{Conc}}_{\underline{B}} \subset \underline{\text{Conc}}_{\underline{T}}$ (in MC^{ν}).

Furthermore Def. 66.1 yields $\alpha \in \underline{\text{Conc}}_{\underline{T}} \supset^{\cap} \alpha \subseteq \underline{\text{El}}$, hence

(171) $\vdash \alpha \in \underline{\text{Conc}}_{\underline{T}} \wedge \underline{u} \in \alpha \supset {}^{\cap}\underline{u} \in \underline{\text{El}}$.

N67. The necessity operator L_{α} corresponding to the conceivability property α. Cases connected with Fey's theory T and S4.

For the sake of simplicity we define the α-necessity operator $\underline{L}_{\alpha}$ — for every property α of the same logical type as El.

DEF. 67.1. $\underline{L}_{\alpha} \underline{p} =_D (\underline{v}) [\underline{v} \in \alpha \supset \diamond (\mathsf{I}_{\underline{v}} \underline{p})]$

where $\underline{v}$ is the first variable that makes $\mathsf{I}_{\underline{v}}$ well-formed and has no free occurrences in the matrix $\underline{p}$.

This abbreviating definition is interesting only when $\alpha \in \underline{\text{Conc}}_{\underline{T}}$ holds. In this case $\underline{L}_{\alpha} \underline{p}$ is generally a contingent proposition, and intuitively Def. 67.1 says that $\underline{L}_\alpha \underline{p}$ holds in the elementary possible case $\underline{u}$ iff it holds in every $\underline{v}$ of these cases, which is conceivable in $\underline{u}$ ($\underline{v}\epsilon\alpha$).

We now show that if α is a $\underline{T}$-conceivability property, then $\underline{L}_\alpha$ fulfills the axioms for the necessity operator $\underline{L}$ in Fey's system $\underline{T}$—cf. [18], p. 31.

THEOR. 67.1. For all matrices $\underline{p}$ and $\underline{q}$ in ML^ν (possibly containing free occurrences of the variable α) we have

(I) $\alpha \epsilon \underline{\mathrm{Conc}}_{\underline{T}} \vdash \underline{L}_\alpha \underline{p} \supset \underline{p}$,

(II) $\alpha \epsilon \underline{\mathrm{Conc}}_{\underline{T}} \vdash \underline{L}_\alpha(\underline{p} \supset \underline{q}) \supset (\underline{L}_\alpha \underline{p} \supset \underline{L}_\alpha \underline{q})$.

Proof. (I) Start with (a) $\alpha \epsilon \underline{\mathrm{Conc}}_{\underline{T}}$ [Def. 66.1] and $\underline{L}_\alpha \underline{p}$.

From $(83)_1$ in N48 and rule C with $\underline{u}$ we have $|_{\underline{u}}$, where $\underline{u}$ is supposed not to occur free in $\underline{p}$.

By Def. 66.1, (a) and $|_{\underline{u}}$ yield $\underline{u}\epsilon\alpha$. This, $|_{\underline{u}}$, and Def. 67.1 yield (b) $\diamond (|_{\underline{u}}\underline{p})$. From this and $(\overline{89})$ in N49 we deduce $|_{\underline{u}} \supset {}^\cap\underline{p}$, which together with $|_{\underline{u}}$ yields $\underline{p}$.

Since $\underline{u}$ is not free in $\underline{p}$ or (a) $\wedge \underline{L}_\alpha \underline{p}$, we conclude [Theor. 33.1] that (a), $\underline{L}_\alpha \underline{p} \vdash \underline{p}$, so that by Theor. 33.2 (I) holds.

(II) We start with (a), (c) $\underline{L}_\alpha(\underline{p} \supset \underline{q})$, $\underline{L}_\alpha \underline{p}$, and (d) $\underline{v}\epsilon\alpha$, where $\underline{v}$ does not occur free in $\underline{p} \wedge \underline{q}$. From (c), $\underline{L}_\alpha \underline{p}$, (d), and Def. 67.1 we deduce (e) $\diamond [|_{\underline{v}}(\underline{p} \supset \underline{q}) \wedge \diamond (|_{\underline{v}}\underline{p})]$.

By $(\overline{89}) \vdash \diamond (|_{\underline{v}}\underline{p}) \supset {}^\cap(|_{\underline{v}} \supset {}^\cap \underline{p})$ so that from (e) we deduce $\diamond [|_{\underline{v}}(\underline{p}\supset\underline{q}) \wedge (|_{\underline{v}} \supset \underline{p})]$, which yields $\diamond [|_{\underline{v}}\underline{p}(\underline{p}\supset\underline{q})]$. Thence we have $\diamond (|_{\underline{v}}\underline{q})$.

We conclude that (a), (c), $\underline{L}_\alpha \underline{p}$, (d) $\vdash \diamond (|_{\underline{v}}\underline{q})$, so that by Theors. 33.2 and 32.2 we have (a), (c), $\underline{L}_\alpha \underline{p} \vdash (\underline{v}) [\underline{v}\epsilon\alpha \supset \diamond (|_{\underline{v}}\underline{q})]$. Thence by Def. 67.1 and two uses of Theor. 33.2 we have (II).

QED

Now we want to show that if α is an S4-conceivability property, then α fulfills the axioms for $\underline{L}$ in S4—cf. [18], p. 46. To this end it suffices to remember (170) and Theor. 67.1 and to prove the following:

THEOR. 67.2. $\alpha \in \underline{\mathrm{Conc}}_{S4} \vdash \underline{L}_{\alpha}\underline{p} \supset \underline{L}_{\alpha}\underline{L}_{\alpha}\underline{p}$ for every matrix in ML^{ν}.

Proof. We start with (a) $\alpha \in \underline{\mathrm{Conc}}_{S4}$, $\underline{L}_{\alpha}\underline{p}$, (b) $\underline{u} \in \alpha$, and (c) $\diamond(\mathsf{I}_{\underline{u}}\underline{v} \in \alpha)$, where $\underline{u}$ and $\underline{v}$ are distinct variables which make $\mathsf{I}_{\underline{u}} \wedge \mathsf{I}_{\underline{v}}$ well-formed and which do not occur free in $\underline{p}$.

From (a), (b), (c), and Def. 66.2 we deduce $\underline{v} \in \alpha$. This, $\underline{L}_{\alpha}\underline{p}$, and Def. 67.1 yield (d) $\diamond(\mathsf{I}_{\underline{v}}\underline{p})$.

By Theor. 33.2 we conclude that (a), $\underline{L}_{\alpha}\underline{p}$, (b) $\vdash$ (c) $\supset$ (d), so that by Theor. 32.2,

(172) (a), $\underline{L}_{\alpha}\underline{p}$, $\underline{u} \in \alpha \vdash \underline{q}$ where $\underline{q} \equiv_D (\underline{v})[\diamond(\mathsf{I}_{\underline{u}}\underline{v} \in \alpha) \supset \diamond(\mathsf{I}_{\underline{v}}\underline{p})]$.

The remaining part of the proof can be considered as a proof of Theor. 67.2 where the result (172) is used as a lemma. Now we begin again, but with only (a), $\underline{L}_{\alpha}\underline{p}$, and (b) $\underline{u} \in \alpha$.

From (a), (b), and $\underline{L}_{\alpha}\underline{p}$ we deduce $\underline{q}$ by (172), and $\diamond(\mathsf{I}_{\underline{u}}\underline{p})$ by Def. 67.1.

Thence we deduce $\diamond \mathsf{I}_{\underline{u}}$ and also $\diamond[\mathsf{I}_{\underline{u}}(\underline{u} \in \alpha \vee \underline{u} \notin \alpha)]$. By $(\overline{92})$ in Theor. 49.3, for $\phi(\underline{p}_1, \ldots, \underline{p}_n)$ equal to $\underline{p}_1 \vee \underline{p}_2$ this yields (e) $\diamond(\mathsf{I}_{\underline{u}}\underline{v} \in \alpha) \vee \diamond(\mathsf{I}_{\underline{u}}\underline{v} \notin \alpha)$ and (e′) $\underline{u} \in \underline{\mathrm{El}}$.

From $\underline{q}$—cf. $(\overline{172})_2$—we easily deduce $\diamond(\mathsf{I}_{\underline{u}}\underline{v} \in \alpha) \supset \diamond[\mathsf{I}_{\underline{u}} \diamond(\mathsf{I}_{\underline{v}}\underline{p})]$, which yields (f) $\diamond(\mathsf{I}_{\underline{v}}\underline{v} \in \alpha) \supset \diamond\{\mathsf{I}_{\underline{u}}[\underline{v} \in \alpha \supset \diamond(\mathsf{I}_{\underline{v}}\underline{p})]\}$.

Furthermore obviously $\vdash \diamond(\mathsf{I}_{\underline{u}}\underline{v} \notin \alpha) \supset \diamond\{\mathsf{I}_{\underline{u}}[\underline{v} \in \alpha \supset \diamond(\mathsf{I}_{\underline{v}}\underline{p})]\}$.

This, (f), and (e) yield $\diamond\{\mathsf{I}_{\underline{u}}[\underline{v} \in \alpha \supset \diamond(\mathsf{I}_{\underline{v}}\underline{p})]\}$. We conclude that

(172′) (a), $\underline{L}_{\alpha}\underline{p}$, $\underline{u} \in \alpha \vdash \diamond(\mathsf{I}_{\underline{u}}\underline{r})$ where $\underline{r} \equiv_D \underline{v} \in \alpha \supset \diamond(\mathsf{I}_{\underline{v}}\underline{p})$.

Since $\underline{v}$ does not occur free in (a), $\underline{L}_{\alpha}\underline{p}$, or $\underline{u}\in\alpha$, and rule $\underline{C}$ has not been used, by rule $\underline{G}$ we deduce $(\underline{v}) \diamond (\mathord{\mid}_{\underline{u}}\underline{r})$ (from (a), $\underline{L}_{\alpha}\underline{p}$, and $\underline{u}\in\alpha$). By $(\overline{92})$ in Theor. 49.3 for $\phi(\underline{r})$ equal to $(\underline{v})\underline{r}$, this and (e) yield $\diamond [\mathord{\mid}_{\underline{u}}(\underline{v})\underline{r}]$, which by Def. 67.1 is $\diamond (\mathord{\mid}_{\underline{u}}\underline{L}_{\alpha}\underline{p})$—cf. (172′).

By Theor. 33.2 we conclude that (a), $\underline{L}_{\alpha}\underline{p} \vdash \underline{u}\in\alpha \supset \diamond (\mathord{\mid}_{\underline{u}}\underline{L}_{\alpha}\underline{p})$. Hence by Theor. 32.2, (a), $\underline{L}_{\alpha}\underline{p} \vdash (\underline{u})[\underline{u}\in\alpha \supset \diamond (\mathord{\mid}_{\underline{u}}\underline{L}_{\alpha}\underline{p})]$.

By Def. 67.1 and Theor. 33.2 this yields Theor. 67.2.

QED

N68. The α-possibility operator $\underline{M}_{\alpha}$. $\underline{L}_{\alpha}$ in connection with the Brouwerian system B and S5.

As a preliminary we introduce the α-possibility operator $\underline{M}_{\alpha}$:

DEF. 68.1. $\underline{M}_{\alpha}\underline{p} \equiv_{D} \sim\underline{L}_{\alpha} \sim\underline{p}$.

THEOR. 68.1. Let $\underline{p}$ be a matrix in ML^{ν} and let $\underline{v}$ be a variable which makes $\mathord{\mid}_{\underline{v}}$ well formed and which does not occur free in $\underline{p}$. Then

(I) $\underline{L}_{\alpha}\underline{p} \equiv^{\cap} \sim\underline{M}_{\alpha} \sim\underline{p}$

(II) $\alpha\in\underline{Conc}_{\underline{T}} \vdash \underline{M}_{\alpha}\underline{p} \equiv^{\cap} (\exists\underline{u})\, [\underline{u}\in\alpha \diamond (\mathord{\mid}_{\underline{u}}\underline{p})]$.

Proof. (I) Def. 68.1 yields (I).

(II) We start with (a) $\alpha\in\underline{Conc}_{\underline{T}}$. By (171) this yields (b) $(\underline{v})\,(\underline{v}\in\alpha \supset \underline{v}\in\underline{El})$. From Defs. 68.1 and 67.1 we deduce $\underline{M}_{\alpha}\underline{p} \equiv^{\cap} (\exists\underline{u})\, [\underline{u}\in\alpha \sim\diamond (\mathord{\mid}_{\underline{u}} \sim\underline{p})]$.

By $(\overline{92})$ in Theor. 49.3 for $\phi(\underline{p}_1)$ equal to $\sim\underline{p}_1$, this and (b) yield $\underline{M}_{\alpha}\underline{p} \equiv^{\cap} (\exists\underline{u})\, [\underline{u}\in\alpha \diamond (\mathord{\mid}_{\underline{v}}\underline{p})]$.

We conclude that (II) holds.

QED

We now show that if α is a $\underline{B}$-conceivability property [Def. 66.3], then $\underline{L}_{\alpha}$ fulfills the axioms for $\underline{L}$ of $\underline{B}$—cf. [18], p. 57. In order to do this we remember $(170)_2$ and Theor. 67.1, and we prove the following:

THEOR. 68.2. $\alpha \epsilon \underline{\text{Conc}}_{\underline{B}} \vdash \underline{p} \supset \underline{L}_\alpha \underline{M}_\alpha \underline{p}$ for every matrix $\underline{p}$ in ML^ν.

Proof. We start with (a) $\alpha \epsilon \underline{\text{Conc}}_{\underline{B}}$, $\underline{p}$, and (b) $\underline{v} \epsilon \alpha$ where $\underline{v}$ is a variable not occurring free in $\underline{p} \wedge \overline{|}_{\underline{u}}$.

By rule $\underline{C}$, $(83)_1$ in N48 yields $\overline{|}_{\underline{u}}$, where the variable $\underline{u}$ does not occur in $\underline{p}$. From (a), (b), $\overline{|}_{\underline{u}}$, and Def. 66.3 we obtain (c) $\diamond(\overline{|}_{\underline{v}} \underline{u} \epsilon \alpha)$. From $\overline{|}_{\underline{u}}$ and $\underline{p}$ we deduce $\diamond(\overline{|}_{\underline{u}} \underline{p})$, which by (c) yields (d) $\diamond[\overline{|}_{\underline{v}} \underline{u} \epsilon \alpha \diamond (\overline{|}_{\underline{u}} \underline{p})]$.

Thence we deduce $(\exists \underline{u})$ (d), whence $\diamond\{\overline{|}_{\underline{v}} (\exists \underline{u})[\underline{u} \epsilon \alpha \diamond (\overline{|}_{\underline{u}} \underline{p})]\}$. By Theor. 68.1(II) this yields (e) $\diamond(\overline{|}_{\underline{v}} \underline{M}_\alpha \underline{p})$.

By Theor. 33.1 we conclude that (a), $\underline{p}$, (b) $\vdash$ (e). Then by Theors. 33.2 and 32.2, (a), $\underline{p} \vdash (\underline{v})\, [\underline{v} \epsilon \alpha \supset \diamond(\overline{|}_{\underline{v}} \underline{M}_\alpha \underline{p})]$. By Def. 67.1 and Theor. 33.2 this yields Theor. 68.2.

QED

Let us remember that on the one hand the axioms of the Brouwerian system $\underline{B}$ together with those of S4 are equivalent with the axioms of S5. On the other hand Def. 66.4, (170), and Theors. 67.1, 66.2, and 68.2 imply the following theorem, where "$\underline{p} \vdash \underline{q}_1, \ldots, \underline{q}_n$" stands for "$\underline{p} \vdash \underline{q}_1, \ldots, \underline{p} \vdash \underline{q}_n$":

THEOR. 68.3. For arbitrary matrices $\underline{p}$ and $\underline{q}$ in ML^ν

$$\alpha \epsilon \underline{\text{Conc}}_{S5} \vdash \begin{cases} \underline{L}_\alpha \underline{p} \supset \underline{p}, \; \underline{L}_\alpha(\underline{p} \supset \underline{q}) \supset (\underline{L}_\alpha \underline{p} \supset \underline{L}_\alpha \underline{q}), \\ \underline{L}_\alpha \underline{p} \supset \underline{L}_\alpha \underline{L}_\alpha \underline{p}, \; \underline{p} \supset \underline{L}_\alpha \underline{M}_\alpha \underline{p}. \end{cases}$$

We conclude that <u>for every S5-conceivability property α, L_α fulfills the axioms of S5 for L.</u>

N69. On the Barcan formula and its converse.

In the first part of this section we show that if α is a $\underline{T}$-conceivability property, then the Barcan formula—briefly $\underline{BF}$—holds for $\underline{L}_{\alpha}$—see Theor. 69.1 below; in the second part we discuss $\underline{BF}$ and in the third its converse.

THEOR. 69.1. $\alpha \in \underline{\mathrm{Conc}}_{\underline{T}} \vdash (\underline{x})\, \underline{L}_{\alpha} \underline{p} \supset \underline{L}_{\alpha} (\underline{x}) \underline{p}$ for every matrix $\underline{p}$ in ML^{ν}.

Proof. We start with (a) $\alpha \in \underline{\mathrm{Conc}}_{\underline{T}}$, (b) $(\underline{x})\, \underline{L}_{\alpha} \underline{p}$, and (c) $\underline{v} \in \alpha$ where $\underline{v}$ does not occur in $\underline{p}$.

By (171), (a) and (c) yield $\underline{v} \in \underline{\mathrm{El}}$. Thence by $(\overline{92})$ in Theor. 49.3, for $\phi(\underline{p}_1, \ldots, \underline{p}_{\underline{n}})$ equal to $(\underline{x})\underline{p}_1$ we deduce (d) $(\underline{x}) \diamond (\,|_{\underline{v}} \underline{p}) \equiv \diamond [\,|_{\underline{v}} (\underline{x}) \underline{p}]$.

By Def. 67.1, (b) implies $(\forall \underline{x}, \underline{v}) [\underline{v} \in \alpha \supset \diamond (\,|_{\underline{v}} \underline{p})]$, which yields $(\underline{v}) [\underline{v} \in \alpha \supset (\underline{x}) \diamond (\,|_{\underline{v}} \underline{p})]$. From this and (c) we deduce $(\underline{x}) \diamond (\,|_{\underline{v}} \underline{p})$, which by (d) yields (e) $\diamond [\,|_{\underline{v}} (\underline{x}) \underline{p}]$.

By Theor. 33.2 we conclude that (a), (b) $\vdash \underline{v} \in \alpha \supset \diamond [\,|_{\underline{v}} (\underline{x}) \underline{p}]$, so that by Theor. 32.2 and Def. 67.1 (a), (b) $\vdash \underline{L}_{\alpha} (\underline{x}) \underline{p}$.

Thence Theor. 69.1 follows by Theor. 33.2.

QED

As is well known—cf. [18], p. 143—the converse of $\underline{BF}$ follows from the axioms of $\underline{T}$ and those of the $\underline{LPC}$, or the lower predicate calculus. In accordance with this, under the condition $\alpha \in \mathrm{Conc}_{\underline{T}}$, $(\underline{x})$ and L_{α} "commute" in MC^{ν}.

Let us add that in connection with MC^{ν} the known objections to $\underline{BF}$—cf. [18], p. 170—do not hold, in my opinion, for the following reasons. They are based on sentences of ordinary language that on the one hand are false and on the other hand are usually schematized into instances of $\underline{BF}$, or briefly into $\underline{BF}$s. These sentences implicitly include quasi-absolute concepts [N24] and they can be satisfactorily schematized within ML^{ν}, not into BFs (where N or L_{α} is used) but into matrices that are similar to BFs and which will be called generalized BFs. Lastly, generalized BFs are not generally L-true in (the semantical system for) ML^{ν} [N9]: hence they are not provable in MC^{ν}.

Let us illustrate the satisfactory situation just described by means of an example based on the following sentence of ordinary language

(a) If every watch (existing at the instant τ) cannot be damaged by water, then necessarily every watch (existing at τ) is not damaged by water.

This sentence is false because it is possible to make the antecedent in (a) true by destroying all nonwaterproof watches. Furthermore the consequent in (a) is obviously false. Hence (a) is false.

In order to translate (a) into ML^{ν} we state that a constant in ML^{ν}, W, denotes the quasi-absolute (proper) concept of a watch (existing at the instant τ), and that another, DW, denotes the extensional property of being damaged by water (at some time).

We assume—cf. Defs. 24.3, 6.11:

$$(173) \qquad \vdash W \in QAbs, \quad \vdash \sim a^* \in W, \quad \vdash DW \in Ext.$$

Incidentally the concept W is proper in that $(173)_2$ holds.

Following is a correct translation of (a) into ML^{ν}:

$$(b) \quad (x)\,[x \in W \supset L_{\alpha}(\sim x \in DW \vee x = a^*)] \supset L_{\alpha}(x)\,(x \in W \supset \sim x \in DW \, \widetilde{\vee\, x = a^*})$$

where by $(173)_2$ the part of (b) under the sign $\frown$ can be deleted (and where $\vee$ is more cohesive than $\supset$).

Incidentally, following is a translation of (b) into ordinary language more straightforward than (a):

(a′) If every watch is not damaged by water in any possible case where it exists, then necessarily no watch is damaged by water.

Obviously (b) is false for a suitable model $\mathcal{M}$ for ML^{ν} [N6], which mirrors the intuitive meanings of $\underline{W}$ and $\underline{DW}$ explained above. Hence (b) cannot be proved in the calculus MC^{ν} supplemented with axioms (173).

This conclusion is essentially related to the fact that (b) is not an ordinary $\underline{BF}$. However we can present (b) as a generalized $\underline{BF}$ as follows. We state the following conventions, where $\underline{x}$ is a variable not occurring free in the predicate α, and where $\underline{p}$ and $\underline{q}$ are sentences:

(174) $$(\forall \underline{x}\underline{q})\underline{p} \equiv_D (\underline{x})(\underline{q} \supset \underline{p}), \quad (\forall \underline{x} \in \alpha)\underline{p} \equiv_D (\underline{x})(\underline{x} \in \alpha \supset \underline{p}).$$

Furthermore we introduce in ML^{ν} the extensional predicate $\underline{NDW'}$, which holds for $\underline{x}$ iff either $\underline{x}$ is not damaged by water or $\underline{x}$ does not exist:

(175) $$\underline{x} \in \underline{NDW'} \equiv_D \sim \underline{x} \in \underline{DW} \vee \underline{x} = \underline{a}^*.$$

By $(174)_2$ and (175) we can turn (b) into

(b′) $$(\forall \underline{x} \in \underline{W})\, \underline{L}_{\alpha}\, \underline{x} \in \underline{NDW'} \supset \underline{L}_{\alpha}(\forall \underline{x} \in \underline{W})\, \underline{x} \in \underline{NDW'},$$

which looks like a $\underline{BF}$ and may be called a generalized $\underline{BF}$.

Let us add that (b′) is not an ordinary but a generalized $\underline{BF}$ essentially in that in (b′) the universally quantified variable $\underline{x}$ is restricted and this occurs by means of a quasi-absolute attribute, $\underline{W}$.

Let us now identify the universe of discourse with the set of watches (existing at τ). Then instead of "watch (. . .)" we can say "thing." However it is important to remark that "thing" is to be used as expressing a quasi-absolute concept.

It would be in accordance with common practice to schematize (a) into something like

(b″) $$(\underline{x})\, \underline{L}_{\alpha}\underline{x} \in \underline{NDW}' \supset \underline{L}_{\alpha}(\underline{x})\underline{x} \in \underline{NDW}'$$

in that $(\underline{x})\underline{p}$ may look like a schematization of "$\underline{p}$ holds for every thing $\underline{x}$." In some formal languages this use of $(\underline{x})\underline{p}$ is correct, but it is not so in ML^{ν} where the values of variables are individual concepts. Hence not even for thing = watch (. . .) can (b″) be accepted in ML^{ν}; it must be replaced by (b) where, besides $(173)_{1,2}$, $\underline{W}$ fulfills the axiom

(173′) $$(\underline{x})\underline{x} \in \underline{W}^{(\underline{e})}$$ [Def. 18.9],

which is $\underline{L}$-true in ML^{ν} provided we assume $\underline{W}$ to be denumerable, for the sake of simplicity. (This assumption would be superfluous if suitable semantics hinted at in N25 were used.)

In connection with more general choices of the universe of discourse, the concept of thing is to be considered as the union of several quasi-absolute concepts, and our conclusions will still hold. Thus what we see from (b′) appears to have a rather general validity.

In addition the consideration of (b′) shows, on the one hand, why in presenting the known objections to $\underline{BF}$, assertion (a) is usually schematized into an ordinary $\underline{BF}$; on the other hand, from (b′) we see that this schematization is not (completely) acceptable at least in connection with MC^{ν}.

As a consequence, the objections to $\underline{BF}$ do not hold (at least) in connection with MC^{ν}.

It is possible to develop some considerations on (and objections to) the converse of $\underline{BF}$—briefly $\underline{CBF}$—which are similar to those just presented in connection with $\underline{BF}$. In accordance with this, in some special version of S5, due to Kripke, $\underline{CBF}$ is not provable—cf. [18], p. 182.

We want to show that in our theory of $\underline{L}_{\alpha}$—which is the same as the one used for $\underline{BF}$—there are generalized $\underline{CBF}$s that mirror the facts above. To this end we consider the following sentence of ordinary language:

(c) If, necessarily, all kings are rich, then every (man who is a) king is necessarily rich.

The antecedent in (c) is reasonably true; the consequent is false. Hence (c) is false. We can satisfactorily schematize (c) within ML^{ν} into the following generalized $\underline{CBF}$:

(c′) $\quad \underline{L}_{\alpha}(\forall \underline{x} \in \underline{Man} \cap \underline{King})\underline{x} \in \underline{Rich} \supset (\forall \underline{x} \in \underline{Man} \cap \underline{King})\underline{L}_{\alpha}\underline{x} \in \underline{King}$,

where convention $(174)_2$ is used, where $\underline{Man}$, $\underline{King}$, and $\underline{Rich}$ stand for suitable constants in ML^{ν}, and where the following axioms on them are understood:

(176) $\quad \underline{Man} \in \underline{QAbs} \qquad \underline{King}, \underline{Rich} \in \underline{Ext}$.

Sentence (c′) is not $\underline{L}$-implied by (176) in ML^{ν}; hence it cannot be deduced from (176), which is satisfactory.

N70. Conclusions about our way of carrying out Lemmon's suggestions. Hints on identity, descriptions, and classes in connection with L_{α}.

The theory of α-conceivability properties developed so far can be used to carry out Lemmon's suggestions in the following general way. Let $n_1, \ldots, n_4$ be positive integers and let c_{ri} ($i = 1, \ldots, n_r$, $r = 1, \ldots, 4$) be distinct constants in ML^{ν}, of the same type as El. Then let us postulate

A70.1. $c_{ri} \in Conc_r$ $\qquad$ ($i = 1, \ldots, n_r$; $r = 1, \ldots, 4$)

where $Conc_r$ is the r-th of the classes $Conc_T$ to $Conc_{S5}$ introduced in N66. Thus in a same calculus based on MC^{ν}, there are necessity operators, $L_{ri} = L_{c_{ri}}$, of all the kinds considered in (our version) of Lemmon's suggestions, and there may be several of them for each kind. Furthermore, in particular, L_{4i} can be chosen in such a way that it is nontrivial, i.e. $L_{4i}p$ is not strictly equivalent with Np for all p's.

In this connection let us remark that—as is said in [18], p. 167—in dealing with S5 according to Kripke it is unimportant whether the accessibility relation R is universal or not. In other words it is unimportant whether the conceivability property α coincides with El or not. Instead, for the S5-necessity operators L_{4i} ($i = 1, \ldots, n_4$) introduced into MC^{ν} above, this question is essential in that whether or not L_{4i} coincides with N depends on it.

Our way of carrying out Lemmon's suggestion seems to me of a general character because of the considerations above and the fact that the operators L_{ri} can be used together with contingent identity, descriptions, and attributes from the same general point of view as in ML^{ν}. As far as I know something like this has not yet been done in connection with any of the modal systems T, B,

S4, and S5, supplemented with $\underline{LPC}$ and $\underline{BF}$. (These systems are substantially those denoted in [18] by $\underline{T}+\underline{LPC}+\underline{BF}$, $\underline{T}+\underline{LPC}+\underline{BF}$, $S4+\underline{LPC}+\underline{BF}$, and $S5+\underline{LPC}$.)

In order to illustrate the (general character of the) aforementioned fact concerning the operators $\underline{L}_{ri}$, we assume that $\alpha \in \underline{Conc}_{\underline{T}}$ holds. Then we can realize that $\underline{L}_{\alpha}$ fulfills the analogues of all axioms in MC^{ν} [Ns12, 25, 45] except AS12.13 for substitution. In particular $\underline{L}_{\alpha}$ fulfills the analogues of the following axioms (or axiom schemes): AS12.8 on specification, ASs12.10–12 on (contingent) identity, the intensionality axioms ASs12.14, 12.15 for attributes and functions, the existence axioms ASs12.16, 12.17 for attributes and functions, AS12.18 for descriptions, the strong (new) axiom AS12.19, the Zermelo axiom AS12.20, and also ASs12.21, 12.22 which mirror some conventions on the nonexisting object $\underline{a}^*_{\underline{t}}$ for $\underline{t} = (\underline{t}_1, \ldots, \underline{t}_{\underline{n}})$ and $\underline{t} = (\underline{t}_1, \ldots, \underline{t}_{\underline{n}} : \underline{t}_0)$, AS45.1 on natural numbers, and AS25.1.

Regarding AS12.13, let us add that unless $\underline{L}_{\alpha}$ is trivial, i.e. equivalent to $\underline{N}$, its analogue

(177) $\underline{L}_{\alpha}\underline{x} = \underline{y} \supset \Phi(\underline{x}) \equiv \Phi(\underline{y})$ where $\Phi(\underline{x})$ and $\Phi(\underline{y})$ are as in AS12.13

is false for a suitable choice of the matrix $\Phi(\underline{x})$.

At this point it seems natural and interesting to consider the question below. As a preliminary let us define (in MC^{ν}) $\underline{R}$ as an α-admissible attribute ($\alpha \in \underline{Ad}$) of the type $\underline{t} = (\underline{t}_1, \ldots, \underline{t}_{\underline{n}})$:

DEF. 70.1. $\underline{R} \in \alpha\text{-}\underline{Ad} \equiv_D (\forall \underline{x}_1, \underline{y}_1, \ldots, \underline{x}_{\underline{n}}, \underline{y}_{\underline{n}}) \{\underline{L}_{\alpha} \bigwedge_{\underline{i}=1}^{\underline{n}} \underline{x}_{\underline{i}} = \underline{y}_{\underline{i}} \supset [\underline{R}(\underline{x}_1, \ldots, \underline{x}_{\underline{n}}) \equiv \underline{R}(\underline{y}_1, \ldots, \underline{y}_{\underline{n}})]\}$

where for $\underline{i} = 1, \ldots, \underline{n}$, $\underline{x}_{\underline{i}}$ and $\underline{y}_{\underline{i}}$ are the first two variables of type $\underline{t}_{\underline{i}}$ which do not occur in $\underline{R}$ and are distinct from $\underline{x}_1$ to $\underline{x}_{\underline{i}-1}$ and $\underline{y}_1$ to $\underline{y}_{\underline{i}-1}$.

Let $\Phi(\underline{x})$ be a matrix built up out of $\underline{m}$ attributes, $\underline{R}_1, \ldots, \underline{R}_m$, variables, connectives, parentheses, commas, all signs, and $\underline{L}_\alpha$. Furthermore let $\underline{R}^{(\alpha)}$ be the matrix $\underline{R}_1 \in \alpha\text{-}\underline{Ad} \wedge \ldots \wedge \underline{R}_m \in \alpha\text{-}\underline{Ad}$.

It is natural to ask which conditions can assure us that (177) is $\underline{L}$-implied [N9] by $\underline{R}^{(\alpha)}$. I conjecture that $\Vdash \alpha \in \underline{Conc}_{S5}$ is one of the (sufficient) conditions above, that condition $\Vdash \alpha \in \underline{Conc}_T$ is not in general such a sufficient condition if taken alone, but that it is so in case $\underline{L}_\alpha$ does not occur in $\underline{p}$, i.e. the modal degree μ of $\underline{p}$—cf. [18], p. 50—is zero.

In case $\mu > 0$ it is natural to try to replace $\underline{R}^{(\alpha)}$ with a suitable stronger sentence of modal degree 1 and to take into account the reduction properties of modalities—cf. e.g. [18], p. 48. The problem just hinted at as an example of a possible extension of the considerations developed in Ns65–70 is open, as far as I know.

Notes to Memoir 3

1. The following considerations concern the comparison of our point of view on the formal translation of factual sentences with Meredith and Prior's—cf. [27]; furthermore they motivate this point of view semantically and syntactically.

On the one hand, from (I) and (IV), i.e. the special postulates (i) and (iv) in [27], it appears that, for every matrix $\underline{p}$, $\vdash \underline{p}$ is equivalent to $\vdash I_\rho \supset {}^\cap\underline{p}$. Then asserting $\underline{p}$ (or $I_\rho \supset {}^\cap\underline{p}$) is in effect equivalent with asserting the validity of $\underline{p}$ in the real case ρ. In this [27] is similar to usual languages; however it differs from them in that e.g. both of the sentences (a) and (a′) are translated into $\underline{p}_a$.

On the other hand, in MC^{ν}_{ρ} there are contingent propositions $\underline{p}$ for which $\vdash \underline{p}$ is not equivalent to $\vdash \mathfrak{R}\underline{p}$—see the remarks following (95). Furthermore asserting $\underline{p}$ [$\mathfrak{R}\underline{p}$ or $|_{\rho} \supset ^{\cap}\underline{p}$] is equivalent to asserting that $\underline{p}$ holds in the arbitrary (elementary) possible case γ (the real case ρ). This makes MC^{ν}_{ρ} appear less similar than [27] to usual languages when e.g. the factual sentence (a) is translated into $\mathfrak{R}\underline{p}_{\underline{a}}$ (or $|_{\rho} \supset ^{\cap}\underline{p}_{\underline{a}}$), but more similar than [27] when the different sentences (a) and (a′) are translated into the different formal sentences $\underline{p}_{\underline{a}}$ and $\mathfrak{R}\underline{p}_{\underline{a}}$ (or $|_{\rho} \supset ^{\cap}\underline{p}_{\underline{a}}$).

On the basis of our semantics—cf. Memoir 1—our point of view on factual sentences is quite natural. In particular, on the one hand our understanding every matrix $\underline{p}$ in MC^{ν}_{ρ}, which is not modally closed, as $\underline{p}$ holds in the elementary possible case γ (and not necessarily in the real case ρ) appears natural when one remembers that the semantical theory for MC^{ν}_{ρ} is obtained by means of slight additions from the semantical theory for the calculus MC^{ν}, which was constructed to solve an axiomatization problem concerning classical physics, a science for which all possible cases are on a par.

On the other hand, our formal translation of the factual sentence (a) into $\mathfrak{R}\underline{p}_{\underline{a}}$ can be explained by the following considerations of pragmatics. If in 1946 a man said "there is a world war," then this sentence is to be translated into a non-pragmatic extensional language by something such as "In 1946 there is a world war." Likewise if a man asserts the sentence (a) above, then he asserts (a) in the real case. So this sentence is to be translated into a non-pragmatic modal language such as MC^{ν}_{ρ} by something like "(a) holds in the real case," e.g. by $\mathfrak{R}\underline{p}_{\underline{a}}$.

As far as syntax is concerned, let us observe that in Memoir 2 a general deduction theory was developed for MC^{ν}; in particular

Theor. 33.1 on the use of rules G and C was proved. By the requirement (d) on modal C-steps in Def. 33.1, the condition that the (non-logical) axioms of a theory, $\mathscr{T}$, based on (MC^{ν}_{ρ}, hence on) MC^{ν} should be totally (hence modally) closed is very useful to apply rules G and C. In particular let $\mathscr{A}$ be a (possible) set of (factual) axioms of astronomy in every-day English. According to our views the formal translation of $\mathscr{A}$ into MC^{ν}_{ρ} is a set $\mathscr{A}'$ of totally closed sentences (of the form $\mathfrak{R}p'$). According to [27] the formal translation $\mathscr{A}''$ of $\mathscr{A}$ into MC^{ν}_{ρ} is a set of sentences that are not modally closed. Let $\mathscr{T}'$ [$\mathscr{T}''$] be the theory based on MC^{ν}_{ρ} which $\mathscr{A}'$ [$\mathscr{A}''$] gives rise to. All uses of rules G and C, which are possible in $\mathscr{T}''$, are also possible in $\mathscr{T}'$; furthermore there are many instances of a deduction in $\mathscr{T}'$, say $p'_1, \ldots, p'_n$ for which (1) a step p'_i is an element of $\mathscr{A}''$ and constitutes a modal C-step—see requirement (d) in Def. 33.1—and (2) the sequence $p'_1, \ldots, p''_n$ is not a deduction in the theory $\mathscr{T}''$ in that p''_i does not fulfill requirement (d) in Def. 33.1. So the theory $\mathscr{T}'$, constructed according to our point of view, is syntactically more efficient than $\mathscr{T}''$.

Appendix A

A Semantical Invariance Property of Logical Constants in ML^{ν}

This appendix, which may be interesting in itself, serves as a preliminary for appendixes B to D, where a natural problem concerning the definition of mass [Ns19, 20] is considered.

In connection with our semantical system $\langle \underline{D}_1, \ldots, \underline{D}_{\nu}, \Gamma, \underline{a}^{\nu} \rangle$ for ML^{ν} [N6], for each $\gamma \in \Gamma$ let $\alpha_{\gamma}^{(\underline{r})}$ be a permutation of $\underline{D}_{\underline{r}}$—i.e. a bijection of $\underline{D}_{\underline{r}}$ onto itself—leaving the "nonexisting object" invariant, i.e. with $\underline{a}_{\underline{r}}^{\nu}(\gamma) = \alpha_{\gamma}^{(\underline{r})}[\underline{a}_{\underline{r}}^{\nu}(\gamma)]$ $(\gamma \in \Gamma; \underline{r} = 1, \ldots, \nu)$, and let ϕ be the family of all these permutations: $\phi = \{\alpha_{\gamma}^{(1)}, \ldots, \alpha_{\gamma}^{(\nu)}\}_{\gamma \in \Gamma}$. Furthermore let β be a permutation of Γ.

DEF. A1. For $\xi \in \underline{QI}_{\underline{t}}$ with $\underline{t} \in \overline{\tau}^{\nu}$ [N2, N6, (8), (9)] we define recursively the ϕ-transform $\phi\xi$ [the β-transform $\beta\xi$] of ξ by means of the following conditions, where for the sake of brevity λ is used within our metalanguage:

(a) $\underline{t} \in \{1, \ldots, \nu\}$ and $\phi\xi = (\lambda\gamma)\alpha_{\gamma}^{(\underline{t})}[\xi(\gamma)]$ $[\beta\xi = (\lambda\gamma)\xi[\beta^{-1}(\gamma)]]$,

(b) $\underline{t} = 0$ (hence $\xi \subseteq \Gamma$) and $\phi\xi = \xi$ $[\beta\xi = (\lambda\gamma)(\beta^{-1}\gamma \in \Gamma)]$,

(c) $\underline{t} = (\underline{t}_1, \ldots, \underline{t}_{\underline{n}})$ and $\phi\xi = (\lambda\eta_1, \ldots, \eta_{\underline{n}}, \gamma)\xi(\phi^{-1}\eta_1, \ldots, \phi^{-1}\eta_{\underline{n}})$ $[\beta\xi = (\lambda\eta_1, \ldots, \eta_{\underline{n}}\gamma)\xi(\eta_1, \ldots, \eta_{\underline{n}}, \beta^{-1}\gamma)]$,

(d) $\underline{t} = (\underline{t}_1, \ldots, \underline{t}_{\underline{n}} : \underline{t}_0)$ and $\phi\xi = (\lambda\eta_1, \ldots, \eta_{\underline{n}})\phi[\xi(\phi^{-1}\eta_1, \ldots, \phi^{-1}\eta_{\underline{n}})]$ $[\beta\xi = (\lambda\eta_1, \ldots, \eta_{\underline{n}})\beta[\xi(\beta^{-1}\eta_1, \ldots, \beta^{-1}\eta_{\underline{n}})]]$.

Now let $\xi = \widetilde{\text{des}}_{\mathcal{M}\mathcal{V}}(\Delta)$ and $\xi_1 = \widetilde{\text{des}}_{\mathcal{M}_1\mathcal{V}_1}(\Delta)$ hold, where Δ is a designator in ML^ν, $\mathcal{M}_1$ is $\phi\mathcal{M}[\beta\mathcal{M}]$, and $\mathcal{V}_1$ is $\phi\mathcal{V}[\beta\mathcal{V}]$ in an obvious sense. Then by the designation rules (δ_1) to (δ_8) in N8, the rule (δ_9) on description [N11], and the assumed conditions $\underline{a}_{\underline{r}}^\nu(\gamma) = \alpha_\gamma^{(\underline{r})}[\underline{a}_{\underline{r}}^\nu(\gamma)]$ $(\gamma \in \Gamma;\ \underline{r} = 1, \ldots, \nu)$, we have $\xi_1 = \phi\xi\ [\xi_1 = \beta\xi]$.

Hence if ξ is independent of $\mathcal{M}$ and $\mathcal{V}$, then $\xi = \phi\xi = \beta\xi$ holds for all ϕ and β of the preceding kind. Thus we have proved the following:

THEOR. A1. (I) Let Δ be a closed designator in ML^ν without constants, so that $\xi = \widetilde{\text{des}}_{\mathcal{M}\mathcal{V}}(\Delta)$ is independent of $\mathcal{M}$ and $\mathcal{V}$. Then ξ is invariant $(\xi = \phi\xi = \beta\xi)$ for the arbitrary family ϕ of permutations of $\underline{D}_1, \ldots, \underline{D}_\nu$, and the arbitrary permutation β of Γ mentioned above.

(II) If in addition Δ is a sentence, then $\xi = \Gamma$ or $\xi = \Lambda$.

Incidentally let us set [Defs. 48.1, 48.2]

DEF. A2. $\underline{\text{Real}}_{\underline{u}} =_D (\lambda\rho)\ [\rho \in \underline{\text{Real}}_0 \vee {}_{\underline{u}}\rho \in \underline{\text{Real}}_0^{(e)}]$.

Then in MC^ν we have

(A1) $\underline{u} \notin \underline{\text{El}} \vdash \underline{\text{Real}}_{\underline{u}} = {}^\frown\underline{\text{Real}}_0,\ \underline{u} \in \underline{\text{El}} \vdash \sim {}_{\underline{u}}\underline{\text{Real}}_{\underline{u}} = \underline{\text{Real}}_0 \vee {}_{\underline{u}}\underline{\text{Real}}_{\underline{u}} = \underline{\text{Real}}^{(e)}$,

so that the following can be asserted in MC_ρ^ν [N52]:

(A2) $\quad \mathfrak{R}\ \underline{\text{Real}}_\rho = \underline{\text{Real}}_0^{(e)},\ \vdash \sim {}_\rho \supset {}^\frown\underline{\text{Real}}_\rho = \underline{\text{Real}}_0$.

Hence if $\underline{u} \in \underline{\text{El}}$ holds in the Γ-case γ at the admissible model $\mathcal{M}$ and the value assignment $\mathcal{V}$, then the $\underline{\text{QI}}$ assigned to $\underline{\text{Real}}_{\underline{u}}$ by $\mathcal{M}$ and $\mathcal{V}$, say ξ, certainly fails to fulfill the invariance conditions mentioned in Theor. A1. In particular this holds for $\underline{\text{Real}}_\rho$ in ML_ρ^ν. So ξ cannot be the $\underline{\text{QI}}$ of a logical constant in ML^ν; furthermore $\underline{\text{Real}}_\rho$ <u>is a logical constant in ML_ρ^ν which cannot be replaced by any logical constant definable in ML^ν</u>.

Appendix B

A Generalization of Theor. 19.1

The aim of appendixes B to D is to show that the use of the privileged concepts of (modally prefixed) real numbers and mass points made in N19 to define mass is practically essential, as we asserted in N20 (which is devoted to the same topic). In N20 we denoted those privileged concepts by $\underline{\text{Real}}_0$ and $\underline{\text{MP}}_0$, whereas in N19 and the main part of this monograph the simpler notations $\underline{\text{Real}}$ and $\underline{\text{MP}}$ were used for the same purposes. Now we shall use $\underline{\text{Real}}_0$ and $\underline{\text{MP}}_0$ as in N20. Hence now formulas (48), $(49)_3$, (50), and (51) are not assumed, but the results obtained from them by substituting $\underline{\text{Real}}_0$ for $\underline{\text{Real}}$ and $\underline{\text{MP}}_0$ for $\underline{\text{MP}}$ are.

Unlike Ns19, 20 appendixes B to D have a syntactical character. Thus proving theorems is often simpler, and stronger results are obtained. These syntactical results trivially imply the semantical theorems that interest us in connection with Ns19, 20.

It is useful to set

(B1) $\Phi(\rho, \underline{\text{Real}}) \equiv_D \rho \in \underline{\text{Real}} \diamond \underline{\text{Exp}}(\underline{M}, \rho)$, $\Psi(\rho, \underline{\text{Real}}) \equiv_D \Phi(\rho, \underline{\text{Real}})$ $(\forall \rho') [\underline{\text{Exp}}(\underline{M}, \rho') \supset^\frown \rho' = \rho]$.

Formulas (52) and $(53)_1$ were understood in N19 as the semantical analogues of the first two of the assertions

(B2) $\vdash \underline{M} \in \underline{\text{MP}}_0 \supset^\frown (\exists \rho)\, \Psi(\rho, \underline{\text{Real}}_0)$, $\vdash \underline{M} \in \underline{\text{MP}}_0 \supset^\frown (\exists_1 \rho)\, \Phi(\rho, \underline{\text{Real}}_0)$, $\vdash (\exists \underline{M})\, \underline{M} \in \underline{\text{MP}}_0$.

Formula (52) was considered as an axiom and $(53)_1$ as a theorem based on (52) and (47) [N19].

The syntactical analogue of the above considerations consists of assuming $(B2)_1$ and deducing $(B2)_2$ from $(B2)_1$ and the syntactical analogue of (47), which is

(B3) $\vdash \underline{\text{Real}}_0 \in \underline{\text{Abs}}, \ \vdash \underline{\text{MP}}_0 \in \underline{\text{Abs}}$—cf. Theor. 47.1(VII).

This deduction is straightforward on the basis of (the proof of) the semantical theorem 19.1. Therefore we consider this as done.

We regard $(B2)_3$ as an obvious axiom.

The semantical analogue of Theor. B1 below embodies the generalization of Theor. 19.1 asserted in N20. More particularly the conditions (1) to (3) considered in N20 are related in order to parts (I) to (III) of Theor. B1.

As a preliminary we write the following syntactical analogues of (47′) and (47‴):

(B4) $\vdash \underline{\text{Real}}_0 \ \underline{\subseteq}^{\cap} \underline{\text{Real}}, \ \vdash \underline{\text{Real}} \ \underline{\subseteq}^{\cap} \underline{\text{Real}}_2, \ \vdash \underline{\text{MP}} =^{\cap} \underline{\text{MP}}_0,$

(B5) $\vdash (\forall \underline{M}, \rho) \, [\underline{\text{Exp}} \, (\underline{M}, \rho) \supset^{\cap} \underline{M} \in \underline{\text{MP}}_0^{(e)}], \ \vdash (\forall \underline{M}, \rho) \, [\underline{\text{Exp}} \, (\underline{M}, \rho)$
$\supset^{\cap} \rho \in \underline{\text{Real}}_0^{(e)}], \ \vdash \underline{\text{Exp}} \in^{\cap} \underline{\text{Ext}}.$

THEOR. B1. (I) Conditions $(B4)_1$ and (B5) [or $(B5)_{2,3}$] imply the equivalence of axiom $(B2)_1$ on $\underline{\text{Real}}_0$ with its analogue for $\underline{\text{Real}}$, i.e. with the first of the assertions—cf. (B1)

(B6) $\vdash \underline{M} \in \underline{\text{MP}}_0 \supset^{\cap} (\exists \rho) \Psi (\rho, \underline{\text{Real}}), \ \vdash \underline{M} \in \underline{\text{MP}}_0 \supset^{\cap} (\exists_1 \rho) \Phi (\rho, \underline{\text{Real}}$

(II) Conditions $(B4)_{1,2}$ and $(B5)_{2,3}$, and either of axioms $(B2)_1$ and $(B6)_1$ yield besides theorem $(B2)_2$ its analogue for $\underline{\text{Real}}$, i.e. $(B6)_2$.

(III) Conditions $(B4)_{1,2}$ and $(B5)_{2,3}$ and either of $(B2)_1$ and (B6 yield

(B7) $\vdash \underline{M} \in \underline{MP}_0 \supset^{\cap} (\imath\rho)\Phi(\rho, \underline{Real}) = (\imath\rho)\Phi(\rho, \underline{Real}_0)$,

so that, for $\underline{Exp} = {}^{\cap}\underline{Exp}_1$ [N19] and $\underline{M} \in \underline{MP}_0$, the mass $\mu(\underline{M}) =_D (\imath\rho)\Phi(\rho, \underline{Real}_0)$ defined on the basis of $\underline{Real}_0$ (and $\underline{MP}_0$) according to the procedure presented in N19 equals the analogue for $\underline{Real}$.

Proof. (I) We assume $(B2)_1$, $(B4)_1$, and $(B5)_2$. By (B1) $(B4)_1$ yields (a) $\vdash (\rho)[\Phi(\rho, \underline{Real}_0) \supset^{\cap} \Phi(\rho, \underline{Real})]$ and (a′) $\vdash (\rho)[\Psi(\rho, \underline{Real}_0) \supset^{\cap} \Psi(\rho, \underline{Real})]$. Then $(B2)_1$ yields $(B6)_1$.

Now we conversely assume $(B4)_1$, $(B5)_2$, and $(B6)_1$. We start with (b) $\underline{M} \in \underline{MP}_0$, which by $(B6)_1$ yields (c) $(\exists_\rho)\Psi(\rho, \underline{Real})$; furthermore, in order to prove (B8) below, also to be used in parts (II) and (III), we also assume (d) $\Phi(\rho, \underline{Real})$, whence $\rho \in \underline{Real}_2$ by $(B1)_1$ and $(B4)_2$. By Def. 20.2 and rule C with ρ_0, this yields (e) $\rho_0 \in \underline{Real}$ and (e′) $\underline{N}(\rho = \rho_0 \vee \sim\rho \in \underline{Real}^{(e)})$. By (e′), (d), and $(B1)_1$, $\rho = \rho_0$ holds. Furthermore from (e′) and $(B5)_{2,3}$ we have (f) $\underline{Exp}(\underline{M}, \rho) \equiv^{\cap} \underline{Exp}(\underline{M}, \rho_0)$, so that by $(B1)_1$, (d) and (e) yield (h) $\Phi(\rho_0, \underline{Real}_0)$.

By Theors. 33.3 and 32.2 we easily conclude that

(B8) $\underline{M} \in \underline{MP}_0 \vdash (\forall\rho)\{\Phi(\rho, \underline{Real}) \supset (\exists\rho_0)[\rho = \rho_0 \Phi(\rho_0, \underline{Real}_0)]\}$.

Now we keep assumption (b) but not (d). Then (c) holds. By (c) and rule C with ρ we have $\Psi(\rho, \underline{Real})$ and (d). Hence by the above reasoning, the matrices (e′) and (h), i.e. $\Phi(\rho_0, \underline{Real}_0)$ follow from (b) using rule C with ρ and ρ_0.

By $(B1)_2$, (c) yields $(\rho')[\underline{Exp}(\underline{M}, \rho') \supset \rho' = \rho]$. Furthermore $(B5)_2$ and (e′) yield $(\rho')[\underline{Exp}(\underline{M}, \rho)\rho = \rho' \supset^{\cap} \rho' = \rho_0]$. Then $(\rho')[\underline{Exp}(\underline{M}, \rho') \supset^{\cap} \rho' = \rho_0]$ holds. By $(B1)_2$ this and (h) yield $\Psi(\rho_0, \underline{Real}_0)$, hence (i) $(\exists\rho)\Psi(\rho, \underline{Real}_0)$. By Theor. 33.1, (b) $\vdash$ (i). This implies $(B2)_1$ by Theor. 33.2. Thus $(B4)_1$, $(B5)_2$, and $(B6)_1$ imply $(B2)_1$, which completes the proof of part (I).

(II) We assume $(B4)_{1,2}$ and $(B5)_{2,3}$ and either of $(B2)_1$ and $(B6)_1$, so that by part (I) both $(B2)_1$ and $(B6)_1$ hold. As we stated above, on the basis of Theor. 19.1 $(B2)_1$ yields $(B2)_2$.

We know that $(B4)_1$ yields (a), whence we obtain (1) $\vdash (\exists\rho)\Phi(\rho, \underline{\text{Real}_0}) \supset (\exists\rho)\Phi(\rho, \underline{\text{Real}})$.

Let us now add the assumption (b) $\underline{M} \in \underline{MP_0}$. Then by (B8) and $(10)_1$ [N34] we easily deduce $(\exists^{(1)}\rho)\Phi(\rho, \underline{\text{Real}_0}) \supset (\exists^{(1)}\rho)\Phi(\rho, \underline{\text{Real}})$, which by (1) yields (m) $(\exists_1\rho)\Phi(\rho, \underline{\text{Real}_0}) \supset (\exists_1\rho)\Phi(\rho, \underline{\text{Real}})$. Hence $\vdash$ (b) $\supset^{\cap}$ $\underline{(\text{m})}$ [Theors. 33.2 and 32.2], which by assertion $(B2)_2$, deduced above, implies $(B6)_2$. Thus part (II) has been proved.

(III) Let the assumptions made in part (III) hold, so that $(B2)_2$, $(B6)_2$, and (B8) hold by part (II). Furthermore $\vdash \underline{MP_0} \in \underline{\text{MConst}}$ by $(B3)_1$ and Def. 18.8. Then (B7) follows easily by AS12.18.

QED

Appendix C

Some Requirements on the Concepts Real and MP in Connection with the Definition of Mass; Consequence concerning MP

The following conditions (a), (C1), and (C2) on Real are requirements on any mathematically satisfactory concept of real number (and the concept Real to be used in a definition of mass such as Def. 19.1 must be of this kind).

(a) Real must be defined on purely logical grounds.

(C1) $$\vdash \underline{\mathrm{Real}}_0 \subseteq^{\frown} \underline{\mathrm{Real}},\ \vdash \underline{\mathrm{Real}}^{(\underline{e})} =^{\frown} \underline{\mathrm{Real}}_0^{(\underline{e})}.$$

(C2) $$\vdash \rho, \sigma \in \underline{\mathrm{Real}} \supset^{\frown} (\rho, \sigma \in \underline{\mathrm{Real}} \supset^{\frown} \rho \pm \sigma,\ \rho\sigma \in \underline{\mathrm{Real}}).$$

By requirements (C1) and (a),

(C3) $$\rho_0, \rho_0' \in \underline{\mathrm{Real}},\ \ \Diamond \underline{p} \Diamond \sim\underline{p},\ \ \Diamond(\rho_1 \in \underline{\mathrm{Real}}\ \rho_1 = \rho_0),\ \ \Diamond(\rho_1 \neq \rho_0 \wedge \rho_1 \in \underline{\mathrm{Real}}_0^{(\underline{e})}) \vdash (\exists \rho_1')[\Diamond(\underline{p}\,\rho_1' \in \underline{\mathrm{Real}}\ \rho_1' = \rho_0') \Diamond (\sim\underline{p} \wedge \rho_1' \neq \rho_0' \wedge \rho_1' \in \underline{\mathrm{Real}}_0^{(\underline{e})})],$$

where (ρ_0, ρ_1, ρ_0', and ρ_1' are distinct variables and) $\underline{p}$ is any matrix not containing ρ_0' or ρ_1'.

To show that requirement (C3) on Real holds we now sketch a proof of its semantical analogue: the L.H.S. of $\vdash$ in (C3) L-implies its R.H.S.

Let the L.H.S. of $\vdash$ in (C3) hold. Then, speaking intuitively, we have $\sigma \in \underline{\text{Real}}_0$ with $\sigma =_D \rho_0' - \rho_0$. So, setting $\rho_1' = \rho_1 + \sigma$, by $(\text{C1})_1$ and the part of (C2) that concerns $\rho + \sigma$, the third and fourth matrices in the L.H.S. of (C3) yield $\diamond\,(\rho_1' \in \underline{\text{Real}}\, \rho_1' = \rho_0')$ and $\diamond\,(\rho_1' \neq \rho_0' \wedge \rho_1' \in \underline{\text{Real}}_0^{(e)})$.

Furthermore, by requirement (a) and Theor. A1, the $\underline{\text{QI}}$ of $\underline{\text{Real}}$ must be invariant for all permutations β of Γ. Now it is easy to see that the R.H.S. of $\vdash$ in (C3) is $\underline{\text{L}}$-implied by the corresponding L.H.S.

Incidentally, by requirement (C2), condition $(\text{C1})_1$ is equivalent to (and not more restrive than) the condition $\vdash \underline{\text{Real}}_0^+ \subseteq {}^\cap\underline{\text{Real}}$ where e.g. $\underline{\text{Real}}^+$ is $(\lambda\rho)(\rho \in \underline{\text{Real}}\,\rho > 0)$.

By Theor. A1 and requirement (a) the following requirement holds.

(a′) The $\underline{\text{QI}}$ ξ of $\underline{\text{Real in any admissible model fulfills the invariance conditions}}$ $(\xi = \phi\xi = \beta\xi)$ $\underline{\text{mentioned in Theor. A1.}}$

Therefore at least in defining mass we cannot identify $\underline{\text{Real}}$ with $\underline{\text{Real}}_{\underline{u}}$ (for $\underline{u} \in \underline{\text{El}}$) or $\underline{\text{Real}}_\rho$ [Def. A2].

Let us add that even if one bases Mach and Painlevé's definition of mass on ML^ν_ρ [N52], so that $\underline{\text{Real}}_\rho$ is defined on purely logical grounds, it is unsatisfactory to identify $\underline{\text{Real}}$ with $\underline{\text{Real}}_\rho$, because in ML^ν_ρ the elementary possible case ρ is privileged, whereas the definition of $\underline{\text{Real}}$ must be compatible with the permutability of these cases. Incidentally an experimentalist may ignore which of them is the real one (in that he does not know which assertions of an extensional language are true in the real case).

More generally, if the concepts $\underline{\text{Real}}$ and $\underline{\text{MP}}$ are used for physical purposes, such as to deal with Mach and Painlevé's definition of mass, then the requirements (C3) and (b) to (e) are reasonable:

(b) The concept MP is understandable to any physicist, even if he ignores the values of mass for many (or all) mass points and is unable to distinguish several (or all) elementary possible cases from one another.

(c) The use of Real and MP allows us to define mass; more precisely the analogue for MP (and Real) of axiom $(B2)_1$ or $(B6)_1$ and the analogue of the uniqueness property $(B2)_2$ or $(B6)_2$ hold:

(C4) $\vdash \underline{M} \in \underline{MP} \supset (\exists \rho)\Psi(\rho, \underline{Real})$, $\vdash \underline{M} \in \underline{MP} \supset (\exists_1 \rho)\Phi(\rho, \underline{Real})$, where (B1) and $\vdash \underline{Exp} = \underline{Exp}_1$ [N19] are understood. Furthermore (C4) must allow us to define mass correctly, so that in particular (C4) must be consistent with (B2).

(d) The concepts MP and Real are acceptable to experimentalists who take into account the possibility of defining, in the same way as mass, other physical magnitudes which may be presently unknown, so that (C4) must hold for other (presently unknown) determinations of Exp.

(e) If $\rho \in \underline{Real}$ holds in the Γ-case γ and $\rho \in \underline{Real}^{(e)}$ holds in the Γ-case γ', then $\rho \in \underline{Real}$ holds in γ', i.e. syntactically,

$$\text{(C5)} \qquad \vdash \rho \in \underline{Real} \supset^{\frown} (\rho \in \underline{Real}^{(e)} \supset^{\frown} \rho \in \underline{Real});$$

and the analogue can be said of MP.

In order to motivate (e) let $\rho \in \underline{Real} \wedge \underline{M} \in \underline{MP}$ hold in γ and $\rho \in \underline{Real}^{(e)} \wedge \underline{M} \in \underline{MP}^{(e)}$ hold in γ'. This may be important in that $\underline{Exp}(\underline{M}, \rho)$ may occur in γ', so that it is natural to be willing to use this fact to show the truth of axiom $(B2)_1$ or any of its generalizations $(B6)_1$ and $(C4)_1$. The usefulness of the same fact reasonably requires the possibility of speaking of it in γ' itself, and hence that $\rho \in \underline{\mathrm{Real}}$ and $\underline{\mathrm{M}} \in \underline{\mathrm{MP}}$ should hold in γ'. Then requirement (e), and in particular (C5), is reasonable especially in connection with requirement (d).

It is clear that if $\underline{\mathrm{Real}}$ satisfies conditions (B4) and the invariance conditions mentioned in Theor. A1, then $\underline{\mathrm{Real}}$ and $\underline{\mathrm{MP}}$ fulfill all requirements considered in this section. Some instances of $\underline{\mathrm{Real}}$, e.g. $\underline{\mathrm{Real}}_1$ and $\underline{\mathrm{Real}}_2$ [Defs. 20.1 and 20.2], are of this type and are distinct from $\underline{\mathrm{Real}}_0$. However they are similar to $\underline{\mathrm{Real}}_0$ and rather awkward—cf. N20.

Let us now remark that $\underline{\mathrm{Real}}$ and $\underline{\mathrm{MP}}$ have different roles in $(C4)_1$. On the one hand, by (B1) $(C4)_1$ involves $\underline{\mathrm{Real}}$ only in that in $(C4)_1$ one asserts the existence of a suitable ρ for which $\rho \in \underline{\mathrm{Real}}$ and other conditions hold. On the other hand $(C4)_1$ involves $\underline{\mathrm{MP}}$ in a way that implies that for <u>every</u> $\underline{\mathrm{M}} \in \underline{\mathrm{MP}}$ certain conditions hold.

Since axiom $(C4)_1$ and (theorem) $(C4)_2$ must enable us to define mass correctly, one can easily convince oneself—cf. (B1)—that $\vdash \underline{\mathrm{MP}}_0 \subseteq {}^\cap \underline{\mathrm{MP}}$ and $\vdash \underline{\mathrm{MP}}_0^{(\underline{e})} = {}^\cap \underline{\mathrm{MP}}^{(\underline{e})}$ must hold.

Now let us assume $\diamond\, \underline{\mathrm{MP}}_0 \neq \underline{\mathrm{MP}}$ as a hypothesis for reductio ad absurdum. Then, for some $\underline{\mathrm{M}}_0$ and $\underline{\mathrm{M}}$, (α) $\underline{\mathrm{M}}_0 \in \underline{\mathrm{MP}}_0 \wedge \underline{\mathrm{M}} \in \underline{\mathrm{MP}} \wedge \underline{\mathrm{M}} = \underline{\mathrm{M}}_0$ holds in a Γ-case, γ_1, whereas $\underline{\mathrm{M}} \neq \underline{\mathrm{M}}_0$ holds in another Γ-case, γ_2. An experimentalist may believe that the real case is γ_1. Furthermore it is reasonable to think that a presently unknown physical magnitude, $\mathscr{M}$, can be defined for $\underline{\mathrm{M}}_0$ in the same way as mass, in connection with a suitable determination of $\underline{\mathrm{Exp}}$, and that (β) $\rho \in \underline{\mathrm{Real}}_0 \wedge \underline{\mathrm{Exp}}(\underline{\mathrm{M}}_0, \rho)$ may hold in γ_2. If (γ) $(\exists \rho)\, \underline{\mathrm{Exp}}(\underline{\mathrm{M}}, \rho)$ does not hold in γ_2, then our physicist is disappointed because γ_2 is not available to assert in γ_1 $\rho_0 \in \underline{\text{Real}} \diamond \underline{\text{Exp}}$ $(\underline{\text{M}}, \rho_0)$ where ρ_0 is the correct value of $\mathscr{M}$ for $\underline{\text{M}}$ in γ_1. Incidentally (γ) might also fail to hold in the other Γ-cases where $\underline{\text{Exp}}(\underline{\text{M}}, \rho_0)$ holds, which would invalidate the existence requirement $(\text{C4})_1$. If (γ) holds in γ_2, then for some ρ_0 the same can be said of (δ) ρ_0 $\in \underline{\text{Real}}_0 \wedge \underline{\text{Exp}}(\underline{\text{M}}, \rho)$. Let our physicist ignore the value of ρ_0. Then he fears that $\rho \neq \rho_0$, may hold (in γ_1), which by (α), (β), and (δ) would contradict the uniqueness property $(\text{C4})_2$.

Since requirement (d) holds, so that in particular the acceptance of the concept MP should precede the knowledge of the experiments characterizing the various determinations of $\underline{\text{Exp}}$ to be considered, under the hypothesis $\diamond\ \underline{\text{MP}}_0 \neq \underline{\text{MP}}$ our physicist would reject the use of the concept $\underline{\text{MP}}$. We conclude that $\Vdash \underline{\text{MP}} =^{\cap} \underline{\text{MP}}_0$ is required to hold, which justifies $(\text{B4})_3$.

Appendix D

A Property of Real Following from the Preceding Requirements

The following theorem constitutes a converse of Theor. B1 (in a generalized sense because e.g. $\vdash \underline{MP} = \underline{MP}_0$ is not assumed).

THEOR. D1. Let the requirements on Real and MP considered in appendix C hold. Then assertions (B4) are justified from the semantical point of view.

Proof. Thesis $(B4)_1$ is requirement $(C1)_1$ and the validity of $(B4)_3$ was justified at the end of appendix C.

Now let $(B4)_2$ not hold, as a hypothesis for reductio ad absurdum. Then we have $\diamond \sim\underline{Real} \subseteq \underline{Real}_2$, so that by Def. 20.2 and $(B4)_1$ we easily deduce, using rule C with ρ_0 and ρ_1,

$$\text{(D1)}\ \ \rho_0 \in \underline{Real}_0, \quad \diamond\, (\rho_1 = \rho_0 \wedge \rho_1 \in \underline{Real}), \quad \diamond\, (\rho_0 \neq \rho_1 \wedge \rho_1 \in \underline{Real}^{(\underline{e})}).$$

By requirement (d) in appendix C, (C4) must be consistent with (B2). Then using rule C with $\underline{M}$ and ρ_0' from $(B2)_{3,1}$ we deduce (a) $\underline{M} \in \underline{MP}_0$ and $\Psi(\rho_0', \underline{Real}_0)$, which by (B1) yields the first two of the matrices

$$\text{(D2)}\ \ \rho_0' \in \underline{Real}_0, \ \diamond\ \underline{Exp}\,(\underline{M}, \rho_0'), \ \rho_0' \in \underline{Real}, \ \underline{N}\,(\exists_1 \rho)\, \Phi\, (\rho, \underline{Real}).$$

By $(B4)_1$, $(D2)_1$ yields $(D2)_3$. Furthermore, by $(B3)_1$ and Defs. 18.8 and 13.2, (a) yields $\underline{NM} \in \underline{MP}_0$. Thence by $(B4)_3$ and $(C4)_2$ we obtain $(D2)_4$—cf. the modal generalization theorem (in Theor. 33.2).

By AS12.23 and $(D2)_2$ there is a contingent matrix $\underline{p}$ which does not contain ρ_0' or ρ_1' and has the form $\underline{q} \wedge \underline{\text{Exp}}(\underline{M}, \rho_0)$. Hence, besides (c) $\diamond \underline{p} \diamond \sim\underline{p}$ we have (d) $\underline{p} \supset {}^\frown\underline{\text{Exp}}(\underline{M}, \rho_0)$.

(C3) holds for the above choice of $\underline{p}$. Hence (c), (D1), $(C1)_2$, and $(D2)_1$ yield the R.H.S. of $\vdash$ in (C3). By rule C with ρ_1' this R.H.S., $(C1)_2$, and (d) yield

(D3) $\diamond [\underline{\text{Exp}}(\underline{M}, \rho_0') \; \rho_1' \in \underline{\text{Real}} \wedge \rho_1' = \rho_0'], \; \diamond (\sim\underline{p}\rho_1' \neq \rho_0' \wedge \rho_1' \in \underline{\text{Real}}^{(\underline{e})})$.

By requirement (C5), $(D3)_{1,2}$ yield (e) $\diamond (\rho_1' \neq \rho_0' \wedge \rho_1' \in \underline{\text{Real}})$. Furthermore by $(B3)_1$, $(D2)_1$ yields $\underline{N}\rho_0' \in \underline{\text{Real}}_0$. From this and $(B4)_1$ we deduce (f) $\underline{N}\rho_0' \in \underline{\text{Real}}$. In addition, by $(B5)_3$, $(D3)_1$ yields $\diamond \underline{\text{Exp}}(\underline{M}, \rho_1')$. This, (f), (e), $(B4)_1$, and $(D3)_1$ yield

(D4) $\diamond [\rho_0', \rho_1' \in \underline{\text{Real}} \wedge \rho_0' \neq \rho_1' \; \diamond \underline{\text{Exp}}(\underline{M}, \rho_0') \diamond \text{Exp}(\underline{M}, \rho_1')]$.

By $(B1)_1$, (D4) contradicts $(D2)_4$. Hence $\diamond \sim\underline{\text{Real}} \subset \underline{\text{Real}}_2$ cannot hold. Thus $(B4)_2$ has also been proved.

QED

Appendix E

On R. Montague's Work, "Pragmatics and Intensional Logic"; General Operators

---

NE1. Introduction

The above work by R. Montague, item [28] in the Bibliography, is denoted hereafter by [M]; Memoir 1 of this volume, is denoted hereafter by [B].

Intensional logic belongs to the subjects of both papers. However in [M], in addition to an intensional language to be denoted here by $\underline{IL}_{\underline{M}}$, pragmatics and belief sentences are also dealt with.

First I want to show that [B] and [M] are also substantially different as far as the bases of intensional logic are concerned, so that the presentation of the results achieved in [B] cannot be shortened by referring to [M]. This occurs especially because the theory set up in [M] has certain inhomogeneities (or is not a unified theory) with substantial consequences, and certain limitations which e.g. cause the specification rule

(E1) $(\underline{x})\Phi(\underline{x}) \supset \Phi(\underline{a})$ where $\underline{a}$ may be a constant or a description,

to hold in connection with no logical type (allowed in [M] for the variable $\underline{x}$).

In [M] possibility is dealt with using a parameter $\underline{i}$ ranging on a set $\underline{A}$, which is the analogue of the parameter γ ranging on the class Γ of the elementary possible cases considered in [B]. Such an analogue is also present in my work [4].

More specifically, I want to show that [M] and [B] are different not only in the set theoretical techniques used but in the following substantial points:

a) On the one hand an aim of [B] is to conform with a unified and homogeneous point of view—cf. requirement (a) in [B], p. 5. Consequently in ML^{ν} we can assert the $\underline{L}$-truth of an axiom system—cf. [B], N12—which is similar to an ordinary axiom system of extensional logic—cf. requirement (b) in [B], p. 5—so that ($\underline{E}$1) holds in ML^{ν}; and which is homogeneous with respect to different types and to variables and constants, so that this axiom system is rather simple. On the other hand the semantical rules in [M] deal in different ways with different types and with variables and constants. Consequently ($\underline{E}$1) does not hold in [M], as we said.

b) It seems strange, but the following matrix holds in $\underline{IL}_{\underline{M}}$ at a suitable model, where $\underline{x}$ is an individual variable:

$$(E2)\ (\exists \underline{x})\underline{x} = {}^{\frown}\underline{a}\,(\exists \underline{x})\underline{x} = {}^{\frown}\underline{b}\ \underline{Na} \neq \underline{b} \diamond \underline{p} \diamond {\sim}\underline{p} \wedge {\sim}(\exists \underline{x})\underline{N}(\underline{px} = \underline{a} \vee {\sim}\underline{px} = \underline{b}).$$

c) In [B] functors are included and infinitely many levels are taken into account for variables and constants. In $\underline{IL}_{\underline{M}}$ functors are excluded and the variables must have one of the levels 0 and 1 whereas the constants are of the levels 0, 1, and 2.

Furthermore variables can express only extensional properties, whereas predicate constants of level 2 can also express nonextensional properties. As a consequence, by the inhomogeneity of [M] mentioned in (a) it is not at all obvious how to generalize $\underline{IL}_{\underline{M}}$ to higher levels.

d) In the event that ζ_1 is an individual constant or a description (incidentally all descriptions in [M] have the level 1) in general we cannot assert in [M]

(E3) $(\exists \zeta_0)\underline{N}\zeta_0 = \zeta_1$ cf. (E2).

e) In [B] an extensional description operator, $\imath$, is used, which is compatible with the axioms ASs12.1–23 in [B]. In [M] an intensional description operator, $\underline{T}$, is used, which is incompatible with ($\underline{E}$1) as we said. In NE3 we want to show that $\underline{T}$ is unsatisfactory not only by inadequacy of the semantical rules, but by the following property asserted in [M], p. 27, for $\underline{T}$:

(E4) $(\exists_1^{\frown} \underline{F})\Psi(\underline{F}) \supset^{\frown} \Psi[(\underline{TF})\Psi(\underline{F})]$ cf. [B], Def. 14.

In connection with point (e), let us incidentally remark that in Memoir 3 an intensional description operator, $\iota_{\underline{u}}$, depending on the parameter $\underline{u}$ is introduced. The operator $\iota_{\underline{u}}$ is somewhat similar to $\underline{T}$ but it is compatible with the axioms ASs12.1–23 in [B] and in particular with ($\underline{E}$1).

As a preliminary to what we do secondly [NE4] let us remark that in [11], p. 84, Carnap briefly hints at general operators. On the basis of [B], N2, as far as the type system is concerned we say that the operator ω has the type $(t_1, \ldots, t_{\underline{n}}; \theta; \theta')$ with $t_1, \ldots, t_{\underline{n}} \in \tau^{\nu}$ and $\theta, \theta' \in \overline{\tau}^{\nu}$ whenever we consider as full formulas of ω expressions of the form $(\omega \underline{x}_1, \ldots, \underline{x}_{\underline{n}})\Delta$ where $\underline{x}_1, \ldots, \underline{x}_{\underline{n}}$ are distinct variables of the respective types $t_1, \ldots, t_{\underline{n}}$, and Δ is a designator of type θ, and in addition we attribute the type θ' to $(\omega \underline{x}_1, \ldots, \underline{x}_{\underline{n}})\Delta$.

For example $\underline{N}$ and $\diamond$ are operators of the type $(0; 0) = (t_1, \ldots, t_{\underline{n}}; 0; 0)$ for $\underline{n} = 0$. The operators considered in pragmatics in [M] have this type, and on p. 22 Montague starts accommodating such operators within $\underline{IL}_M$, using a suitable correspondence between operators and attributes. So intensional logic is reduced to pragmatics, and on p. 24 this reduction is asserted to be partial by certain limitations. I think that these limitations occur because of the features of $\underline{IL}_M$ mentioned in the points (a) to (e) above.

This reduction of intensional logic to pragmatics seems to me related to the way general operators are introduced. In NE4 below we want to show that we can introduce general operators in MC^ν without the limitations, in particular without the limitations for the description operator $\underline{T}$ in [M] shown in NE3.

NE2. On the set theoretical techniques in [M] and [B].

In [B], Ns8, 11, for every designator Δ in ML^ν the intension $\underline{\widetilde{des}}_{\mathcal{M}\mathcal{V}}(\Delta)$ at the model $\mathcal{M}$ and value assignment $\mathcal{V}$ is recursively defined—cf. rules (δ_1) to (δ_9). In [M] the extension $\underline{Ext}_{ia}(\Delta)$ of Δ (in the possible interpretation $\underline{a}$ and at the reference point $\underline{i}$) is defined by means of Def. III on p. 10 for a pragmatic language, and by means of Def. XII on pp. 19–20 for $\underline{IL}_M$; and the intension $\underline{Int}_a(\Delta)$ (in $\underline{a}$) is defined by means of Def. XII on p. 20, for $\underline{IL}_M$. Then according to Montague's way of speaking, we could say that the extension $\underline{Ext}_\gamma(\underline{p})$ of $\underline{p}$ in the Γ-case γ is the class

(E6) $$(\lambda\mathcal{V})\,\gamma \in \underline{\widetilde{des}}_{\mathcal{M}\mathcal{V}}(\underline{p}).$$

This gives an indication of the relation between the set theoretical techniques used in [M] and [B].

In many cases in which $\underline{p}$ may be thought of as belonging both to ML^ν and to $\underline{IL}_M$, the class (E6) does not coincide with $\underline{Ext}_{\gamma a}(\underline{p})$ according to [M], i.e. the semantical systems constructed in [M] and [B] are substantially different—cf. NE3 below.

NE3. On the intensional language $\underline{IL}_M$ considered in [M].

The formation rules for $\underline{IL}_M$ are in [M], p. 14. Only variables of the levels 0 and 1 occur in $\underline{IL}_M$.

As to semantics, Montague uses, instead of the value assignment $\mathscr{V}$, a certain system $\underline{x}[\underline{x}_{n,k}]$—see [M], p. 18—which substantially assigns extensions to individual variables and intensions to attribute variables—so in [M] different levels are dealt with differently.

In accord with the above assertion, in [M], Def. IX, p. 19—where semantical rules for $\underline{IL}_M$ are given—the extension $\underline{Ext}_{\underline{ia}}(\underline{v}_{tn})$ of the individual variable $\underline{v}_{tn}$ is independent of the reference point $\underline{i}$—see Def. IX (2). Furthermore by Def. IX (1) the extension of a constant at $\underline{i}$ can depend on $\underline{i}$. Then by Def. IX (6) and (12) the matrix (E3) cannot be asserted in $\underline{IL}_M$ in general, in case ζ_1 is an individual constant. As a consequence the specification principle (1) cannot be asserted in $\underline{IL}_M$ in case x is an individual variable. Def. IX—in particular (E1) and (E2)—implies (E2).

By Def. IX (7) attribute variables [of level 1] can denote only extensional properties, so that for a non-extensional choice of the matrix $\underline{p}$,

$$\text{(E7)} \qquad (\exists \underline{F})(\forall \underline{x}_1, \ldots, \underline{x}_n)[\underline{F}(\underline{x}_1, \ldots, \underline{x}_n) \equiv {}^{\frown}\underline{p}]$$

does not hold in $\underline{IL}_M$ whereas it holds in ML^{ν}.

As for descriptions in $\underline{IL}_M$—$(\underline{TG})\underline{p}$—the only criterium for $\underline{T}$ in [M] is the validity of our criterium (Π′) on p. 17, in connection with attributes only. The same limitation affects rule (4) in Def. IX, where only descriptions of the form $(\underline{TG}_{kn})\underline{p}$, where $\underline{G}_{kn}$ is the $\underline{k}$-th $\underline{n}$-place attribute variable, are concerned.

Montague deduces from rule (4) that, using the symbolism of [B], we substantially have—cf. [M], p. 21—

(E8) $\Vdash \underline{P}[(\underline{TG})\underline{Q}(\underline{G})] \equiv (\exists_1^\cap \underline{G})\underline{Q}(\underline{G})(\exists\underline{G})[\underline{Q}(\underline{G})\underline{P}(\underline{G})] \vee \sim(\exists_1^\cap \underline{G})[\underline{Q}(\underline{G})\underline{P}(\Lambda)]$.

where $\underline{P}$ and $\underline{Q}$ are predicate constants of level 2. Then, since $\Vdash (\exists\underline{G})\underline{G} \neq \Lambda$, it is easy to deduce

(E9) $$\underline{Q} \in \underline{Ext} \supset \underline{P}[(\underline{TG})\underline{Q}(\underline{G})] \equiv \underline{P}(\Lambda),$$

so that the descriptor T is useless in connection with extensional properties.

Note that, in contrast with (E8), by Theor. 14.3 $\underline{N}(\exists_1^\cap \underline{x})\Phi(\underline{x})$—hence $(\exists_1^\cap \underline{x})\Phi(\underline{x})$—does not imply $\Phi[(\iota\underline{x})\Phi(\underline{x})]$ for some choices of $\Phi(\underline{x})$, e.g. for

(E10) $\Phi(\underline{x}) \equiv_D \underline{p}_1\underline{x} =^\cap \underline{M}_1 \vee \sim\underline{p}\underline{x} =^\cap \underline{M}_2$ where $\Vdash \diamond \underline{p} \diamond \sim\underline{p}$ and $\Vdash \underline{NM}_1 \neq \underline{M}_2$.

Let us assume (E8) in accordance with [M]. Then

(E11) $$(\exists_1^\cap \underline{x})\Phi(\underline{x}) \supset^\cap \Phi(\xi) \quad \text{where} \quad \xi =_D (\underline{Tx})\Phi(\underline{x}).$$

From (E10) it is natural to deduce $\underline{N}(\exists_1^\cap \underline{x})\Phi(\underline{x})$; hence $\underline{N}\Phi(\xi)$ by $(E11)_1$. From $\underline{N}\Phi(\xi)$ we easily deduce by $(E10)_{1,2}$

(E12) $$\underline{p} \supset^\cap \xi =^\cap \underline{M}_1, \quad \sim\underline{p} \supset^\cap \xi =^\cap \underline{M}_1, \quad \xi =^\cap \underline{M}_i \ (\underline{i} = 1, 2).$$

By the axiom $(\underline{x})\underline{K}(\underline{x})$ where $\underline{K}(\underline{x})$ is $\underline{x} =^\cap \underline{M}_1 \wedge \underline{x} =^\cap \underline{M}_2 \supset \underline{M}_1 =^\cap \underline{M}_2$, and by the instantiation principle $(\underline{x})\underline{K}(\underline{x}) \supset \underline{K}(\xi)$ we easily obtain from $(E12)_3$, $\underline{NM}_1 = \underline{M}_2$, which contradicts $(E10)_3$. It may be concluded that the kind of treatment proposed for descriptions in [M] is not satisfactory.

NE4. Operators of a general kind for MC^ν.

In NE1 we intuitively hinted at a generic operator ω of the general type $\underline{t}$ where

(E13) $\underline{t} = (\underline{t}_1, \ldots, \underline{t}_{\underline{n}}; \theta, \theta')$ with $\underline{t}_1, \ldots, \underline{t}_{\underline{n}} \in \tau^\nu$ and $\theta, \theta' \in \bar{\tau}^\nu$.

To consider an example of such an operator let us put ${}_0\int^1 (\underline{x}^2 + \underline{y})\, \underline{dx}$ into the form of an operator full formula: $({}_0\int^1 \underline{x})(\underline{x}^2 + \underline{y})$. In mathematics ${}_0\int^1 \underline{f}(\underline{x})\, \underline{dx}$ is often conceived of as a function (or functional) from functions to real numbers. We can mirror this in the writing:

(E14) $$({}_0\int^1 \underline{x})(\underline{x}^2 + \underline{y}) = {}_0\int^1 (\underline{x}^2 + \underline{y})\, \underline{dx} = {}_0\int^1 [(\lambda \underline{x})(\underline{x}^2 + \underline{y})].$$

Let ρ be the logical type of real numbers. Then in (E14) "${}_0\int^1$" is used first as an operator of type $(\rho; \rho; \rho)$, then in the usual way, and lastly as a functor of type $((\rho : \rho) : \rho)$. This suggests the following identification of any operator full formula $(\omega \underline{x}_1, \ldots, \underline{x}_{\underline{n}})\Delta$ of type $\underline{t}$—see (E13)—for $\underline{n} \neq 0$ and $\theta' \neq 0$:

(E15) $$(\omega \underline{x}_1, \ldots, \underline{x}_{\underline{n}})\Delta = {}^\frown\omega[(\lambda \underline{x}_1, \ldots, \underline{x}_{\underline{n}})\Delta] \quad (\underline{n} > 0, \theta' \neq 0).$$

This in turn suggests—cf. [B], (2)—

(E16) $$\underline{t} = (\underline{t}_1, \ldots, \underline{t}_{\underline{n}}; \theta; \theta') = (\langle \underline{t}_1, \ldots, \underline{t}_{\underline{n}}, \theta \rangle : \theta') \quad (\underline{n} > 0, \theta' \neq 0).$$

By analogy it is natural to assume, for $\underline{n} > 0$ and $\theta' = 0$,

(E15′) $$(\omega \underline{x}_1, \ldots, \underline{x}_{\underline{n}})\Delta \equiv {}^\frown\omega[(\lambda \underline{x}_1, \ldots, \underline{x}_{\underline{n}})\Delta] \quad (\underline{n} > 0, \theta' = 0)$$

and to put

(E16′) $$\underline{t} = (\underline{t}_1, \ldots, \underline{t}_{\underline{n}}; \theta; 0) = (\langle \underline{t}_1, \ldots, \underline{t}_{\underline{n}}, \theta \rangle) \quad (\underline{n} > 0).$$

Now we consider the case $\underline{n} = 0$. As an example we take the operator $\underline{N}$ into account. Intuitively we may write $\underline{N}(\underline{p})$ instead of $\underline{Np}$.

However in MC^ν propositional variables or predicates of propositions are not available. So as a preliminary, we now define in MC^{ν} proposition (Prop) and property of proposition (PProp) as follows:

DEF. E1. $\underline{F} \in \underline{Prop} \equiv_D (\forall \underline{v}_{(1)1}) [\underline{v}_{(1)1} \in \underline{F} \supset {}^{\cap}\underline{v}_{(1)1} = {}^{\cap}\Lambda]$.

DEF. E2. $\Phi \in \underline{PProp} \equiv_D (\forall \underline{v}_{((1))1}) [\underline{v}_{((1))1} \in \Phi \supset \underline{v}_{((1))1} \in \underline{Prop}]$.

The (obviously) modally constant property $\underline{Prop}$ is in a one-to-one correspondence with the class of intensions of propositions. Technically, in MC^{ν} we can prove

(E17) $\vdash \underline{Prop} \in \underline{MConst}, \vdash (\exists_1^{\cap} \underline{F}) [\underline{F} \in \underline{Prop} \wedge (\Lambda \in \underline{F} \equiv {}^{\cap}\underline{p})]$.

Now we can extend the definition Def. 4.3 in [B] of $(\lambda \underline{x}_1, \ldots, \underline{x}_n)\underline{p}$ where $\underline{p}$ is any matrix in ML^{ν}, to the case $\underline{n} = 0$:

DEF. E3. $\lambda \underline{p} =_D (\lambda)\underline{p} =_D (\imath \underline{F}) [\underline{F} \in \text{Prop} \wedge (\Lambda \in \underline{F} \equiv \underline{p})]$.

From Defs. E1, E3, we easily deduce in MC^{ν} that

(E18) $\vdash \underline{F} = {}^{\cap}\lambda \underline{p} \equiv \underline{F} \in \underline{Prop} (\Lambda \in \underline{F} \equiv {}^{\cap}\underline{p})$.

Now it is clear that we can replace the form $\underline{N}(\underline{p})$, intuitively considered for $\underline{Np}$ and not belonging to ML^{ν}, with the full formula $\mathcal{N}(\lambda \underline{p})$ in ML^{ν}, of a suitable constant $\mathcal{N}$ which has the type (((1))) and fulfills the condition $\mathcal{N} \in \underline{Prop}$.

The above considerations suggest we assume

(E19) $\omega \underline{p} \equiv {}^{\cap}\omega(\lambda \underline{p})$ for $\underline{n} = 0, \theta = 0$,

and for $\theta' = 0$, and in addition to set

(E20) $(0; \theta') = \langle ((1)), \theta' \rangle$ for $\underline{n} = 0, \theta = 0$,

and for $\theta' = 0$.

It is natural to assume that (E20) and an analogue of (E19) should hold for $\theta' \neq 0$. This analogue of (E19) is

(E19′) $\omega \underline{p} = {}^{\cap}\omega(\lambda \underline{p})$ $(\underline{n} = 0, \theta = 0)$.

Let us remark that under condition (E13) we have considered the analogues of (E15) and (E16) in all cases except for $\underline{n} = 0$ and $\theta \neq 0$. With regard to this case we now observe that for $\theta = 0$ we took $\lambda \underline{p}$ to be something different from $\underline{p}$ only because in MC^{ν} property variables and properties of propositions are not available. However $\lambda \underline{p}$ substantially coincides with $\underline{p}$—cf. (E18).

Furthermore for $\theta \neq 0$ ($\theta \in \tau^{\nu}$), variables of the types θ and (θ) are available. So it is natural to identify $\lambda \Delta$ with Δ, to assume

(E21) $\omega\Delta \equiv^{\cap} \omega(\Delta)$ for $\theta \neq 0 = \theta'$, and $\omega\Delta =^{\cap} \omega(\Delta)$ for $\theta \neq 0 \neq \theta'$,

and to set

(E22) $(\theta; \theta') = \langle\theta, \theta'\rangle$—cf. [B], (2)—for $\theta \neq 0$.

Then ω is a predicate of type θ for $\theta' = 0$ and a functor of type $(\theta : \theta')$ for $\theta' \neq 0$. So the case $\theta' \neq 0$ is of very little interest.

The preceding considerations are rather intuitive with an exception for Defs. E1–3, (E17), and (E18), and the definitions (E16), (E16′), (E20), and (E22).

Now we suggest two ways of putting the above considerations into technical form. First we can consider (E15), (E15′), (E19), (E19′), and (E21) as metalinguistic definitions. So nothing has to be added. Second we can consider the L.H.S.s of $=^{\cap}$ or $\equiv^{\cap}$ in (E15), (E15′), (E19), (E19′), and (E21) as additional designators. So by adding a formation rule to ($\underline{f}_1$) to ($\underline{f}_9$) in [B], N3, say ($\underline{f}_{10}$), we obtain an extended ML^{ν}, say ML^{ν}_{ω}.

Then we must add to the designation rules (δ_1) to (δ_9) for ML^{ν}—see [B], Ns8, 11—a rule, say (δ_{10}), by which (E15), (E15′), (E19), (E19′), and (E21) hold. This condition determines (δ_{10}).

Then we have to extend MC^{ν} into say MC^{ν}_{ω}, by adding that (E15), (E15′), (E19′), and (E21) should hold as axioms.

Note that by this second procedure to introduce operators in MC^{ν}, no primitive symbol is added, but the designators having an operator type—i.e. a type $\underline{t}$ of the form expressed by (E13), (E16), (E16′), (E20), or (E22)—have attribute full formulas of two kinds: ordinary ones and operator full formulas.

By this procedure it is quite natural to use operators in argument positions.

Bibliography

[1] Austin, J. L. Philosophical Papers, chap. 6, "A Plea for Excuses." New York: Oxford University Press, 1961.

[2] Barcan, R. "Functional Calculus of the First Order Based on Strict Implication." Journal of Symbolic Logic 11 (1946): 1.

[3] Barcan, R. "Deduction Theorem in a Functional Calculus of the First Order Based on Strict Implication." Journal of Symbolic Logic 11 (1946): 115.

[4] Bressan, A. "Metodo di assiomatizzazione in senso stretto della Meccanica classica. Applicazione di esso ad alcuni problemi di assiomatizzazione non ancora copletamente risolti." Rend. Sem. Mat. Univ. di Padova 32 (1962): 55–212.

[5] Bressan, A. "Cinematica dei sistemi continui in relatività generale." Annali di Mat. pura ed applicata, Serie IV, 62 (1963): 99–148.

[6] Bressan, A. "Termodinamica e magneto-visco-elasticità con deformazioni finite in relatività generale." Rend. Sem. Mat. Univ. di Padova 34 (1964): 1–73.

[7] Bressan, A. "Una teoria di relatività generale includente, oltre all'elettromagnetismo e alla termodinamica, le equazioni costitutive dei materiali ereditari. Sistemazione assiomatica." Rend. Sem. Mat. Univ. di Padova 34 (1964): 74–109.

[8] Bressan, A. "Sui fluidi capaci di elettro-magneto-strizione dai punti di vista classico e relativistico." Annali di Mat. pura ed applicata 74 (1966): 317–44.

[9] Bressan, A. "Elasticità con elettro-magneto-strizione." Annali di Mat. pura ed applicata 74 (1966): 383–99.

[10] Burks, A. W. "The logic of Causal Propositions." Mind 60 (1951): 363.

[11] Carnap, R. Introduction to Symbolic Logic and Its Applications. New York: Dover Publications, 1958.

[12] Carnap, R. Meaning and Necessity. Chicago: University of Chicago Press, 1956.

[13] Carnap, R. Replies and Systematic Expositions. The Philosophy of R. Carnap, edited by Paul A Schilpp, pp. 859–999. New York: Tudor Publishing Co., Library of Living Philosophers, 1963.

[14] Curry, H. B. Foundations of Mathematical Logic. New York: McGraw Hill, Inc., 1963.

[15] Hermes, H. Eine Axiomatisierung der allgemeinen Mechanik. Forschungen zur Logik und zur Grundlegung der exakten Wissenschaften. Vol. 3, Leipzig: Verlag von Hirzel, 1938.

[16] Hermes, H. "Zur Axiomatisierung der Mechanik." Proceedings of the International Symposium on the Axiomatic Method, Berkeley, 1957–58. Amsterdam: North-Holland Publishing Co., 1959, p. 250.

[17] Hermes, H. "Modal Operators in an Axiomatization of Mechanics." Proceedings of the Colloque International sur la méthode axiomatique classique et moderne, Paris, 1959, pp. 29–36.

[18] Hughes, G. E., and Creswell, M. J. An Introduction to Modal Logic. London: Methuen and Co., Ltd., 1968.

[19] Hutten, E. H. The Language of Modern Physics. London: George Allen & Unwin; New York: The Macmillan Co., 1956.

[20] Kleene, S. C. Introduction to Metamathematics. Amsterdam: North Holland Publishing Co., 1952.

[21] Kripke, S. A. "Semantical Analysis of Modal Logic I, Normal, Propositional Calculi." Zeitschrift für mathematische Logik und Grundlagen der Mathematik 9 (1963): 67–96.

[22] Lemmon, E. J. "Is There Only One Correct System of Modal Logic?" Proceedings of the Aristotelian Society, Suppl. vol. 33 (1959): 23–40.

[23] McKinsey, J. C. C.; Sugar, A. C.; Suppes, P. "Axiomatic Foundations of Classical Particle Mechanics." Journal of Rational Mechanics and Analysis 2 (1953): 253–72.

[24] McKinsey, J. C. C., and Suppes, P. "Philosophy and the Axiomatic Foundations of Physics." Proceedings of the 11th International Congress of Philosophy 6 (1953): 49–53.

[25] McKinsey, J. C. C., and Tarsky, A. "Some Theorems about the Sentential Calculi of Lewis and Heyting." Journal of Symbolic Logic 13 (1948): 1–15.

[26] Meredith, C. A. "A Single Axiom of Positive Logic." Journal of Computing Systems 1 (1953): 169–70.

[27] Meredith, C. A., and Prior, A. N. "Investigations into Implicational S5." Zeitschrift für mathematische Logik und Grundlagen der Mathematik 10 (1964): 203–20.

[28] Montague, R. "Pragmatics and Intensional Logic." Mimeographed.

[29] Painlevé, P. Les Axiomes de la Mechanique. Paris: Gauthier-Villars Editeur, 1922.

[30] Prior, A. N. Formal Logic. Oxford: Clarendon Press, 1962.

[31] Rosser, J. B. "Review of H. Hermes." Journal of Symbolic Logic 3 (1938): 119–20.

[32] Rosser, J. B. Logic for Mathematicians. New York: McGraw-Hill Inc., 1953.

[33] Signorini, A. Meccanica Razionale. Vol. 2, chap. 10. 2d ed. Rome: Perrella, 1954, p. 1.

[34] Thomason, H. R. "Modal Logic and Metaphysics." In The Logical Way of Doing Things, edited by K. Lambert, pp. 119–46. New Haven: Yale University Press, 1969.

[35] Thomason, H. R., and Stalnaker, R. C. "Modality and Reference." Nous 2 (1968): 359–72.

[36] Thomason, H. R., and Stalnaker, R. C. "Some Completeness Results for Modal Predicate Calculi." Mimeographed.

[37] Thomason, H. R., and Stalnaker, R. C. "A Semantical Analysis of Conditional Logic." Mimeographed.

[38] Von Wright, G. H. An Essay in Modal Logic. Studies in Logic and the Foundations of Mathematics. Amsterdam: North Holland Publishing Co., 1951.

[39] Wittgenstein, L. Tractatus Logico-Philosophicus. London, 1922.

List of Symbols

$(\underline{f}_1)$ to $(\underline{f}_9)$, formation rules for ML^{ν}, Def. 3.1.

$\mathscr{E}_{\underline{t}}$, Def. 3.1, p. 12.

$\sim$, $\wedge$, $\forall$, $\underline{N}$, N3, p. 11.

ι (description operator), N3, AS12.18.

$=$ (contingent identity), N3, ASs12.10–12, 14, 15, and 17.

τ^{ν}, $\bar{\tau}^{\nu}$, Def. 2.1, (1).

$\diamond$, p. 14.

λ, Defs. 4.4, 5.

$\underline{D}_{\underline{i}}$, p. 18.

$\underline{O}_{\underline{t}}$ $(\underline{t}\epsilon\tau^{\nu})$, N6, p. 19.

$\underline{v}_{\underline{tn}}$, $\underline{n}$-th variable of type $\underline{t}$, N3.

$\underline{c}_{\underline{tn}}$, $\underline{n}$-th constant of type $\underline{t}$, N3.

$\underline{lev}$, (3) in N2.

Γ, the class of possible elementary cases, N6.

Γ-case, element of Γ.

$\underline{a}^{\nu}$, $\underline{a}^{\nu}_{\underline{t}}$, N6.

$\underline{QI}$ (quasi intension), $\underline{QI}_{\underline{t}}$, N6.

$\underline{QI}'_{\underline{t}}$ (new quasi intension of type $\underline{t}$), N25.

$\underline{V}$, $\mathfrak{M}$, Def. 6.1.

$\mathscr{V}$, $\mathscr{M}$, Def. 6.2.

$\mathscr{V}\begin{pmatrix} \underline{x}_1, \ldots, \underline{x}_{\underline{n}} \\ \tilde{\underline{x}}_1, \ldots, \tilde{\underline{x}}_{\underline{n}} \end{pmatrix}$, (11) in N6.

$(\underline{T}_1)$ to $(\underline{T}_9)$, i.e. rules of extensional translation, p. 57.

$\underline{\text{MConst}}$, Def. 13.2.

$\underline{\text{MSep}}$, Def. 18.7.

$\underline{\text{Abs}}$, Def. 18.8.

$\overline{\text{F}}$, $\underline{F} \cap \underline{G}$, $\underline{F} \cup \underline{G}$, $\underline{F} \subseteq \underline{G}$, $\underline{F} \subset \underline{G}$, Defs. 18.1–5.

κ is $\underline{v}_{\underline{t}1}$ for $\underline{t} = \nu + 1$.

Λ, Def. 18.6.

$\underline{F}^{(\underline{e})}$, Def. 18.9.

$\{\underline{x}_1, \ldots, \underline{x}_n\}$, Def. 18.12.

$\{\underline{x}_1, \ldots, \underline{x}_n\}^{(\underline{i})}$, Def. 18.13.

$\mu(\underline{M})$, $\mu^{(\underline{e})}(\overline{\underline{M}})$, Defs. 19.1, 2.

$\mathfrak{m}$, pp. 24, 71.

$\underline{\text{MP}}$ (mass point), $\underline{\text{MP}}_0$, Ns19, 20, Apps. C and D.

$\underline{\text{Real}}$ (real number), $\underline{\text{Real}}_0$, Ns19, 20.

$\underline{\text{Real}}_1$, $\underline{\text{Real}}_2$, Defs. 20.1, 2.

$\underline{\text{QMConst}}$, Def. 24.1, for properties see N41.

$\underline{\text{QMSep}}$, Def. 24.2, N41.

$\underline{\text{QAbs}}$, Def. 24.3, N41.

$\underset{\sim}{\cap}$, $\underset{\sim}{\cup}$, Defs. 26.1, 2.

$\underline{\text{Her}}$, Def. 26.3.

$\underline{\text{clos}}$, Def. 26.4, Theor. 44.2.

0, 1, Defs. 27.1, 2.

$\alpha + \beta$, Def. 27.3.

$\underline{\text{Nn}}$, Def. 27.5, N45.

$\underline{\text{LPC}}$ (lower predicate calculus), Ns30, 31.

$\overline{\text{EL}}^{\nu+1}$, N30, p. 125.

$\underline{\text{LPC}}\,(\overline{\text{EL}}^{\nu+1})$, p. 127.

$(\text{ML}^{\nu})^{\eta}$, $\text{LPC}(\text{ML}^{\nu})^{\eta}$, N30, p. 125.

$\underline{A}(\underline{x})$, $\underline{A}(\Delta)$, see conventions (a) and (b) in N30.

$\Phi(\underline{x})$, $\Phi(\underline{y})$, see convention (c) in N30.

$\vdash_{\underline{c}}$ (deduction including rules $\underline{G}$ and $\underline{C}$), Def. 33.2, Theor. 33.1.

$\exists^{(1)}$, Def. 34.1.

$\exists^{(1)\frown}$, Def. 34.2.

$(\underline{ext}\ \underline{x}_1, \ldots, \underline{x}_{\underline{n}})\Delta$, Defs. 35.1, 2.

$\underline{ElR}$ (elementary range), Def. 47.3.

$\underline{El}$ (absolute elementary range), Def. 48.1.

$\underline{El}'$ (analogue of $\underline{El}$ used for the translation $\Delta \to \Delta^*$ of $\underline{EL}^{\nu+1}$ into ML^{ν}), (100′) in N56.

$|_{\underline{u}}$ (the absolute range $\underline{u}$ actually holds), Def. 48.2.

$|'_{\underline{u}}$ (analogue of $|_{\underline{u}}$ for the translation $\Delta \to \Delta^*$), Def. 56.2.

$\iota_{\underline{u}}$ (intensional description operator), Def. 50.1, N54.

ρ (real elementary range), A52.1.

$\mathfrak{R}$, Def. 52.1.

$\underline{MC}^{\nu}_{\rho}$, Ns52, 53.

$\underline{t}^*$ for $\underline{t} \epsilon \bar{\tau}^{\nu}$, Def. 56.1.

$\underline{v}^*_{\underline{t}}$ and $\underline{c}^*_{\underline{t}}$ for $\underline{t} \epsilon \tau^{\nu}$, (97) and (98) on p. 229.

Δ^* (translation into ML^{ν} of the designator Δ in $\underline{EL}^{\nu+1}$ or $\underline{EL}^{\nu}$), N57.

$\underline{I}_{\underline{t}}(\bar{\underline{y}}, \underline{y}, \underline{v}_{(1)1}, \ldots, \underline{v}_{(\nu)1}, \underline{v}_{\nu+1*5})$ or $\underline{I}_{\underline{t}}(\bar{\underline{y}}, \underline{y})$, N59.

$(\underline{T}^*_1)$ to $(\underline{T}^*_9)$, translation rules of $\underline{EL}^{\nu+1}$ into ML^{ν}, N57.

$\underline{Conc}_{\underline{T}}$, $\underline{Conc}_{S4}$, $\underline{Conc}_{\underline{B}}$, $\underline{Conc}_{S5}$, Defs. 66.1–4.

$\underline{L}_{\alpha}$ (necessity operator corresponding to the conceivability property α), N67.

$\underline{M}_{\alpha}$ (possibility operator corresponding to α), N68.

S4, S5, Chap. 13.

$\underline{BF}$ (Barcan formula), N69.

$\underline{T}$ (Fey's system), Chap. 13.

Index of Subjects